T0298013

Structure-Preserving Algorithms for Oscillatory Differential Equations II

Xinyuan Wu · Kai Liu · Wei Shi

Structure-Preserving Algorithms for Oscillatory Differential Equations II

 Science Press
Beijing

 Springer

Xinyuan Wu
Department of Mathematics
Nanjing University
Nanjing
China

Wei Shi
Nanjing Tech University
Nanjing
China

Kai Liu
Nanjing University of Finance
and Economics
Nanjing
China

ISBN 978-3-662-48155-4   ISBN 978-3-662-48156-1   (eBook)
DOI 10.1007/978-3-662-48156-1

Jointly published with Science Press, Beijing, China
ISBN: 978-7-03-043918-5 Science Press, Beijing

Library of Congress Control Number: 2015950922

Springer Heidelberg New York Dordrecht London

Printed on acid-free paper

Springer-Verlag GmbH Berlin Heidelberg is part of Springer Science+Business Media
(www.springer.com)

This monograph is dedicated to Prof. Kang Feng on the thirtieth anniversary of his pioneering study on symplectic algorithms.

His profound work, which opened up a rich new field of research, is of great importance to numerical mathematics in China, and the influence of his seminal contributions has spread throughout the world.

2014 Nanjing Workshop on Structure-Preserving Algorithms for Differential Equations (Nanjing, November 29, 2014)

Preface

Numerical integration of differential equations, as an essential tool for investigating the qualitative behaviour of the physical universe, is a very active research area since large-scale science and engineering problems are often modelled by systems of ordinary and partial differential equations, whose analytical solutions are usually unknown even when they exist. Structure preservation in numerical differential equations, known also as geometric numerical integration, has emerged in the last three decades as a central topic in numerical mathematics. It has been realized that an integrator should be designed to preserve as much as possible the (physical/geometric) intrinsic properties of the underlying problem. The design and analysis of numerical methods for oscillatory systems is an important problem that has received a great deal of attention in the last few years. We seek to explore new efficient classes of methods for such problems, that is high accuracy at low cost. The recent growth in the need of geometric numerical integrators has resulted in the development of numerical methods that can systematically incorporate the structure of the original problem into the numerical scheme. The objective of this sequel to our previous monograph, which was entitled "Structure-Preserving Algorithms for Oscillatory Differential Equations", is to study further structure-preserving integrators for multi-frequency oscillatory systems that arise in a wide range of fields such as astronomy, molecular dynamics, classical and quantum mechanics, electrical engineering, electromagnetism and acoustics. In practical applications, such problems can often be modelled by initial value problems of second-order differential equations with a linear term characterizing the oscillatory structure. As a matter of fact, this extended volume is a continuation of the previous volume of our monograph and presents the latest research advances in structure-preserving algorithms for multi-frequency oscillatory second-order differential equations. Most of the materials of this new volume are drawn from very recent published research work in professional journals by the research group of the authors.

Chapter 1 analyses in detail the matrix-variation-of-constants formula which gives significant insight into the structure of the solution to the multi-frequency and multidimensional oscillatory problem. It is known that the Störmer–Verlet formula is a very popular numerical method for solving differential equations. Chapter 2 presents novel improved multi-frequency and multidimensional Störmer–Verlet formulae. These methods are applied to solve four significant problems. For structure-preserving integrators in differential equations, another related area of increasing importance is the computation of highly oscillatory problems. Therefore, Chap. 3 explores improved Filon-type asymptotic methods for highly oscillatory differential equations. In recent years, various energy-preserving methods have been developed, such as the discrete gradient method and the average vector field (AVF) method. In Chap. 4, we consider efficient energy-preserving integrators based on the AVF method for multi-frequency oscillatory Hamiltonian systems. An extended discrete gradient formula for multi-frequency oscillatory Hamiltonian systems is introduced in Chap. 5. It is known that collocation methods for ordinary differential equations have a long history. Thus, in Chap. 6, we pay attention to trigonometric Fourier collocation methods with arbitrary degrees of accuracy in preserving some invariants for multi-frequency oscillatory second-order ordinary differential equations. Chapter 7 analyses the error bounds for explicit ERKN integrators for systems of multi-frequency oscillatory second-order differential equations. Chapter 8 contains an analysis of the error bounds for two-step extended Runge–Kutta–Nyström-type (TSERKN) methods. Symplecticity is an important characteristic property of Hamiltonian systems and it is worthwhile to investigate higher order symplectic methods. Therefore, in Chap. 9, we discuss high-accuracy explicit symplectic ERKN integrators. Chapter 10 is concerned with multi-frequency adapted Runge–Kutta–Nyström (ARKN) integrators for general multi-frequency and multidimensional oscillatory second-order initial value problems. Butcher's theory of trees is widely used in the study of Runge–Kutta and Runge–Kutta–Nyström methods. Chapter 11 develops a simplified tricoloured tree theory for the order conditions for ERKN integrators and the results presented in this chapter are an important step towards an efficient theory of this class of schemes. Structure-preserving algorithms for multi-symplectic Hamiltonian PDEs are of great importance in numerical simulations. Chapter 12 focuses on general approach to deriving local energy-preserving integrators for multi-symplectic Hamiltonian PDEs.

The presentation of this volume is characterized by mathematical analysis, providing insight into questions of practical calculation, and illuminating numerical simulations. All the integrators presented in this monograph have been tested and verified on multi-frequency oscillatory problems from a variety of applications to observe the applicability of numerical simulations. They seem to be more efficient than the existing high-quality codes in the scientific literature.

The authors are grateful to all their friends and colleagues for their selfless help during the preparation of this monograph. Special thanks go to John Butcher of The University of Auckland, Christian Lubich of Universität Tübingen, Arieh Iserles of University of Cambridge, Reinout Quispel of La Trobe University, Jesus Maria Sanz-Serna of Universidad de Valladolid, Peter Eris Kloeden of Goethe-Universität, Elizabeth Louise Mansfield of University of Kent, Maarten de Hoop of Purdue University, Tobias Jahnke of Karlsruher Institut für Technologie (KIT),

Achim Schädle of Heinrich Heine University Düsseldorf and Jesus Vigo-Aguiar of Universidad de Salamanca for their encouragement.

The authors are also indebted to many friends and colleagues for reading the manuscript and for their valuable suggestions. In particular, the authors take this opportunity to express their sincere appreciation to Robert Peng Kong Chan of The University of Auckland, Qin Sheng of Baylor University, Jichun Li of University of Nevada Las Vegas, Adrian Turton Hill of Bath University, Choi-Hong Lai of University of Greenwich, Xiaowen Chang of McGill University, Jianlin Xia of Purdue University, David McLaren of La Trobe University, Weixing Zheng and Zuhe Shen of Nanjing University.

Sincere thanks also go to the following people for their help and support in various forms: Cheng Fang, Peiheng Wu, Jian Lü, Dafu Ji, Jinxi Zhao, Liangsheng Luo, Zhihua Zhou, Zehua Xu, Nanqing Ding, Guofei Zhou, Yiqian Wang, Jiansheng Geng, Weihua Huang, Jiangong You, Hourong Qin, Haijun Wu, Weibing Deng, Rong Shao, Jiaqiang Mei, Hairong Xu, Liangwen Liao and Qiang Zhang of Nanjing University, Yaolin Jiang of Xi'an Jiao Tong University, Yongzhong Song, Jinru Chen and Yushun Wang of Nanjing Normal University, Xinru Wang of Nanjing Medical University, Mengzhao Qin, Geng Sun, Jialin Hong, Zaijiu Shang and Yifa Tang of Chinese Academy of Sciences, Guangda Hu of University of Science and Technology Beijing, Jijun Liu, Zhizhong Sun and Hongwei Wu of Southeast University, Shoufo Li, Aiguo Xiao and Liping Wen of Xiang Tan University, Chuanmiao Chen of Hunan Normal University, Siqing Gan of Central South University, Chengjian Zhang and Chengming Huang of Huazhong University of Science and Technology, Shuanghu Wang of the Institute of Applied Physics and Computational Mathematics, Beijing, Yuhao Cong of Shanghai University, Hongjiong Tian of Shanghai Normal University, Yongkui Zou of Jilin University, Jingjun Zhao of Harbin Institute of Technology, Qin Ni and Chunwu Wang of Nanjing University of Aeronautics and Astronautics, Guoqing Liu, and Hao Cheng of Nanjing Tech University, Hongyong Wang of Nanjing University of Finance and Economics, Theodoros Kouloukas of La Trobe University, Anders Christian Hansen, Amandeep Kaur and Virginia Mullins of University of Cambridge, Shixiao Wang of The University of Auckland, Qinghong Li of Chuzhou University, Yonglei Fang of Zaozhuang University, Fan Yang, Xianyang Zeng and Hongli Yang of Nanjing Institute of Technology, Jiyong Li of Hebei Normal University, Bin Wang of Qufu Normal University, Xiong You of Nanjing Agricultural University, Xin Niu of Hefei University, Hua Zhao of Beijing Institute of Tracking and Tele Communication Technology, Changying Liu, Lijie Mei, Yuwen Li, Qihua Huang, Jun Wu, Lei Wang, Jinsong Yu, Guohai Yang and Guozhong Hu.

The authors would like to thank Kai Hu, Ji Luo and Tianren Sun for their help with the editing, the editorial and production group of the Science Press, Beijing and Springer-Verlag, Heidelberg.

The authors also thank their family members for their love and support throughout all these years.

The work on this monograph was supported in part by the Natural Science Foundation of China under Grants 11271186, by NSFC and RS International Exchanges Project under Grant 113111162, by the Specialized Research Foundation for the Doctoral Program of Higher Education under Grant 20100091110033 and 20130091110041, by the 985 Project at Nanjing University under Grant 9112020301, and by the Priority Academic Program Development of Jiangsu Higher Education Institutions.

Nanjing, China   Xinyuan Wu
                                                                                          Kai Liu
                                                                                          Wei Shi

Contents

Chapter 1
Matrix-Variation-of-Constants Formula

The first chapter presents the matrix-variation-of-constants formula which is fundamental to structure-preserving integrators for multi-frequency and multidimensional oscillatory second-order differential equations in the current volume and the previous volume [23] of our monograph since the formula makes it possible to incorporate the special structure of the multi-frequency oscillatory problems into the integrators.

1.1 Multi-frequency and Multidimensional Problems

Oscillatory second-order initial value problems constitute an important category of differential equations in pure and applied mathematics, and in applied sciences such as mechanics, physics, astronomy, molecular dynamics and engineering. Among traditional and typical numerical schemes for solving these kinds of problems is the well-known Runge-Kutta-Nyström method [13], which has played an important role since 1925 in dealing with second-order initial value problems of the conventional form

$$\begin{cases} y'' = f(y, y'), & x \in [x_0, x_{\text{end}}], \\ y(x_0) = y_0, & y'(x_0) = y_0'. \end{cases} \tag{1.1}$$

However, many systems of second-order differential equations arising in applications have the general form

$$\begin{cases} y'' + My = f(y, y'), & x \in [x_0, x_{\text{end}}], \\ y(x_0) = y_0, & y'(x_0) = y_0', \end{cases} \tag{1.2}$$

where $M \in \mathbb{R}^{d \times d}$ is a positive and semi-definite matrix (not necessarily diagonal nor symmetric, in general) that implicitly contains and preserves the main frequencies of the oscillatory problem. Here, $f : \mathbb{R}^d \times \mathbb{R}^d \to \mathbb{R}^d$, with the position $y(x) \in \mathbb{R}^d$ and

© Springer-Verlag Berlin Heidelberg and Science Press, Beijing, China 2015
X. Wu et al., *Structure-Preserving Algorithms for Oscillatory
Differential Equations II*, DOI 10.1007/978-3-662-48156-1_1

the velocity $y'(x)$ as arguments. The system (1.2) is a multi-frequency and multidimensional oscillatory problem which exhibits pronounced oscillatory behaviour due to the linear term My. Among practical examples we mention the damped harmonic oscillator, the van der Pol equation, the Liénard equation (see [10]) and the damped wave equation. The design and analysis of numerical integrators for nonlinear oscillators is an important problem that has received a great deal of attention in the last few years.

It is important to observe that the special case $M = 0$ in (1.2) reduces to the conventional form of second-order initial value problems (1.1). Therefore, each integrator for the system (1.2) is applicable to the conventional second-order initial value problems (1.1). Consequently, this extended volume of our monograph focuses only on the general second-order oscillatory system (1.2).

When the function f does not contain the first derivative y', (1.2) reduces to the special and important multi-frequency oscillatory system

$$
\begin{cases}
y'' + My = f(y), & x \in [x_0, x_{\text{end}}], \\
y(x_0) = y_0, & y'(x_0) = y_0'.
\end{cases}
\tag{1.3}
$$

If M is symmetric and positive semi-definite and $f(y) = -\nabla U(y)$, then with $q = y$, $p = y'$, (1.3) becomes identical to a multi-frequency and multidimensional oscillatory Hamiltonian system of the form

$$
\begin{cases}
p' = -\nabla_q H(p, q), & p(x_0) = p_0, \\
q' = \nabla_p H(p, q), & q(x_0) = q_0,
\end{cases}
\tag{1.4}
$$

with the Hamiltonian

$$
H(p, q) = \frac{1}{2} p^\mathsf{T} p + \frac{1}{2} q^\mathsf{T} M q + U(q),
\tag{1.5}
$$

where $U(q)$ is a smooth potential function. The solution of the system (1.4) exhibits nonlinear oscillations. Mechanical systems with a partitioned Hamiltonian function yield examples of this type. It is well known that two fundamental properties of Hamiltonian systems are:

(i) the solutions preserve the Hamiltonian H, i.e., $H(p(x), q(x)) \equiv H(p_0, q_0)$ for any $x \geq x_0$;

(ii) the corresponding flow is symplectic, i.e., it preserves the differential 2-form
$$\sum_{i=1}^{d} \mathrm{d} p_i \wedge \mathrm{d} q_i.$$

It is true that great advances have been made in the theory of general-purpose methods for the numerical solution of ordinary differential equations. However, the numerical implementation of a general-purpose method cannot respect the qualitative behaviour of a multi-frequency and multidimensional oscillatory problem. It turns out that structure-preserving integrators are required in order to produce the qualitative properties of the true flow of the multi-frequency oscillatory problem. This new volume represents an attempt to extend our previous volume [23] and presents the very recent advances in Runge-Kutta-Nyström-type (RKN-type) methods for multi-frequency oscillatory second-order initial value problems (1.2). To this end, the following matrix-variation-of-constants formula is fundamental and plays an important role.

1.2 Matrix-Variation-of-Constants Formula

The following matrix-variation-of-constants formula gives significant insight into the structure of the solution to the multi-frequency and multidimensional problem (1.2), which has motivated the formulation of multi-frequency and multidimensional adapted Runge-Kutta-Nyström (ARKN) schemes, and multi-frequency and multidimensional extended Runge-Kutta-Nyström (ERKN) integrators, as well as classical RKN methods.

Theorem 1.1 (Wu et al. [21]) *If $M \in \mathbb{R}^{d \times d}$ is a positive semi-definite matrix and $f : \mathbb{R}^d \times \mathbb{R}^d \to \mathbb{R}^d$ in (1.2) is continuous, then the exact solution of (1.2) and its derivative satisfy*

$$
\begin{cases}
y(x) = \phi_0\big((x - x_0)^2 M\big)y_0 + (x - x_0)\phi_1\big((x - x_0)^2 M\big)y_0' \\
\qquad + \displaystyle\int_{x_0}^{x} (x - \zeta)\phi_1\big((x - \zeta)^2 M\big)\hat{f}(\zeta)\mathrm{d}\zeta, \\
y'(x) = -(x - x_0)M\phi_1\big((x - x_0)^2 M\big)y_0 + \phi_0\big((x - x_0)^2 M\big)y_0' \\
\qquad + \displaystyle\int_{x_0}^{x} \phi_0\big((x - \zeta)^2 M\big)\hat{f}(\zeta)\mathrm{d}\zeta,
\end{cases} \tag{1.6}
$$

for $x_0, x \in (-\infty, +\infty)$, where

$$
\hat{f}(\zeta) = f\big(y(\zeta), y'(\zeta)\big)
$$

and the matrix-valued *functions $\phi_0(M)$ and $\phi_1(M)$ are defined by*

$$
\phi_0(M) = \sum_{k=0}^{\infty} \frac{(-1)^k M^k}{(2k)!}, \quad \phi_1(M) = \sum_{k=0}^{\infty} \frac{(-1)^k M^k}{(2k+1)!}. \tag{1.7}
$$

We notice that these matrix-valued functions reduce to the ϕ-functions used for Gautschi-type trigonometric or exponential integrators in [4, 7] when M is a symmetric and positive semi-definite matrix.

With regard to algorithms for computing the matrix-valued functions $\phi_0(M)$ and $\phi_1(M)$, we refer the reader to [1] and references therein.

Taking the importance of the matrix-variation-of-constants formula into account, a brief proof is presented in a self-contained way.

Proof Multiplying both sides of the equation in formula (1.2) (using ζ for the independent variable) by $(x - \zeta)\phi_1((x - \zeta)^2 M)$ and integrating from x_0 to x yields

$$\int_{x_0}^{x}(x - \zeta)\phi_1((x - \zeta)^2 M)y''(\zeta)d\zeta + \int_{x_0}^{x}(x - \zeta)\phi_1((x - \zeta)^2 M)My(\zeta)d\zeta$$
$$= \int_{x_0}^{x}(x - \zeta)\phi_1((x - \zeta)^2 M)\hat{f}(\zeta)d\zeta.$$

Applying integration by parts twice to the first term on the left-hand side gives the first equation of (1.6) on noticing that

$$\left((x - \zeta)\phi_1((x - \zeta)^2 M)\right)' = -\phi_0((x - \zeta)^2 M),$$

and

$$\left(\phi_0((x - \zeta)^2 M)\right)' = (x - \zeta)\phi_1((x - \zeta)^2 M)M.$$

Likewise, multiplying both sides of the equation in formula (1.2) by $\phi_0((x - \zeta)^2 M)$ and integrating from x_0 to x yields

$$\int_{x_0}^{x}\phi_0((x - \zeta)^2 M)y''(\zeta)d\zeta + \int_{x_0}^{x}\phi_0((x - \zeta)^2 M)My(\zeta)d\zeta = \int_{x_0}^{x}\phi_0((x - \zeta)^2 M)\hat{f}(\zeta)d\zeta.$$

Integration by parts twice in the first term on the left-hand side gives the second equation of (1.6). □

Here and in the remainder of this chapter, the integral of a matrix function is understood componentwise.

We notice that a conventional form of variation-of-constants formula in the literature (see, e.g. [3] by García-Archilla et al., [4, 5] by Hairer et al., and [7] by Hochbruck et al.) for the special oscillatory system (1.3) is expressed by

$$\left[\begin{array}{l} y(x) = \cos\left((x - x_0)\Omega\right)y_0 + \Omega^{-1}\sin\left((x - x_0)\Omega\right)y_0' \\ \qquad + \int_{x_0}^{x}\Omega^{-1}\sin\left((x - \tau)\Omega\right)\hat{g}(\tau)d\tau, \\ y'(x) = -\Omega\sin\left((x - x_0)\Omega\right)y_0 + \cos\left((x - x_0)\Omega\right)y_0' \\ \qquad + \int_{x_0}^{x}\cos\left((x - \tau)\Omega\right)\hat{g}(\tau)d\tau, \end{array}\right. \qquad (1.8)$$

where $M = \Omega^2$ and $\hat{g}(\tau) = f(y(\tau))$.

It is very clear that the matrix-variation-of-constants formula (1.6) for general oscillatory systems (1.2) does not involve the decomposition of M and is different from the conventional one for the oscillatory system (1.3) in which f is independent of y'. It is also important to avoid matrix decompositions in an integrator for multi-frequency and multidimensional oscillatory systems as M is not necessarily diagonal nor symmetric in (1.2) and the decomposition $M = \Omega^2$ is not always feasible. Therefore, the matrix-variation-of-constants formula (1.6) without the decomposition of matrix M has wider applications in general.

In the particular case $M = \omega^2 I_d$, (1.3) becomes

$$\begin{cases} y'' + \omega^2 y = f(y), & x \in [x_0, x_{\text{end}}], \\ y(x_0) = y_0, & y'(x_0) = y_0', \end{cases} \tag{1.9}$$

and the corresponding variation-of-constants formula is well known. That is,

$$\begin{cases} y(x) = \cos\big((x - x_0)\omega\big)y_0 + \omega^{-1}\sin\big((x - x_0)\omega\big)y_0' \\ \qquad + \displaystyle\int_{x_0}^{x} \omega^{-1}\sin\big((x - \tau)\omega\big)\hat{g}(\tau)\mathrm{d}\tau, \\ y'(x) = -\omega\sin\big((x - x_0)\omega\big)y_0 + \cos\big((x - x_0)\omega\big)y_0' \\ \qquad + \displaystyle\int_{x_0}^{x} \cos\big((x - \tau)\omega\big)\hat{g}(\tau)\mathrm{d}\tau. \end{cases} \tag{1.10}$$

If M is symmetric and positive semi-definite, it is easy to see that (1.8) can be straightforwardly obtained by formally replacing ω with $\Omega = M^{1/2}$ in the formula (1.10) as pointed out in Sect. XIII.1.2 in [5] by Hairer et al. However, it is required to show (1.6) by a rigorous proof as M is not necessarily symmetric or diagonal and f depends on both y and y' in (1.2).

It follows from formula (1.6) of Theorem 1.1 that, for any $x, \mu, h \in \mathbb{R}$ with $x, x + \mu h \in [x_0, x_{\text{end}}]$, the solution to (1.2) satisfies the following integral equations:

$$\begin{cases} y(x + \mu h) = \phi_0(\mu^2 V)y(x) + h\mu\phi_1(\mu^2 V)y'(x) \\ \qquad + h^2 \displaystyle\int_0^{\mu} (\mu - \zeta)\phi_1\big((\mu - \zeta)^2 V\big)\hat{f}(x + \zeta h)\,\mathrm{d}\zeta, \\ y'(x + \mu h) = \phi_0(\mu^2 V)y'(x) - \mu h M\phi_1(\mu^2 V)y(x) \\ \qquad + h \displaystyle\int_0^{\mu} \phi_0\big((\mu - \zeta)^2 V\big)\hat{f}(x + \zeta h)\,\mathrm{d}\zeta, \end{cases} \tag{1.11}$$

where $V = h^2 M$, and the $d \times d$ matrix-valued functions $\phi_0(V)$ and $\phi_1(V)$ are defined by (1.7).

It can be observed that the matrix-variation-of-constants formula (1.11) subsumes the structure of the internal stages and updates for an extended and improved RKN-type integrator. In fact, if $y(x_n)$ and $y'(x_n)$ are prescribed, it follows from (1.11) that

$$
\begin{cases}
\begin{aligned}
y(x_n + \mu h) = {} & \phi_0(\mu^2 V)y(x_n) + \mu h \phi_1(\mu^2 V)y'(x_n) \\
& + h^2 \int_0^\mu (\mu - \zeta)\phi_1\big((\mu - \zeta)^2 V\big)\hat{f}(x_n + h\zeta)\mathrm{d}\zeta, \\
y'(x_n + \mu h) = {} & -\mu h M \phi_1(\mu^2 V)y(x_n) + \phi_0(\mu^2 V)y'(x_n) \\
& + h \int_0^\mu \phi_0\big((\mu - \zeta)^2 V\big)\hat{f}(x_n + h\zeta)\mathrm{d}\zeta,
\end{aligned}
\end{cases}
\tag{1.12}
$$

for $0 < \mu < 1$, and

$$
\begin{cases}
\begin{aligned}
y(x_n + h) = {} & \phi_0(V)y(x_n) + h\phi_1(V)y'(x_n) \\
& + h^2 \int_0^1 (1 - \zeta)\phi_1\big((1 - \zeta)^2 V\big)\hat{f}(x_n + h\zeta)\mathrm{d}\zeta, \\
y'(x_n + h) = {} & -h M \phi_1(V)y(x_n) + \phi_0(V)y'(x_n) \\
& + h \int_0^1 \phi_0\big((1 - \zeta)^2 V\big)\hat{f}(x_n + h\zeta)\mathrm{d}\zeta,
\end{aligned}
\end{cases}
\tag{1.13}
$$

for $\mu = 1$.

The formulae (1.12) and (1.13) suggest clearly the structure of the internal stages and updates for an improved RKN-type integrator when they are applied to multi-frequency and multidimensional oscillatory systems (1.2), respectively.

As a simple example, the trapezoidal discretization of the integrals in formula (1.13) with a fixed stepsize h gives the implicit scheme

$$
\begin{cases}
\begin{aligned}
y_{n+1} = {} & \phi_0(V)y_n + h\phi_1(V)y_n' + \frac{1}{2}h^2\phi_1(V)f(y_n, y_n'), \\
y_{n+1}' = {} & -h M \phi_1(V)y_n + \phi_0(V)y_n' \\
& + \frac{1}{2}h\big(\phi_0(V)f(y_n, y_n') + f(y_{n+1}, y_{n+1}')\big).
\end{aligned}
\end{cases}
\tag{1.14}
$$

If the function f does not depend on the first derivative y', formula (1.14) reduces to an explicit scheme

$$
\begin{cases}
y_{n+1} = \phi_0(V)y_n + h\phi_1(V)y_n' + \frac{1}{2}h^2\phi_1(V)f(y_n), \\
y_{n+1}' = -h M \phi_1(V)y_n + \phi_0(V)y_n' + \frac{1}{2}h\big(\phi_0(V)f(y_n) + f(y_{n+1})\big),
\end{cases}
\tag{1.15}
$$

which was first given in [2] for the initial value problem

$$
y'' + \omega^2 y = f(y), \quad y(x_0) = y_0, \quad y'(x_0) = y_0',
$$

where $\omega > 0$. This means that the integrator (1.14) reduces to the Deuflhard method in the particular case $M = \omega^2 I$.

As another example, we consider to apply the variation-of constants formula (1.6) to high-dimensional nonlinear Hamiltonian wave equations:

$$\begin{cases} u_{tt}(X,t) = f\big(u(X,t)\big) + a^2 \Delta u(X,t), & X \in \Re \subseteq \mathbb{R}^d,\ t > t_0, \\ u(X,t_0) = \varphi_1(X),\ u_t(X,t_0) = \varphi_2(X), & X \in \Re \cup \partial\Re, \end{cases} \tag{1.16}$$

where $f(u) = -\dfrac{\mathrm{d}G\big(u(X,t)\big)}{\mathrm{d}u}$ and

$$\Delta = \sum_{i=1}^{d} \frac{\partial^2}{\partial x_i^2}.$$

Applying the variation-of-constants formula (1.6) to (1.16) gives an analytical expression for the solution of (1.16):

$$\begin{cases} u(X,t) = u(X,t_0) + (t - t_0)u_t(X,t_0) \\ \qquad + \displaystyle\int_{t_0}^{t} (t - \zeta)\Big(f\big(u(X,\zeta)\big) + a^2 \Delta u(X,\zeta)\Big)\mathrm{d}\zeta, \\ u_t(X,t) = u_t(X,t_0) + \displaystyle\int_{t_0}^{t} \Big(f\big(u(X,\zeta)\big) + a^2 \Delta u(X,\zeta)\Big)\mathrm{d}\zeta, \\ X \in \Re \cup \partial\Re, \end{cases} \tag{1.17}$$

i.e.,

$$\begin{cases} u(X,t) = \varphi_1(X) + (t - t_0)\varphi_2(X) \\ \qquad + \displaystyle\int_{t_0}^{t} (t - \zeta)\Big(f\big(u(X,\zeta)\big) + a^2 \Delta u(X,\zeta)\Big)\mathrm{d}\zeta, \\ u_t(X,t) = \varphi_2(X) + \displaystyle\int_{t_0}^{t} \Big(f\big(u(X,\zeta)\big) + a^2 \Delta u(X,\zeta)\Big)\mathrm{d}\zeta, \\ X \in \Re \cup \partial\Re. \end{cases} \tag{1.18}$$

Under suitable assumptions it can be proved that (1.18) is consistent with Dirichlet boundary conditions, Neumann boundary conditions, and Robin boundary conditions, respectively.

Formula (1.17) for the purpose of numerical simulation can be written as

$$\begin{cases} u(X, t_n + h) = u(X, t_n) + hu_t(X, t_n) \\ \qquad + h^2 \int_0^1 (1 - \zeta)\Big(f\big(u(X, t_n + \zeta h)\big) + a^2 \Delta u(X, t_n + \zeta h)\Big)\mathrm{d}\zeta, \\ u_t(X, t_n + h) = u_t(X, t_n) \\ \qquad + h \int_0^1 \Big(f\big(u(X, t_n + \zeta h)\big) + a^2 \Delta u(X, t_n + \zeta h)\Big)\mathrm{d}\zeta, \\ n = 0, 1, \dots, \qquad X \in \Re \cup \partial\Re. \end{cases} \qquad (1.19)$$

1.3   Towards Classical Runge-Kutta-Nyström Schemes

It is well known that Nyström [13] proposed a direct approach to solving second-order initial value problems (1.1) numerically. To show this point clearly, from the matrix-variation-of-constants formula (1.11) with $M = 0$, we first give the following integral formulae for second-order initial value problems (1.1):

$$\begin{cases} y(x_n + \mu h) = y(x_n) + \mu h y'(x_n) + h^2 \int_0^\mu (\mu - \zeta)\varphi(x_n + h\zeta)\,\mathrm{d}\zeta, \\ y'(x_n + \mu h) = y'(x_n) + h \int_0^\mu \varphi(x_n + h\zeta)\,\mathrm{d}\zeta, \end{cases} \qquad (1.20)$$

for $0 < \mu < 1$, and

$$\begin{cases} y(x_n + h) = y(x_n) + h y'(x_n) + h^2 \int_0^1 (1 - \zeta)\varphi(x_n + h\zeta)\,\mathrm{d}\zeta, \\ y'(x_n + h) = y'(x_n) + h \int_0^1 \varphi(x_n + h\zeta)\,\mathrm{d}\zeta, \end{cases} \qquad (1.21)$$

for $\mu = 1$, where $\varphi(v) := f\big(y(v), y'(v)\big)$.

Formulae (1.20) and (1.21) contain and show clearly the structure of the internal stages and updates of a Runge-Kutta-type integrator for solving (1.1), respectively. This suggests the classical Runge-Kutta-Nyström scheme in a quite simple and natural way in comparison to the original idea (that is, with the block vector $(y^\mathsf{T}, y'^\mathsf{T})^\mathsf{T}$ considered as a new variable, (1.1) can be transformed into a first-order differential equation of doubled dimension. Then apply Runge-Kutta methods to the first-order differential equation, together with some simplifications). In fact, approximating the integrals in (1.20) and (1.21) by using suitable quadrature formulae straightforwardly yields the classical Runge-Kutta-Nyström scheme (see, e.g. [6, 13]) given by the following definition.

Definition 1.1 An s-stage *Runge-Kutta-Nyström (RKN) method* for the initial value problem (1.1) is defined by

$$
\begin{cases}
Y_i = y_n + c_i h y'_n + h^2 \sum_{j=1}^{s} \bar{a}_{ij} f(Y_j, Y'_j), \quad i = 1, \ldots, s, \\[2mm]
Y'_i = y'_n + h \sum_{j=1}^{s} a_{ij} f(Y_j, Y'_j), \quad i = 1, \ldots, s, \\[2mm]
y_{n+1} = y_n + h y'_n + h^2 \sum_{i=1}^{s} \bar{b}_i f(Y_i, Y'_i), \\[2mm]
y'_{n+1} = y'_n + h \sum_{i=1}^{s} b_i f(Y_i, Y'_i),
\end{cases}
\tag{1.22}
$$

or expressed equivalently by the conventional form

$$
\begin{cases}
k_i = f\left(y_n + c_i h y'_n + h^2 \sum_{j=1}^{s} \bar{a}_{ij} k_j, \; y'_n + h \sum_{j=1}^{s} a_{ij} k_j\right), \; i = 1, \ldots, s, \\[2mm]
y_{n+1} = y_n + h y'_n + h^2 \sum_{i=1}^{s} \bar{b}_i k_i, \\[2mm]
y'_{n+1} = y'_n + h \sum_{i=1}^{s} b_i k_i,
\end{cases}
\tag{1.23}
$$

where $\bar{a}_{ij}$, a_{ij}, $\bar{b}_i$, b_i, c_i for $i, j = 1, \ldots, s$ are real constants.

Conventionally, the RKN method (1.22) can be expressed by the following partitioned Butcher tableau:

$$
\begin{array}{c|c|c}
c & \bar{A} & A \\ \hline
& \bar{b}^{\mathsf{T}} & b^{\mathsf{T}}
\end{array}
=
\begin{array}{c|ccc|ccc}
c_1 & \bar{a}_{11} & \cdots & \bar{a}_{1s} & a_{11} & \cdots & a_{1s} \\
\vdots & \vdots & \ddots & \vdots & \vdots & \ddots & \vdots \\
c_s & \bar{a}_{s1} & \cdots & \bar{a}_{ss} & a_{s1} & \cdots & a_{ss} \\ \hline
& \bar{b}_1 & \cdots & \bar{b}_s & b_1 & \cdots & b_s
\end{array}
$$

where $\bar{b} = (\bar{b}_1, \ldots, \bar{b}_s)^{\mathsf{T}}$, $b = (b_1, \ldots, b_s)^{\mathsf{T}}$ and $c = (c_1, \ldots, c_s)^{\mathsf{T}}$ are s-dimensional vectors, and $\bar{A} = (\bar{a}_{ij})$ and $A = (a_{ij})$ are $s \times s$ constant matrices.

1.4 Towards ARKN Schemes and ERKN Integrators

1.4.1 ARKN Schemes

Inheriting the internal stages of the classical RKN methods and approximating the integrals in (1.13) by a suitable quadrature formula to modify the updates of the classical RKN methods gives the ARKN methods for the multi-frequency and multidimensional oscillatory system (1.2).

Definition 1.2 (*Wu et al.* [22]) An s-stage ARKN method for numerical integration of the multi-frequency and multidimensional oscillatory system (1.2) is defined as

$$\left\{ \begin{aligned} & Y_i = y_n + h c_i y_n' + h^2 \sum_{j=1}^{s} \bar{a}_{ij}\big(f(Y_j, Y_j') - MY_j\big), \qquad i = 1, \dots, s, \\ & Y_i' = y_n' + h \sum_{j=1}^{s} a_{ij}\big(f(Y_j, Y_j') - MY_j\big), \qquad i = 1, \dots, s, \\ & y_{n+1} = \phi_0(V) y_n + h\phi_1(V) y_n' + h^2 \sum_{i=1}^{s} \bar{b}_i(V) f(Y_i, Y_i'), \\ & y_{n+1}' = \phi_0(V) y_n' - h M \phi_1(V) y_n + h \sum_{i=1}^{s} b_i(V) f(Y_i, Y_i'), \end{aligned} \right. \tag{1.24}$$

where $\bar{a}_{ij}, a_{ij}, c_i$ for $i, j = 1, \dots, s$ are real constants, and the weight functions $\bar{b}_i(V), b_i(V)$ for $i = 1, \dots, s$ in the updates are matrix-valued functions of $V = h^2 M$.

The ARKN scheme (1.24) can also be denoted by the Butcher tableau

$$\begin{array}{c|c} c & \bar{A} \quad A \\ \hline & \bar{b}^{\mathsf{T}}(V) \;\, b^{\mathsf{T}}(V) \end{array} = \begin{array}{c|ccc|ccc} c_1 & \bar{a}_{11} & \cdots & \bar{a}_{1s} & a_{11} & \cdots & a_{1s} \\ \vdots & \vdots & \ddots & \vdots & \vdots & \ddots & \vdots \\ c_s & \bar{a}_{s1} & \cdots & \bar{a}_{ss} & a_{s1} & \cdots & a_{ss} \\ \hline & \bar{b}_1(V) & \cdots & \bar{b}_s(V) & b_1(V) & \cdots & b_s(V) \end{array}$$

It should be noticed that the internal stages of an ARKN method are exactly the same as the classical RKN methods, but the updates have been revised in light of the matrix-variation-of-constants formula (1.13).

A detailed analysis for a six-stage ARKN method of order five will be presented in Chap. 10 for the case of general oscillatory second-order initial value problems (1.2).

1.4.2 ERKN Integrators

Explicitly, the matrix-variation-of-constants formula (1.6) with $\hat{f}(\zeta) = f\big(y(\zeta)\big)$ can be easily applied to the special oscillatory system (1.3), as in (1.12) and (1.13). Then, approximating the integrals in (1.12) and (1.13) using suitable quadrature formulae leads to the following ERKN integrator for the oscillatory system (1.3).

Definition 1.3 (*Wu et al.* [21]) An s-stage ERKN integrator for the numerical integration of the oscillatory system (1.3) is defined by

$$
\left\{
\begin{aligned}
Y_i &= \phi_0(c_i^2 V)y_n + hc_i\phi_1(c_i^2 V)y_n' + h^2 \sum_{j=1}^{s} \bar{a}_{ij}(V)f(Y_j), \qquad i = 1,\ldots,s, \\
y_{n+1} &= \phi_0(V)y_n + h\phi_1(V)y_n' + h^2 \sum_{i=1}^{s} \bar{b}_i(V)f(Y_i), \\
y_{n+1}' &= -hM\phi_1(V)y_n + \phi_0(V)y_n' + h \sum_{i=1}^{s} b_i(V)f(Y_i),
\end{aligned}
\right.
$$

$$(1.25)$$

where c_i for $i = 1,\ldots,s$ are real constants, $b_i(V)$, $\bar{b}_i(V)$ for $i = 1,\ldots,s$, and $\bar{a}_{ij}(V)$ for $i, j = 1,\ldots,s$ are matrix-valued functions of $V = h^2 M$, which are assumed to have series expansions of real coefficients

$$
\bar{a}_{ij}(V) = \sum_{k=0}^{\infty} \frac{\bar{a}_{ij}^{(2k)}}{(2k)!} V^k, \quad \bar{b}_i(V) = \sum_{k=0}^{\infty} \frac{\bar{b}_i^{(2k)}}{(2k)!} V^k, \quad b_i(V) = \sum_{k=0}^{\infty} \frac{b_i^{(2k)}}{(2k)!} V^k.
$$

The scheme (1.25) can also be denoted by the Butcher tableau as

$$
\begin{array}{c|c}
c & \bar{A}(V) \\
\hline
& \bar{b}^{\mathsf{T}}(V) \\
& b^{\mathsf{T}}(V)
\end{array}
\quad = \quad
\begin{array}{c|ccc}
c_1 & \bar{a}_{11}(V) & \cdots & \bar{a}_{1s}(V) \\
\vdots & \vdots & \ddots & \vdots \\
c_s & \bar{a}_{s1}(V) & \cdots & \bar{a}_{ss}(V) \\
\hline
& \bar{b}_1(V) & \cdots & \bar{b}_s(V) \\
& b_1(V) & \cdots & b_s(V)
\end{array}
$$

Here, it should be noted that both the internal stages and the updates of an ERKN method are revised in light of the matrix-variation-of-constants formula (1.12) and (1.13) with $\hat{f}(\zeta) = f(y(\zeta))$.

An important observation is that when $M = 0$, both the ARKN scheme and the ERKN integrator reduce to the classical RKN method.

An error analysis for explicit ERKN integrators for the multi-frequency oscillatory second-order differential equation (1.3) will be presented in Chap. 7.

With regard to the order conditions for ERKN integrators, a simplified tri-coloured tree theory will be introduced in Chap. 11.

1.5 Towards Two-Step Multidimensional ERKN Methods

We first consider a consequence of (1.11):

$$
\begin{aligned}
y(x_n + \mu h) &= \phi_0(\mu^2 V)y(x_n) + \mu h\phi_1(\mu^2 V)y'(x_n) \\
&\quad + h^2 \int_0^{\mu} (\mu - z)\phi_1\big((\mu - z)^2 V\big)\hat{f}(x_n + hz)\,dz,
\end{aligned}
$$

$$(1.26)$$

where $V = h^2 M$, $\hat{f}(\xi) = f(y(\xi))$, and the matrix-valued functions $\phi_0(V)$ and $\phi_1(V)$ are defined by (1.7).

Theorem 1.2 *The exact solution of the problem* (1.3) *satisfies*

$$
y(x_n + c_i h) = (1 + c_i)\phi_1^{-1}(V)\phi_1\big((1 + c_i)^2 V\big) y(x_n) - c_i\phi_1^{-1}(V)\phi_1(c_i^2 V) y(x_n - h)
$$
$$
+ h^2 \int_0^1 c_i (1 - z)\Big(\phi_1^{-1}(V)\phi_1(c_i^2 V)\phi_1\big((1 - z)^2 V\big)\hat{f}(x_n - hz)
$$
$$
+ c_i\phi_1\big(c_i^2(1 - z)^2 V\big)\hat{f}(x_n + c_i hz)\Big)\, dz,
$$

$$(1.27)$$

and

$$
y(x_n + h) = 2\phi_0(V) y(x_n) - y(x_n - h)
$$
$$
+ h^2 \int_0^1 (1 - z)\phi_1\big((1 - z)^2 V\big)\big(\hat{f}(x_n + hz) + \hat{f}(x_n - hz)\big)\, dz.
$$

$$(1.28)$$

Approximating the integrals in (1.27) and (1.28) with suitable quadrature formulae leads to the following two-step multidimensional extended Runge-Kutta-Nyström method.

Definition 1.4 (*Li et al.* [11]) An s-stage two-step multidimensional extended Runge-Kutta-Nyström (TSERKN) method for the initial value problem (1.3) is defined by

$$
\begin{cases}
Y_i = (1 + c_i)\phi_1^{-1}(V)\phi_1\big((1 + c_i)^2 V\big) y_n - c_i\phi_1^{-1}(V)\phi_1(c_i^2 V) y_{n-1} \\
\quad + h^2 \sum_{j=1}^s a_{ij}(V) f(Y_j), \qquad i = 1, \ldots, s, \\
\\
y_{n+1} = 2\phi_0(V) y_n - y_{n-1} + h^2 \sum_{i=1}^s b_i(V) f(Y_i),
\end{cases}
$$

$$(1.29)$$

where c_i for $i = 1, \ldots, s$ are real constants, and $a_{ij}(V), b_i(V)$ for $i, j = 1, \ldots, s$ are matrix-valued functions of $V = h^2 M$, which are assumed to have power series with real coefficients

$$
a_{ij}(V) = \sum_{k=0}^\infty \frac{a_{ij}^{(2k)}}{(2k)!} V^k, \quad b_i(V) = \sum_{k=0}^\infty \frac{b_i^{(2k)}}{(2k)!} V^k.
$$

Based on the *SEN-tree theory*, order conditions for the TSERKN methods are derived via the B-series defined on the set *SENT* of trees and the B^f-series defined on the subset $SENT^f$ of $SENT$. We refer the reader to Li et al. [11] for details. Furthermore, an analysis of the error bounds for the TSERKN methods will be presented in Chap. 8.

1.6   Towards AAVF Methods for Multi-frequency Oscillatory Hamiltonian Systems

Consider the initial value problem of the system of multi-frequency oscillatory second-order differential equations

$$
\begin{cases}
\ddot{q} + Mq = f(q), & t \in [t_0, t_{\text{end}}], \\
q(t_0) = q_0, & \dot{q}(t_0) = \dot{q}_0,
\end{cases}
\tag{1.30}
$$

where M is a $d \times d$ symmetric positive semi-definite matrix and $f : \mathbb{R}^d \to \mathbb{R}^d$ is continuous. We assume that $f(q) = -\nabla U(q)$ for a real-valued function $U(q)$. Then, (1.30) can be written as the Hamiltonian system

$$
\begin{cases}
\dot{p} = -\nabla_q H(p, q), \\
\dot{q} = \nabla_p H(p, q),
\end{cases}
\tag{1.31}
$$

with the initial values $q(t_0) = q_0$, $p(t_0) = p_0 = \dot{q}_0$ and the Hamiltonian

$$
H(p, q) = \frac{1}{2} p^\mathsf{T} p + \frac{1}{2} q^\mathsf{T} M q + U(q).
\tag{1.32}
$$

As is well known, one of the characteristic properties of a Hamiltonian system is energy conservation.

By the matrix-variation-of-constants formula (1.6), the solution of (1.30) and its derivative satisfy the following equations:

$$
\begin{cases}
q(t) = \phi_0\big((t - t_0)^2 M\big) q_0 + (t - t_0)\phi_1\big((t - t_0)^2 M\big) p_0 \\
\qquad + \displaystyle\int_{t_0}^{t} (t - \zeta)\phi_1\big((t - \zeta)^2 M\big) \hat{f}(\zeta)\,d\zeta, \\
p(t) = -(t - t_0) M \phi_1\big((t - t_0)^2 M\big) q_0 + \phi_0\big((t - t_0)^2 M\big) p_0 \\
\qquad + \displaystyle\int_{t_0}^{t} \phi_0\big((t - \zeta)^2 M\big) \hat{f}(\zeta)\,d\zeta,
\end{cases}
\tag{1.33}
$$

where t_0, t are any real numbers and $\hat{f}(\zeta) = f\big(q(\zeta)\big)$. These equations suggest the following scheme:

$$
\begin{cases}
q_{n+1} = \phi_0(V)q_n + h\phi_1(V)p_n + h^2 I Q_1, \\
p_{n+1} = -h M \phi_1(V)q_n + \phi_0(V)p_n + h I Q_2,
\end{cases}
\tag{1.34}
$$

where h is the stepsize, $V = h^2 M$, and $I Q_1$, $I Q_2$ are determined by the condition of energy preservation at each time step

$$H(p_{n+1}, q_{n+1}) = H(p_n, q_n).$$

The following theorem (see Wang et al.[14]) gives a sufficient condition for the scheme (1.34) to be energy-preserving.

Theorem 1.3 *If*

$$\begin{cases} IQ_1 = \phi_2(V) \int_0^1 f\big((1-\tau)q_n + \tau q_{n+1}\big)\mathrm{d}\tau, \\ IQ_2 = \phi_1(V) \int_0^1 f\big((1-\tau)q_n + \tau q_{n+1}\big)\mathrm{d}\tau, \end{cases} \tag{1.35}$$

then the scheme (1.34) preserves the Hamiltonian (1.32) exactly, i.e.,

$$H(p_{n+1}, q_{n+1}) = H(p_n, q_n), \ n = 0, 1, \dots . \tag{1.36}$$

Definition 1.5 *An adapted average-vector-field (AAVF) method* for the system (1.30) is defined by

$$\begin{cases} q_{n+1} = \phi_0(V)q_n + h\phi_1(V)p_n + h^2\phi_2(V)\int_0^1 f\big((1-\tau)q_n + \tau q_{n+1}\big)\mathrm{d}\tau, \\ p_{n+1} = -hM\phi_1(V)q_n + \phi_0(V)p_n + h\phi_1(V)\int_0^1 f\big((1-\tau)q_n + \tau q_{n+1}\big)\mathrm{d}\tau, \end{cases} \tag{1.37}$$

where h is the stepsize, $\phi_0(V)$, $\phi_1(V)$ are determined by (1.7) and $\phi_2(V) = \sum_{k=0}^\infty \frac{(-1)^k V^k}{(2k+2)!}$.

In Chap. 4, efficient energy-preserving integrators are analysed in detail based on the AVF method, and in Chap. 5, an extended discrete gradient formula is introduced for multi-frequency oscillatory Hamiltonian systems.

1.7   Towards Filon-Type Methods for Multi-frequency Highly Oscillatory Systems

We consider the particular highly oscillatory second-order linear system

$$\begin{cases} \ddot{q}(t) + Mq(t) = g(t), \quad t \in [t_0, t_{\mathrm{end}}], \\ q(t_0) = q_0, \quad \dot{q}(t_0) = \dot{q}_0. \end{cases} \tag{1.38}$$

Based on the matrix-variation-of-constants formula, we consider the following integrators:

$$\begin{cases} q_{n+1} = \phi_0(V)q_n + h\phi_1(V)\dot{q}_n + \displaystyle\int_0^h (h-z)\phi_1\big((h-z)^2 M\big)g(t_n+z)\mathrm{d}z, \\[2mm] \dot{q}_{n+1} = -hM\phi_1(V)q_n + \phi_0(V)\dot{q}_n + \displaystyle\int_0^h \phi_0\big((h-z)^2 M\big)g(t_n+z)\mathrm{d}z. \end{cases} \tag{1.39}$$

Below, we devote ourselves to deriving efficient methods for computing the following two highly oscillatory integrals:

$$\begin{cases} \tilde{I}_1 := \displaystyle\int_0^h (h-z)\phi_1\big((h-z)^2 M\big)g(t_n+z)\mathrm{d}z, \\[2mm] \tilde{I}_2 := \displaystyle\int_0^h \phi_0\big((h-z)^2 M\big)g(t_n+z)\mathrm{d}z. \end{cases} \tag{1.40}$$

The Filon-type method for highly oscillatory integrals was first introduced in [9]. It is an efficient method for dealing with highly oscillatory systems. Here we apply Filon-type quadratures to the two integrals in (1.40). We interpolate the vector-valued function g by a vector-valued polynomial p

$$p(t) = \sum_{l=1}^{v}\sum_{j=0}^{\theta_l-1} \alpha_{l,j}(t)g^{(j)}(t_n+c_l h), \tag{1.41}$$

such that

$$p^{(j)}(t_n+c_l h) = g^{(j)}(t_n+c_l h), \tag{1.42}$$

for $l = 1, 2, \ldots, v$, and $j = 0, 1, \ldots, \theta_l-1$ for each fixed l. Here $c_1, \ldots, c_v$ can be chosen as any value which satisfies $0 = c_1 < c_2 < \cdots < c_v = 1$. $\theta_1, \ldots, \theta_v$ can be any positive integer and the superscript (j) denotes the jth-derivative with respect to t.

Replacing g in (1.40) by p yields a Filon-type method $\tilde{I}_1^F$, $\tilde{I}_2^F$ for $\tilde{I}_1$, $\tilde{I}_2$:

$$\begin{cases} \tilde{I}_1^F := \displaystyle\sum_{l=1}^{v}\sum_{j=0}^{\theta_l-1} I_1[\alpha_{l,j}](t_n)g^{(j)}(t_n+c_l h), \\[2mm] \tilde{I}_2^F := \displaystyle\sum_{l=1}^{v}\sum_{j=0}^{\theta_l-1} I_2[\alpha_{l,j}](t_n)g^{(j)}(t_n+c_l h), \end{cases} \tag{1.43}$$

where $I_1[\alpha_{l,j}](t_n)$ and $I_2[\alpha_{l,j}](t_n)$ are defined by

$$\begin{cases} I_1[\alpha_{l,j}](t_n) := \displaystyle\int_0^h (h-z)\phi_1\big((h-z)^2 M\big)\alpha_{l,j}(t_n+z)\mathrm{d}z, \\[2mm] I_2[\alpha_{l,j}](t_n) := \displaystyle\int_0^h \phi_0\big((h-z)^2 M\big)\alpha_{l,j}(t_n+z)\mathrm{d}z. \end{cases} \tag{1.44}$$

On the basis of the above analysis and formula (1.39), we present the Filon-type method for the oscillatory linear system (1.38).

Definition 1.6 A Filon-type method for integrating the oscillatory linear system (1.38) is defined as

$$
\begin{cases}
q_{n+1} = \phi_0(V)q_n + h\phi_1(V)\dot{q}_n + \sum_{l=1}^{v}\sum_{j=0}^{\theta_l-1} I_1[\alpha_{l,j}](t_n)g^{(j)}(t_n + c_l h), \\[4mm]
\dot{q}_{n+1} = -hM\phi_1(V)q_n + \phi_0(V)\dot{q}_n + \sum_{l=1}^{v}\sum_{j=0}^{\theta_l-1} I_2[\alpha_{l,j}](t_n)g^{(j)}(t_n + c_l h),
\end{cases}
$$

(1.45)

where h is the stepsize, and I_1, I_2 are defined by (1.44).

An elementary analysis of Filon-type methods for multi-frequency highly oscillatory nonlinear systems is presented by Wang et al. [15]. Further discussions can be found in [16] and will be described in Chap. 3.

1.8 Towards ERKN Methods for General Second-Order Oscillatory Systems

This section turns to the effective integration of the multi-frequency oscillatory second-order initial value problem (1.2).

In order to obtain new RKN-type methods for (1.2), we approximate the integrals in (1.12) and (1.13) with some quadrature formulae. This leads to the following definition.

Definition 1.7 An s-stage extended Runge-Kutta-Nyström method for the numerical integration of the general IVP (1.2) is defined by the following scheme:

$$
\begin{cases}
Y_i = \phi_0(c_i^2 V)y_n + c_i\phi_1(c_i^2 V)hy_n' + h^2\sum_{j=1}^{s}\bar{a}_{ij}(V)f(Y_j, Y_j'), & i = 1, \ldots, s, \\[4mm]
Y_i' = -c_i hM\phi_1(c_i^2 V)y_n + \phi_0(c_i^2 V)y_n' + h\sum_{j=1}^{s}a_{ij}(V)f(Y_j, Y_j'), & i = 1, \ldots, s, \\[4mm]
y_{n+1} = \phi_0(V)y_n + \phi_1(V)hy_n' + h^2\sum_{i=1}^{s}\bar{b}_i(V)f(Y_i, Y_i'), \\[4mm]
y_{n+1}' = -hM\phi_1(V)y_n + \phi_0(V)y_n' + h\sum_{i=1}^{s}b_i(V)f(Y_i, Y_i'),
\end{cases}
$$

(1.46)

where $\bar{a}_{ij}(V), a_{ij}(V), \bar{b}_i(V), b_i(V), i, j = 1, \ldots, s$ are matrix-valued functions which can be expanded in a power series of $V = h^2 M$ with real coefficients, h

is the stepsize, and y_n and y'_n are approximations to the values of $y(x)$ and $y'(x)$ at $x_n = x_0 + nh$, respectively, for $n = 1, 2, \ldots$.

This method can also be represented compactly in Butcher's tableau of coefficients:

$$
\begin{array}{c|c}
c & \bar{A}(V) \mid A(V) \\
\hline
& \bar{b}^{\mathsf{T}}(V) \mid b^{\mathsf{T}}(V)
\end{array}
=
\begin{array}{c|ccc|ccc}
c_1 & \bar{a}_{11}(V) & \cdots & \bar{a}_{1s}(V) & a_{11}(V) & \cdots & a_{1s}(V) \\
\vdots & \vdots & \vdots & \vdots & \vdots & \vdots & \vdots \\
c_s & \bar{a}_{s1}(V) & \cdots & \bar{a}_{ss}(V) & a_{s1}(V) & \cdots & a_{ss}(V) \\
\hline
& \bar{b}_1(V) & \cdots & \bar{b}_s(V) & b_1(V) & \cdots & b_s(V)
\end{array}
$$

We note that for the non-autonomous system $y'' + My = f(x, y, y')$, the ERKN method (1.46) has the form

$$
\begin{cases}
Y_i = \phi_0(c_i^2 V)y_n + c_i\phi_1(c_i^2 V)hy'_n + h^2 \displaystyle\sum_{j=1}^{s} \bar{a}_{ij}(V)f(x_n + c_j h, Y_j, Y'_j), & i = 1, \ldots, s, \\[2mm]
Y'_i = -c_i h M \phi_1(c_i^2 V)y_n + \phi_0(c_i^2 V)y'_n + h \displaystyle\sum_{j=1}^{s} a_{ij}(V)f(x_n + c_j h, Y_j, Y'_j), & i = 1, \ldots, s, \\[2mm]
y_{n+1} = \phi_0(V)y_n + \phi_1(V)hy'_n + h^2 \displaystyle\sum_{i=1}^{s} \bar{b}_i(V)f(x_n + c_i h, Y_i, Y'_i), \\[2mm]
y'_{n+1} = -hM\phi_1(V)y_n + \phi_0(V)y'_n + h \displaystyle\sum_{i=1}^{s} b_i(V)f(x_n + c_i h, Y_i, Y'_i).
\end{cases}
\tag{1.47}
$$

A detailed analysis of the ERKN method for the general second-order oscillatory systems (1.2) can be found in [24].

1.9 Towards High-Order Explicit Schemes for Hamiltonian Nonlinear Wave Equations

Hamiltonian nonlinear wave equations arise in various fields of science and engineering. There is a lot of current interest in high-order explicit schemes with the conservation law of semi-discrete energy for Hamiltonian partial differential equations.

Let us consider Hamiltonian nonlinear wave equations of the form

$$
\frac{\partial^2 u(x, t)}{\partial t^2} - a^2 \frac{\partial^2 u(x, t)}{\partial x^2} = f(u), \qquad (x, t) \in [x_L, x_R] \times [t_0, T], \tag{1.48}
$$

where $u(x, t)$ represents the wave displacement at position x and time t, and $f(u) = -\frac{dG(u(x,t))}{du}$ is a nonlinear function of u, the negative derivative of a potential energy $G(u)$ with respect to u. It is assumed that the Hamiltonian nonlinear wave equation is subject to the following initial conditions:

$$
\begin{cases}
u(x, t_0) = \varphi_1(x), & x \in [x_L, x_R], \\
\dfrac{\partial u}{\partial t}(x, t_0) = \varphi_2(x), & x \in [x_L, x_R],
\end{cases}
\tag{1.49}
$$

and Neumann boundary conditions

$$
\frac{\partial u}{\partial x}(x_L, t) = \frac{\partial u}{\partial x}(x_R, t) = 0, \qquad t \geq t_0.
\tag{1.50}
$$

With the method of lines, the given Hamiltonian nonlinear wave equation can be reduced to a system of second-order ordinary differential equations in time, then ERKN methods can be applied to the semi-discrete system in time. For example, a fourth-order finite difference scheme for discretizing the spatial derivative and a fourth-order multidimensional ERKN integrator for time integration lead to an efficient high-order explicit scheme for solving the Hamiltonian nonlinear wave equation (1.48) with the initial conditions (1.49) and the Neumann boundary conditions (1.50). The conservation law of the semi-discrete energy as well as the convergence and stability of the semi-discrete system can be shown (see [12]).

1.10 Conclusions and Discussions

The matrix-variation-of-constants formula (1.6) for the multi-frequency oscillatory system (1.2) is the lighthouse which guides the authors to a new perspective for RKN-type schemes. Essentially, the matrix-variation-of-constants formula (1.6) is also the fundamental approach to a true understanding of the novel structure-preserving integrators in this extended volume and the previous volume [23] of our monograph. For example, when this formula is combined with the ideas of collocation methods and the local Fourier expansion of the system (1.3), a type of trigonometric Fourier collocation method for (1.3) can be devised which will be presented in Chap. 6 (see, Wang et al. [18]). The formula (1.6) is also important for the error analysis of this kind of improved RKN-type integrators when they are applied to the multi-frequency and multidimensional oscillatory second-order equation (1.2). Readers are referred to Wang et al. [17].

Last but not least, it should be stressed that these integrators which are based on the matrix-variation-of-constants formula (1.6) are especially important when M has large positive eigenvalues such as in the case of discretization of the spatial derivative in wave equations. Another important advantage is that the matrix-variation-of-constants formula (1.6) can be extended to the initial-value problem of the general higher-dimensional nonlinear wave equation;

$$\begin{cases} U_{tt}(X,t) - a^2 \Delta U(X,t) = f\big(U(X,t), U_t(X,t)\big), & X \in \Omega, \ t_0 < t \le T, \\ U(X,t_0) = U_0(X), \\ U_t(X,t_0) = U_1(X), \end{cases}$$

$$(1.51)$$

with the Robin boundary condition:

$$\nabla U \times \mathbf{n} + \lambda U = \beta(X,t), \qquad X \in \partial\Omega, \tag{1.52}$$

where $f(\cdot, \cdot)$ is a function of U and U_t, and $U : \mathbb{R}^d \times \mathbb{R} \to \mathbb{R}$ with $d \ge 1$ representing the wave displacement at position $X \in \mathbb{R}^d$ and time t, $\mathbf{n}$ is the unit outward normal vector at the boundary $\partial\Omega$, and λ is a constant. Robin boundary conditions are a weighted combination of Dirichlet boundary conditions and Neumann boundary conditions. Robin boundary conditions are also called impedance boundary conditions from their application in electromagnetic problems, or convective boundary conditions from their application in heat transfer problems.

In order to model the Robin boundary conditions, we restrict ourselves to the case where Δ is defined on the domain

$$D(\Delta) = \big\{ v \in H^2(\Omega) \mid \nabla v \cdot \mathbf{n} + \lambda v = \beta(x,t), \ x \in \partial\Omega \big\}.$$

In this case, the series

$$\sum_{k=0}^{\infty} \frac{\Delta^k}{(2k+j)!}$$

is unconditionally convergent under the Sobolev norm $\| \cdot \|_{L^2(\Omega) \leftarrow L^2(\Omega)}$ (see, e.g., [8]), and which is denoted by $\phi_j(\Delta)$, namely

$$\phi_j(\Delta) := \sum_{k=0}^{\infty} \frac{\Delta^k}{(2k+j)!}, \qquad j = 0, 1, \dots. \tag{1.53}$$

We then have the following operator-variation-of-constants formula for the initial-value problem of the general higher-dimensional nonlinear wave equation (1.51).

Theorem 1.4 *Let $U_0(X)$, $U_1(X) \in C^\infty(\bar{\Omega})$. If Δ is a Laplacian defined on the domain $D(\Delta)$, and $f(U, U_t)$ in (1.51) is continuous, then the exact solution of (1.51) and its derivative satisfy*

$$\begin{cases} U(X,t) = \phi_0\big((t-t_0)^2 a^2 \Delta\big) U(X,t_0) + (t-t_0)\phi_1\big((t-t_0)^2 a^2 \Delta\big) U_t(X,t_0) \\ \qquad + \displaystyle\int_{t_0}^{t} (t-\varsigma)\phi_1\big((t-\varsigma)^2 a^2 \Delta\big) \tilde{f}(\varsigma)\mathrm{d}\varsigma, \\ U'(X,t) = (t-t_0)a^2 \Delta\phi_1\big((t-t_0)^2 a^2 \Delta\big) U(X,t_0) + \phi_0\big((t-t_0)^2 a^2 \Delta\big) U_t(X,t_0) \\ \qquad + \displaystyle\int_{t_0}^{t} \phi_0\big((t-\varsigma)^2 a^2 \Delta\big) \tilde{f}(\varsigma)\mathrm{d}\varsigma \end{cases}$$

$$(1.54)$$

for $t_0, t \in (-\infty, +\infty)$, where

$$\tilde{f}(\varsigma) = f\big(U(X, \varsigma), U_t(X, \varsigma)\big),$$

and the Laplacian-valued functions ϕ_0 and ϕ_1 are defined by (1.53).

Moreover, the operator-variation-of-constants formula (1.54) is completely consistent with the Robin boundary conditions (1.52) under suitable assumptions. Especially, if $f(U, U_t) = 0$, then (1.51) becomes the homogeneous linear wave equation:

$$\begin{cases} U_{tt} - a^2 \Delta U = 0, \\ U(X, t_0) = U_0(X), \\ U_t(X, t_0) = U_1(X), \end{cases} \tag{1.55}$$

and accordingly, (1.54) reduces to

$$\begin{cases} U(X, t) = \phi_0\big((t - t_0)^2 a^2 \Delta\big) U_0(X) + (t - t_0)\phi_1\big((t - t_0)^2 a^2 \Delta\big) U_1(X), \\ U'(X, t) = (t - t_0)a^2 \Delta\phi_1\big((t - t_0)^2 a^2 \Delta\big) U_0(X) + \phi_0\big((t - t_0)^2 a^2 \Delta\big) U_1(X), \end{cases} \tag{1.56}$$

which exactly integrates the homogeneous linear wave equation (1.55). This means that (1.56) gives a closed-form solution to the higher-dimensional homogeneous wave equation (1.55). In comparison with the well-known D'Alembert, Poisson and Kirchhoff formula, formula (1.56) is independent of the computation of integrals and presents a closed-form solution of (1.55). The details can be found in Wu et al. [20].

This chapter is based on the work of Wu [19].

References

1. Al-Mohy AH, Higham NJ, Relton SD (2015) New algorithms for computing the matrix sine and cosine separately or simultaneously. SIAM J Sci Comput 37:A456–A487
2. Deuflhard P (1979) A study of extrapolation methods based on multistep schemes without parasitic solutions. Z Angew Math Phys 30:177–189
3. García-Archilla B, Sanz-Serna JM, Skeel RD (1998) Long-time-step methods for oscillatory differential equations. SIAM J Sci Comput 20:930–963
4. Hairer E, Lubich C (2000) Long-time energy conservation of numerical methods for oscillatory differential equations. SIAM J Numer Anal 38:414–441
5. Hairer E, Lubich C, Wanner G (2006) Geometric numerical integration: structure-preserving algorithms for ordinary differential equations, 2nd edn. Springer, Berlin
6. Hairer E, Nørsett SP, Wanner G (1993) Solving ordinary differential equations I: nonstiff problems. Springer, Berlin
7. Hochbruck M, Lubich C (1999) A Gautschi-type method for oscillatory second-order differential equations. Numer Math 83:403–426
8. Hochbruck M, Ostermann A (2010) Exponential integrators. Acta Numer 19:209–286
9. Iserles A, Nørsett SP (2005) Efficient quadrature of highly oscillatory integrals using derivatives. Proc R Soc Lond, Ser A, Math Phys Eng Sci 461:1383–1399

10. Jordn DW, Smith P (2007) Nonlinear ordinary differential equations. An introduction for scientists and engineers, 3rd edn. Oxford University Press, Oxford
11. Li J, Wang B, You X, Wu X (2011) Two-step extended RKN methods for oscillatory systems. Comput Phys Commun 182:2486–2507
12. Liu C, Shi W, Wu X (2014) An efficient high-order explicit scheme for solving Hamiltonian nonlinear wave equations. Appl Math Comput 246:696–710
13. Nyström EJ (1925) Ueber die numerische Integration von Differentialgleichungen. Acta Soc Sci Fenn 50:1–54
14. Wang B, Wu X (2013) A new high precision energy-preserving integrator for system of oscillatory second-order differential equations. Phys Lett A 376:1185–1190
15. Wang B, Wu X, Liu K (2013) A Filon-type asymptotic approach to solving highly oscillatory second-order initial value problems. J Comput Phys 243:210–223
16. Wang B, Wu X (2014) Improved Filon-type asymptotic methods for highly oscillatory differential equations with multiple time scales. J Comput Phys 276:62–73
17. Wang B, Wu X, Xia J (2013) Error bounds for explicit ERKN integrators for systems of multi-frequency oscillatory second-order differential equations. Appl Numer Math 74:17–34
18. Wang B, Iserles A, Wu X (2014) Arbitrary order trigonometric Fourier collocation methods for multi-frequency oscillatory systems. Found Comput Math. doi:10.1007/s10208-014-9241-9
19. Wu X (2014) Matrix-variation-of-constants formula with applications. A seminar report of Nanjing University (preprint)
20. Wu X, Mei L, Liu C (2015) An analytical expression of solutions to nonlinear wave equations in higher dimensions with Robin boundary conditions. J Math Anal Appl 426:1164–1173
21. Wu X, You X, Shi W, Wang B (2010) ERKN integrators for systems of oscillatory second-order differential equations. Comput Phys Commun 181:1873–1887
22. Wu X, You X, Xia J (2009) Order conditions for ARKN methods solving oscillatory systems. Comput Phys Commun 180:2250–2257
23. Wu X, You X, Wang B (2013) Structure-preserving integrators for oscillatory ordinary differential equations. Springer, Heidelberg (jointly published with Science Press Beijing)
24. You X, Zhao J, Yang H, Fang Y, Wu X (2014) Order conditions for RKN methods solving general second-order oscillatory systems. Numer Algo 66:147–176

Chapter 2
Improved Störmer–Verlet Formulae with Applications

The Störmer–Verlet formula is a popular numerical integration method which has played an important role in the numerical simulation of differential equations. In this chapter, we analyse two improved multi-frequency Störmer–Verlet formulae with four applications, including time-independent Schrödinger equations, wave equations, orbital problems and the problem of Fermi, Pasta and Ulam. Stability and phase properties of the two improved Störmer–Verlet formulae are analysed. In order to derive the first improved multi-frequency Störmer–Verlet formula, the symplectic conditions for the one-stage explicit multi-frequency ARKN method are investigated in detail. Moreover, the coupled conditions for explicit symplectic and symmetric multi-frequency ERKN integrators are presented.

2.1 Motivation

A good numerical integrator should meet different requirements of the governing differential equation describing physical phenomena of the universe. For a differential equation with a particular structure, it is natural to require numerical algorithms to adapt to the structure of the problem and to preserve as much as possible the intrinsic properties of the true solution to the problem. A good theoretical foundation of structure-preserving algorithms for ordinary differential equations can be found in Feng et al. [8], Hairer et al. [19] and references contained therein. The time-independent Schrödinger equation is frequently encountered and is one of the basic equations of quantum mechanics. In fact, solutions of the time-independent Schrödinger equation are required in the study of atomic and molecular structure, molecular dynamics and quantum chemistry. Many numerical methods have been proposed to solve this type of Schrödinger equation. Readers are referred to [38, 48, 50] for example. In applied science and engineering, the wave equation is an important second-order partial differential equation for the description of waves. Examples of waves in physics are sound waves, light waves and water waves. It also

© Springer-Verlag Berlin Heidelberg and Science Press, Beijing, China 2015
X. Wu et al., *Structure-Preserving Algorithms for Oscillatory Differential Equations II*, DOI 10.1007/978-3-662-48156-1_2

arises in fields like electromagnetics and fluid dynamics. The numerical treatment of
wave equations is fundamental for understanding non-linear phenomena. Besides,
orbital problems also constitute a very important category of differential equations in
scientific computing. For an orbital problem that arises in the analysis of the motion
of spacecraft, asteroids, comets, and natural or man-made satellites, it is important
to maintain the accuracy of a numerical integration to a high degree of accuracy.
In recent years, numerical studies of non-linear effects in physical systems have
received much attention. The Fermi–Pasta–Ulam problem [9] is an important model
for simulating the physics of non-linear phenomena, which reveals highly unexpected
dynamical behaviour. All of the problems described above can be expressed using
the following multi-frequency oscillatory second-order initial value problem:

$$\begin{cases} y'' + My = f(t, y), & t \in [t_0, t_{end}], \\ y(t_0) = y_0, \quad y'(t_0) = y_0', \end{cases} \tag{2.1}$$

where $M \in \mathbb{R}^{d \times d}$ and $f : \mathbb{R} \times \mathbb{R}^d \to \mathbb{R}^d$, $y_0 \in \mathbb{R}^d$, $y_0' \in \mathbb{R}^d$. Problems in the form
(2.1) also arise in mechanics, theoretical physics, quantum dynamics, molecular
biology, etc. In fact, by applying the shooting method to the one-dimensional time-
independent Schrödinger equation, the boundary value problem can be converted
into an initial value problem of the form (2.1). The spatial semi-discretization of a
wave equation with the method of lines is an important source for (2.1). Furthermore,
some orbital problems and the Fermi–Pasta–Ulam problem can also be expressed by
(2.1).

If M is a symmetric and positive semi-definite matrix and $f(t, y) = -\nabla U(y)$,
then with the new variables $q = y$ and $p = y'$ the system (2.1) is simply the following
multi-frequency and multidimensional oscillatory Hamiltonian system

$$\begin{cases} p'(t) = -\nabla_q H(p, q), & p(t_0) = p_0 = y_0', \\ q'(t) = \nabla_p H(p, q), & q(t_0) = q_0 = y_0, \end{cases} \tag{2.2}$$

with the Hamiltonian

$$H(p, q) = \frac{1}{2} p^\mathsf{T} p + \frac{1}{2} q^\mathsf{T} Mq + U(q),$$

(see Cohen et al. [5], for example). In essence, a large number of mechanical systems
with a partitioned Hamiltonian function fit this pattern. Some useful approaches to
constructing Runge-Kutta-Nyström (RKN) type methods for the system (2.1) have
been proposed (see, e.g. [11–14, 16, 23, 40, 51]). Meanwhile, symplectic methods
for the Hamiltonian system (2.2) have been developed, and readers are referred to
[3, 7, 20, 34, 35, 37, 43, 44, 59] for some examples on this topic. Wu et al. [63]
presented the general multi-frequency and multidimensional ARKN methods (RKN
methods adapted to the system (2.1)) and derived the corresponding order conditions

based on the B-series theory. Some concrete multi-frequency and multidimensional ARKN methods are obtained in [56]. Furthermore, Wu et al. [61] formulated a standard form of the multi-frequency and multidimensional ERKN methods (extended RKN methods) for the oscillatory system (2.1) and derived the corresponding order conditions using B-series theory based on the set of ERKN trees (tri-coloured trees). Following this research, two-step ERKN methods and energy-preserving integrators are studied for the oscillatory system (2.1). Readers are referred to [28, 29, 53, 57, 62].

On the other hand, the Störmer–Verlet scheme is a classical technique for the system of second-order differential equations

$$\begin{cases} y''(t) = g\big(t, y(t)\big), & t \in [t_0, t_{end}], \\ y(t_0) = y_0, & y'(t_0) = y'_0. \end{cases} \qquad (2.3)$$

Störmer [42] used higher-order variants for numerical computations of the motion of ionized particles [42], and Verlet (1967) [49] proposed this method for computing problems in molecular dynamics. This method became known as the Störmer–Verlet method and readers are referred to [17] for a survey of this method. The Störmer–Verlet method has become by far the most widely used numerical scheme in this respect. Further examples and references can be found in [18, 30, 46] and references contained therein. Applying the Störmer–Verlet formula to (2.3) gives

$$\begin{cases} Y_1 = y_n + \dfrac{h}{2}y'_n, \\ y_{n+1} = y_n + hy'_n + \dfrac{h^2}{2}g(t_n + \dfrac{1}{2}h, Y_1), \\ y'_{n+1} = y'_n + hg(t_n + \dfrac{1}{2}h, Y_1). \end{cases} \qquad (2.4)$$

The Störmer–Verlet formula (2.4) could also be applied directly to the system (2.1) providing it is rewritten in the form

$$y''(t) = f(t, y(t)) - My(t) \triangleq g(t, y(t)).$$

However, this form does not take account of the specific structure of the oscillatory system (2.1) generated by the linear term My. Both the multi-frequency ARKN scheme and ERKN scheme are formulated from the formula of the integral equations (see Theorem 1.1 in Chap. 1) adapted to the system (2.1). They are expected to have better numerical behaviour than the classical Störmer–Verlet formula. The key point here is that each new multi-frequency and multidimensional Störmer–Verlet formula utilizes a combination of existing trigonometric integrators and symplectic schemes.

2.2 Two Improved Störmer–Verlet Formulae

Two improved multi-frequency and multidimensional Störmer–Verlet formulae for the oscillatory system (2.1) are presented below.

2.2.1 Improved Störmer–Verlet Formula 1

The first improved Störmer–Verlet formula is based on the multi-frequency and multidimensional ARKN schemes and the corresponding symplectic conditions. Taking advantage of the specific structure of (2.1) introduced by the linear term My and revising the updates of RKN methods, we obtain s-stage multi-frequency and multidimensional ARKN methods for (2.1) (see [63])

$$\begin{cases} Y_i = y_n + c_i h y_n' + h^2 \sum_{j=1}^{s} \bar{a}_{ij} \big(f(t_n + c_j h, Y_j) - M Y_j \big), \quad i = 1, 2, \ldots, s, \\[2ex] y_{n+1} = \phi_0(V) y_n + \phi_1(V) h y_n' + h^2 \sum_{i=1}^{s} \bar{b}_i(V) f(t_n + c_i h, Y_i), \\[2ex] h y_{n+1}' = -V \phi_1(V) y_n + \phi_0(V) h y_n' + h^2 \sum_{i=1}^{s} b_i(V) f(t_n + c_i h, Y_i), \end{cases}$$
$$(2.5)$$

where h is the stepsize, $\bar{a}_{ij}$ for $i, j = 1, 2, \ldots, s$ are real constants, $b_i(V)$ and $\bar{b}_i(V)$ for $i = 1, 2, \ldots, s$ are matrix-valued functions of $V = h^2 M$, $\phi_0(V)$ and $\phi_1(V)$ are given by (1.7). In order to derive an improved Störmer–Verlet formula for (2.1), the one-stage explicit multi-frequency and multidimensional ARKN method of the form below is considered:

$$\begin{cases} Y_1 \;\; = y_n + c_1 h y_n', \\ y_{n+1} = \phi_0(V) y_n + \phi_1(V) h y_n' + h^2 \bar{b}_1(V) f(t_n + c_1 h, Y_1), \\ h y_{n+1}' = -V \phi_1(V) y_n + \phi_0(V) h y_n' + h^2 b_1(V) f(t_n + c_1 h, Y_1). \end{cases} \quad (2.6)$$

The symplectic conditions for the one-stage explicit multi-frequency and multidimensional ARKN method (2.6) are examined below.

Theorem 2.1 *Suppose that M is symmetric and positive semi-definite and $f(y) = -\nabla U(y)$ is the negative gradient of the function $U(y)$ with continuous second derivatives with respect to y. If the coefficients of a multi-frequency and multidimensional ARKN method (2.6) satisfy*

$$b_1(V) \phi_0(V) + \bar{b}_1(V) V \phi_1(V) = d_1 I, \; d_1 \in \mathbb{R}, \quad (2.7)$$

$$\bar{b}_1(V)(\phi_0(V) + c_1 V \phi_1(V)) = b_1(V)(\phi_1(V) - c_1 \phi_0(V)), \quad (2.8)$$

where I is the identity matrix, then the method is symplectic.

Proof Following the approach used in [37], we will adopt exterior forms.

First consider the special case where M is a diagonal matrix with nonnegative entries: $M = \text{diag}(m_{11}, m_{22}, \ldots, m_{dd})$. Accordingly, $\phi_0(V)$, $\phi_1(V)$, $b_1(V)$ and $\bar{b}_1(V)$ are all diagonal matrices. Denote $f_1 = f(Y_1)$. Then the ARKN scheme (2.6) becomes

$$
\begin{cases}
Y_1^J = y_n^J + c_1 h y_n'^J, \\
y_{n+1}^J = \phi_0(h^2 m_{JJ}) y_n^J + \phi_1(h^2 m_{JJ}) h y_n'^J + h^2 \bar{b}_1(h^2 m_{JJ}) f_1^J, \\
y_{n+1}'^J = -h m_{JJ} \phi_1(h^2 m_{JJ}) y_n^J + \phi_0(h^2 m_{JJ}) y_n'^J + h b_1(h^2 m_{JJ}) f_1^J,
\end{cases}
\tag{2.9}
$$

where the superscript J ($J = 1, 2, \ldots, d$) denotes the Jth component of a vector. In terms of the above notations and the Hamiltonian system (2.2), symplecticity of the method (2.6) is identical to

$$
\sum_{J=1}^d dy_{n+1}^J \wedge dy_{n+1}'^J = \sum_{J=1}^d dy_n^J \wedge dy_n'^J.
$$

To show this equality, it is required to compute

$$
\begin{aligned}
dy_{n+1}^J \wedge dy_{n+1}'^J = & [\phi_0^2(h^2 m_{JJ}) + h^2 m_{JJ} \phi_1^2(h^2 m_{JJ})] dy_n^J \wedge dy_n'^J \\
& + h[b_1(h^2 m_{JJ}) \phi_0(h^2 m_{JJ}) + \bar{b}_1(h^2 m_{JJ}) h^2 m_{JJ} \phi_1(h^2 m_{JJ})] dy_n^J \wedge df_1^J \\
& + h^2[b_1(h^2 m_{JJ}) \phi_1(h^2 m_{JJ}) - \bar{b}_1(h^2 m_{JJ}) \phi_0(h^2 m_{JJ})] dy_n'^J \wedge df_1^J.
\end{aligned}
$$

It follows from the definition (1.7) that

$$
\begin{aligned}
\phi_0^2(h^2 m_{JJ}) + h^2 m_{JJ} \phi_1^2(h^2 m_{JJ}) &= \cos^2(h\sqrt{m_{JJ}}) + h^2 m_{JJ} \left(\frac{\sin(h\sqrt{m_{JJ}})}{h\sqrt{m_{JJ}}} \right)^2 \\
&= \cos^2(h\sqrt{m_{JJ}}) + \sin^2(h\sqrt{m_{JJ}}) = 1.
\end{aligned}
$$

Thus,

$$
\begin{aligned}
dy_{n+1}^J \wedge dy_{n+1}'^J = & dy_n^J \wedge dy_n'^J \\
& + h[b_1(h^2 m_{JJ}) \phi_0(h^2 m_{JJ}) + \bar{b}_1(h^2 m_{JJ}) h^2 m_{JJ} \phi_1(h^2 m_{JJ})] dy_n^J \wedge df_1^J \\
& + h^2[b_1(h^2 m_{JJ}) \phi_1(h^2 m_{JJ}) - \bar{b}_1(h^2 m_{JJ}) \phi_0(h^2 m_{JJ})] dy_n'^J \wedge df_1^J.
\end{aligned}
\tag{2.10}
$$

Differentiating the first formula of (2.9) yields

$$
dy_n^J = dY_1^J - h c_1 dy_n'^J.
$$

Then, one obtains

$$\mathrm{d}y_n^J \wedge \mathrm{d}f_1^J = \mathrm{d}Y_1^J \wedge \mathrm{d}f_1^J - hc_1\mathrm{d}y_n'^J \wedge \mathrm{d}f_1^J.$$

Inserting this formula into (2.10) gives

$$\begin{aligned}
\mathrm{d}y_{n+1}^J \wedge \mathrm{d}y_{n+1}'^J = {} & \mathrm{d}y_n^J \wedge \mathrm{d}y_n'^J \\
& + h[b_1(h^2m_{JJ})\phi_0(h^2m_{JJ}) + \bar{b}_1(h^2m_{JJ})h^2m_{JJ}\phi_1(h^2m_{JJ})]\mathrm{d}Y_1^J \wedge \mathrm{d}f_1^J \\
& + h^2[b_1(h^2m_{JJ})\phi_1(h^2m_{JJ}) - \bar{b}_1(h^2m_{JJ})\phi_0(h^2m_{JJ}) \\
& - c_1b_1(h^2m_{JJ})\phi_0(h^2m_{JJ}) - c_1\bar{b}_1(h^2m_{JJ})h^2m_{JJ}\phi_1(h^2m_{JJ})]\mathrm{d}y_n'^J \wedge \mathrm{d}f_1^J.
\end{aligned}$$

$$(2.11)$$

Summing over all J yields

$$\begin{aligned}
\sum_{J=1}^{d} \mathrm{d}y_{n+1}^J \wedge \mathrm{d}y_{n+1}'^J = {} & \sum_{J=1}^{d} \mathrm{d}y_n^J \wedge \mathrm{d}y_n'^J \\
& + h \sum_{J=1}^{d}[b_1(h^2m_{JJ})\phi_0(h^2m_{JJ}) + \bar{b}_1(h^2m_{JJ})h^2m_{JJ}\phi_1(h^2m_{JJ})]\mathrm{d}Y_1^J \wedge \mathrm{d}f_1^J \\
& + h^2 \sum_{J=1}^{d}[b_1(h^2m_{JJ})\phi_1(h^2m_{JJ}) - \bar{b}_1(h^2m_{JJ})\phi_0(h^2m_{JJ}) \\
& - c_1b_1(h^2m_{JJ})\phi_0(h^2m_{JJ}) - c_1\bar{b}_1(h^2m_{JJ})h^2m_{JJ}\phi_1(h^2m_{JJ})]\mathrm{d}y_n'^J \wedge \mathrm{d}f_1^J.
\end{aligned}$$

$$(2.12)$$

By (2.7) and $f(y) = -\nabla_y U(y)$, where U has continuous second derivatives with respect to y, one obtains

$$\begin{aligned}
& \sum_{J=1}^{d}[b_1(h^2m_{JJ})\phi_0(h^2m_{JJ}) + \bar{b}_1(h^2m_{JJ})h^2m_{JJ}\phi_1(h^2m_{JJ})]\mathrm{d}Y_1^J \wedge \mathrm{d}f_1^J \\
& = \sum_{J=1}^{d} d_1\mathrm{d}Y_1^J \wedge \mathrm{d}f_1^J = -d_1 \sum_{J,I=1}^{d} (\frac{\partial f^J}{\partial y^I}\mathrm{d}Y_1^I) \wedge \mathrm{d}Y_1^J \\
& = -d_1 \sum_{J,I=1}^{d} (-\frac{\partial^2 U}{\partial y^J \partial y^I})\mathrm{d}Y_1^I \wedge \mathrm{d}Y_1^J = 0.
\end{aligned}$$

According to (2.8), the last term of (2.12) is equal to zero. Therefore,

$$\sum_{J=1}^{d} \mathrm{d}y_{n+1}^J \wedge \mathrm{d}y_{n+1}'^J = \sum_{J=1}^{d} \mathrm{d}y_n^J \wedge \mathrm{d}y_n'^J.$$

For the general case, where M is a symmetric and positive semi-definite matrix, the decomposition of M may be written as follows

$$M = P^{\mathsf{T}}W^2P = \Omega_0^2 \text{ with } \Omega_0 = P^{\mathsf{T}}WP,$$

where P is an orthogonal matrix and W is a diagonal matrix with nonnegative diagonal entries which are the square roots of the eigenvalues of M. Using the variable substitution $z(t) = Py(t)$ the system (2.1) is equivalent to

$$\begin{cases} z'' + W^2 z = Pf(P^\mathsf{T}z), \\ z(t_0) = z_0 = Py_0, \\ z'(t_0) = z'_0 = Py'_0. \end{cases} \tag{2.13}$$

Then symplectic integrators for the diagonal matrix M with nonnegative entries can be applied to the transformed system. Moreover, the methods are invariant under linear transformations. This means that the ARKN method (2.6) with symplecticity conditions (2.7) and (2.8) can be applied to systems with M a symmetric and positive semi-definite matrix.

It may be concluded that the ARKN method (2.6), satisfying symplectic conditions (2.7) and (2.8), is a symplectic integrator for the system (2.1) with symmetric and positive semi-definite M. □

With Theorem 2.1, a one-stage explicit symplectic multi-frequency and multi-dimensional ARKN method can be derived. Regarding c_1, d_1 as parameters, the Eqs. (2.7) and (2.8) may be rewritten as

$$\begin{cases} b_1(V) = d_1\big(\phi_0(V) + c_1 V\phi_1(V)\big), \\ \bar{b}_1(V) = d_1\big(\phi_1(V) - c_1\phi_0(V)\big). \end{cases} \tag{2.14}$$

The choice of $d_1 = 1$ and $c_1 = \dfrac{1}{2}$ gives the following symplectic multi-frequency and multidimensional ARKN formula:

$$\begin{cases} Y_1 = y_n + \dfrac{1}{2}hy'_n, \\ y_{n+1} = \phi_0(V)y_n + \phi_1(V)hy'_n + h^2\big(\phi_1(V) - \dfrac{1}{2}\phi_0(V)\big)f(t_n + \dfrac{1}{2}h, Y_1), \\ hy'_{n+1} = -V\phi_1(V)y_n + \phi_0(V)hy'_n + h^2\big(\phi_0(V) + \dfrac{1}{2}V\phi_1(V)\big)f(t_n + \dfrac{1}{2}h, Y_1). \end{cases} \tag{2.15}$$

It can be verified that this formula is of order two from the order conditions of ARKN methods in Wu et al. [63]. This improved Störmer–Verlet formula (2.15) is denoted by ISV1.

2.2.2 Improved Störmer–Verlet Formula 2

The second improved Störmer–Verlet formula is based on the multi-frequency and multidimensional ERKN integrators and the corresponding symplectic conditions.

Taking advantage of the specific structure given by the linear term My of (2.1), and revising not only the updates but also the internal stages of classical RKN methods leads to the so-called multi-frequency and multidimensional ERKN integrators.

Definition 2.1 An s-stage multi-frequency and multidimensional ERKN integrator for the system (2.1) is defined as in [61]:

$$
\begin{cases}
Y_i = \phi_0(c_i^2 V)y_n + c_i\phi_1(c_i^2 V)hy_n' + h^2 \sum_{j=1}^{s} \bar{a}_{ij}(V)f(t_n + c_j h, Y_j), \quad i = 1, 2, \ldots, s, \\[2mm]
y_{n+1} = \phi_0(V)y_n + \phi_1(V)hy_n' + h^2 \sum_{i=1}^{s} \bar{b}_i(V)f(t_n + c_i h, Y_i), \\[2mm]
hy_{n+1}' = -V\phi_1(V)y_n + \phi_0(V)hy_n' + h^2 \sum_{i=1}^{s} b_i(V)f(t_n + c_i h, Y_i),
\end{cases}
$$

$$(2.16)$$

where c_i for $i = 1, 2, \ldots, s$, are real constants, $b_i(V)$, $\bar{b}_i(V)$ and $\bar{a}_{ij}(V)$ for $i, j = 1, 2, \ldots, s$, are matrix-valued functions of $V = h^2 M$.

Our attention is only focused on the one-stage explicit multi-frequency and multidimensional ERKN method as given below:

$$
\begin{cases}
Y_1 = \phi_0(c_1^2 V)y_n + c_1\phi_1(c_1^2 V)hy_n', \\[2mm]
y_{n+1} = \phi_0(V)y_n + \phi_1(V)hy_n' + h^2\bar{b}_1(V)f(t_n + c_1 h, Y_1), \\[2mm]
hy_{n+1}' = -V\phi_1(V)y_n + \phi_0(V)hy_n' + h^2 b_1(V)f(t_n + c_1 h, Y_1).
\end{cases}
$$

$$(2.17)$$

The symplectic conditions for the one-stage explicit multi-frequency and multidimensional ERKN method (2.17) are presented by the following theorem.

Theorem 2.2 *Under the conditions of Theorem 2.1, the one-stage explicit multi-frequency and multidimensional ERKN method (2.17) is symplectic if its coefficients satisfy*

$$
\begin{cases}
b_1(V)\phi_0(V) + \bar{b}_1(V)V\phi_1(V) = d_1\phi_0(c_1^2 V), \quad d_1 \in \mathbb{R}, \\[2mm]
\bar{b}_1(V)\big(\phi_0(V) + c_1 V\phi_1(V)\phi_0^{-1}(c_1^2 V)\phi_1(c_1^2 V)\big) \\[2mm]
= b_1(V)\big(\phi_1(V) - c_1\phi_0(V)\phi_0^{-1}(c_1^2 V)\phi_1(c_1^2 V)\big).
\end{cases}
$$

$$(2.18)$$

Proof It follows from the Theorem 3.1 in Wu et al. [59]. $\qquad\square$

The one-stage explicit symplectic multi-frequency and multidimensional ERKN method (2.17) can be obtained by solving the equations in (2.18) with c_1, d_1 as parameters. This leads to

$$
\begin{cases}
b_1(V) = d_1\big(\phi_0(V)\phi_0(c_1^2 V) + c_1 V\phi_1(V)\phi_1(c_1^2 V)\big), \\[2mm]
\bar{b}_1(V) = d_1\big(\phi_1(V)\phi_0(c_1^2 V) - c_1 V\phi_0(V)\phi_1(c_1^2 V)\big).
\end{cases}
$$

$$(2.19)$$

Choosing $d_1 = 1$ and $c_1 = \dfrac{1}{2}$ yields the following symplectic multidimensional ERKN formula

$$\begin{cases} Y_1 = \phi_0\left(\dfrac{V}{4}\right)y_n + \dfrac{1}{2}\phi_1\left(\dfrac{V}{4}\right)hy_n', \\[2mm] y_{n+1} = \phi_0(V)y_n + \phi_1(V)hy_n' + \dfrac{h^2}{2}\phi_1\left(\dfrac{V}{4}\right)f\left(t_n + \dfrac{1}{2}h, Y_1\right), \\[2mm] hy_{n+1}' = -V\phi_1(V)y_n + \phi_0(V)hy_n' + h^2\phi_0\left(\dfrac{V}{4}\right)f\left(t_n + \dfrac{1}{2}h, Y_1\right), \end{cases} \quad (2.20)$$

which is denoted by ISV2. From the order conditions of multidimensional ERKN methods presented in Wu et al. [61], this formula can be verified to be of order two. Moreover, this formula is symplectic and symmetric.

Note that the formula ISV1 amends the updates, and the formula ISV2 modifies both the internal stages and the updates to adapt them to the qualitative behaviour of the true solution. Therefore, the two formulae ISV1 and ISV2 are derived by making use of the specific structure of the Eq. (2.1) introduced by the linear term My. An important observation is that as $M \to \mathbf{0}_{d \times d}$, both formulae ISV1 and ISV2 reduce to the well-known classical Störmer–Verlet formula (2.4). In this sense the formulae ISV1 and ISV2 are extensions of the classical Störmer–Verlet formula (2.4).

2.3 Stability and Phase Properties

It is important to analyse the stability and phase properties of an oscillatory integrator, including dispersion and dissipation (see, e.g. [47]) of the numerical method. This section examines the stability and phase properties of the new formulae ISV1 and ISV2.

The following test equation is given in [47, 55] for stability analysis of the new formulae:

$$y''(t) + \omega^2 y(t) = -\varepsilon y(t) \text{ with } \omega^2 + \varepsilon > 0, \quad (2.21)$$

where ω represents an estimation of the dominant frequency λ and $\varepsilon = \lambda^2 - \omega^2$ is the error of the estimation.

Applying the formula ISV1 determined by (2.15) to (2.21) yields

$$\begin{cases} Y_1 = y_n + \dfrac{1}{2}hy_n', \quad z = \varepsilon h^2, \quad V = h^2\omega^2, \\[3mm] y_{n+1} = \phi_0(V)y_n + \phi_1(V)hy_n' - z\left(\phi_1(V) - \dfrac{1}{2}\phi_0(V)\right)Y_1, \\[3mm] hy_{n+1}' = -V\phi_1(V)y_n + \phi_0(V)hy_n' - z\left(\phi_0(V) + \dfrac{1}{2}V\phi_1(V)\right)Y_1. \end{cases} \quad (2.22)$$

It follows from (2.22) that

$$\begin{pmatrix} y_{n+1} \\ hy'_{n+1} \end{pmatrix} = S_1(V, z) \begin{pmatrix} y_n \\ hy'_n \end{pmatrix},$$

where the stability matrix $S_1(V, z)$ of the formula ISV1 is given by

$$S_1(V, z) = \begin{pmatrix} \phi_0(V) - z\big(\phi_1(V) - \frac{1}{2}\phi_0(V)\big) & \phi_1(V) - \frac{1}{2}z\big(\phi_1(V) - \frac{1}{2}\phi_0(V)\big) \\ -V\phi_1(V) - z\big(\phi_0(V) + \frac{1}{2}V\phi_1(V)\big) & \phi_0(V) - \frac{1}{2}z\big(\phi_0(V) + \frac{1}{2}V\phi_1(V)\big) \end{pmatrix}.$$

Likewise, applying the formula ISV2 determined by (2.20) to (2.21) yields

$$\begin{cases} Y_1 = \phi_0\Big(\dfrac{V}{4}\Big)y_n + \dfrac{1}{2}\phi_1\Big(\dfrac{V}{4}\Big)hy'_n, \qquad z = \varepsilon h^2, \qquad V = h^2\omega^2, \\[2mm] y_{n+1} = \phi_0(V)y_n + \phi_1(V)hy'_n - \dfrac{z}{2}\phi_1\Big(\dfrac{V}{4}\Big)Y_1, \\[2mm] hy'_{n+1} = -V\phi_1(V)y_n + \phi_0(V)hy'_n - z\phi_0\Big(\dfrac{V}{4}\Big)Y_1. \end{cases}$$

Thus we have

$$\begin{pmatrix} y_{n+1} \\ hy'_{n+1} \end{pmatrix} = S_2(V, z) \begin{pmatrix} y_n \\ hy'_n \end{pmatrix},$$

where the stability matrix $S_2(V, z)$ of the formula ISV2 is given by

$$S_2(V, z) = \begin{pmatrix} \phi_0(V) - \frac{z}{2}\phi_1\big(\frac{1}{4}V\big)\phi_0\big(\frac{1}{4}V\big) & \phi_1(V) - \frac{z}{2}\phi_1\big(\frac{1}{4}V\big)\big(\frac{1}{2}\phi_1\big(\frac{1}{4}V\big)\big) \\ -V\phi_1(V) - z\phi_0\big(\frac{1}{4}V\big)\phi_0\big(\frac{1}{4}V\big) & \phi_0(V) - z\phi_0\big(\frac{1}{4}V\big)\big(\frac{1}{2}\phi_1\big(\frac{1}{4}V\big)\big) \end{pmatrix}.$$

The spectral radii $\rho\big(S_1(V, z)\big)$ and $\rho\big(S_2(V, z)\big)$ represent the stability of formulae ISV1 and ISV2, respectively. Since each stability matrix $S(V, z)$ depends on the variables V and z, geometrically, the characterization of stability becomes a two-dimensional region in the (V, z)-plane. Accordingly, the following definition of stability for the two new formulae can be used.

Definition 2.2 For the stability matrix $S(V, z)$ of each formula, $S_s = \{(V, z)|\ V > 0\ and\ \rho(S) < 1\}$ is called the stability region of the formula. Accordingly, $S_p = \{(V, z)|\ V > 0,\ \rho(S) = 1\ and\ \mathrm{tr}(S)^2 < 4\det(S)\}$ is called the *periodicity region of the formula.*

The stability regions of the formulae ISV1 and ISV2 are shown in Fig. 2.1.

Definition 2.3 The quantities

$$\phi(H) = H - \arccos\Big(\frac{\mathrm{tr}(S)}{2\sqrt{\det(S)}}\Big), \quad d(H) = 1 - \sqrt{\det(S)}$$

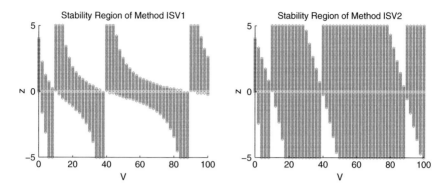

Fig. 2.1 Stability regions for the methods ISV1 (*left*) and ISV2 (*right*)

are called the dispersion error and the dissipation error of the integrator, respectively, where $H = \sqrt{V + z}$. Then, an integrator is said to be dispersive of order q and dissipative of order r, if $\phi(H) = \mathcal{O}(H^{q+1})$ and $d(H) = \mathcal{O}(H^{r+1})$, respectively. If $\phi(H) = 0$ and $d(H) = 0$, the integrator is said to be zero dispersive and zero dissipative, respectively.

With regard to the phase properties of the new formulae ISV1 and ISV2, we have

- ISV1: $\phi(H) = -\dfrac{\varepsilon(\varepsilon + 3\omega^2)H^3}{24(\varepsilon + \omega^2)^2} + \mathcal{O}(H^5)$,

- ISV2: $\phi(H) = -\dfrac{\varepsilon^2 H^3}{24(\varepsilon + \omega^2)^2} + \mathcal{O}(H^5)$.

Since the formulae ISV1 and ISV2 are both symplectic, one has $\det(S_i) = 1$ for $i = 1, 2$ and hence both the formulae ISV1 and ISV2 are zero dissipative.

2.4 Applications

In this section, four applications of the two improved Störmer–Verlet formulae are used to demonstrate their remarkable efficiency. These four types of problems are the one-dimensional time-independent Schrödinger equations, the non-linear wave equations, orbital problems and the Fermi–Pasta–Ulam problem. The numerical methods used for comparison are:

- SV: the classical Störmer–Verlet formula (2.4);
- A: the symmetric and symplectic Gautschi-type method of order two given in [13];
- B: the symmetric Gautschi-type method of order two given in [16];
- ISV1: the improved Störmer–Verlet formula given by (2.15) in this chapter;
- ISV2: the improved Störmer–Verlet formula given by (2.20) in this chapter.

2.4.1 Application 1: Time-Independent Schrödinger Equations

The time-independent Schrödinger equation is one of the basic equations of quantum mechanics. For the one-dimensional case, it may be written in the form (see, e.g. [24])

$$\frac{d^2\Psi}{dx^2} + 2E\Psi = 2V(x)\Psi, \tag{2.23}$$

where E is the energy eigenvalue, $V(x)$ the potential and $\Psi(x)$ the wave function. Consider the Eq. (2.23) with boundary conditions

$$\Psi(a) = 0, \ \Psi(b) = 0, \tag{2.24}$$

and use the shooting scheme in the implementation of the above methods. It is well known that the shooting method converts the boundary value problem to an initial value problem. Here, the boundary value at the end point b is transformed to an initial value $\Psi'(a)$, and the results are independent of $\Psi'(a)$ if $\Psi'(a) \neq 0$. The eigenvalue E is a parameter here and its value making $\Psi(b) = 0$ is the eigenvalue computed.

The two improved Störmer–Verlet formulae are tested against the classical Störmer–Verlet formula and the two Gautschi-type methods A and B through the following two cases.

Case 1: The harmonic oscillator
The potential of the one-dimensional harmonic oscillator is

$$V(x) = \frac{1}{2}kx^2,$$

with the interval of integration $[a, b]$ and boundary conditions $\Psi(a) = \Psi(b) = 0$. With $k = 1$ and $M = 2E$, the eigenvalues are computed as $E = 5.5$ and $E = 20.5$ using the stepsizes $h = 0.01, 0.02, 0.03, 0.04$ on the interval $[-8, 8]$. The errors between the energy eigenvalue E and that calculated by each numerical method ($ERR = |E - E_{\text{calculated}}|$) are shown in Fig. 2.2.

Case 2: Doubly anharmonic oscillator
The potential of the doubly anharmonic oscillator is

$$V(x) = \frac{1}{2}x^2 + \lambda_1 x^4 + \lambda_2 x^6.$$

The boundary conditions for this problem are $\Psi(a) = \Psi(b) = 0$. With $\lambda_1 = \lambda_2 = \frac{1}{2}$, the eigenvalues are computed as $E = 54.2225$ and $E = 145.0317$ on the interval $[-5, 5]$ using the stepsizes $h = 0.005, 0.01, 0.015, 0.02$. The results are shown in Fig. 2.3.

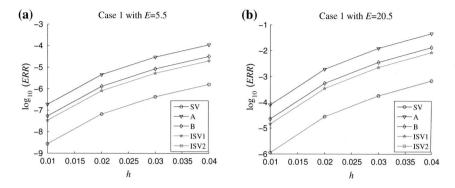

Fig. 2.2 Results for Case 1. The logarithm of the error $ERR = |E - E_{\text{calculated}}|$ against the stepsizes h

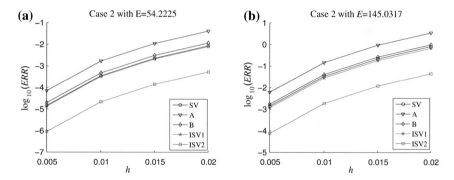

Fig. 2.3 Results for Case 2. The logarithm of the error $ERR = |E - E_{\text{calculated}}|$ against the stepsizes h

2.4.2 Application 2: Non-linear Wave Equations

For each problem in this section and the next two sections, the efficiency curves, i.e. accuracy versus the computational cost measured by the number of function evaluations required by each method, will be shown. If a problem is a Hamiltonian system, the numerical energy conservation is presented for each method. In this section, a non-linear wave equation is studied.

Problem 2.1 Consider the non-linear wave equation (see [11])

$$\begin{cases} \dfrac{\partial^2 u}{\partial t^2} - a(x)\dfrac{\partial^2 u}{\partial x^2} + 92u = f(t, x, u), & 0 < x < 1, \ t > 0. \\ u(0, t) = 0, \quad u(1, t) = 0, \quad u(x, 0) = a(x), \quad u_t(x, 0) = 0, \end{cases}$$

where

$$f(t, x, u) = u^5 - a^2(x)u^3 + \frac{a^5(x)}{4} \sin^2(20t)\cos(10t), \quad a(x) = 4x(1-x).$$

The exact solution is $u(x, t) = a(x)\cos(10t)$. The oscillatory problem represents a vibrating string with angular speed 10. A semi-discretization of the wave equation of the spatial variable by means of second-order symmetric differences yields the following second-order ordinary differential equations in time

$$\begin{cases} \dfrac{d^2 u_i(t)}{dt^2} - a(x_i)\dfrac{u_{i+1}(t) - 2u_i(t) + u_{i-1}(t)}{\Delta x^2} + 92u_i(t) = f\big(t, x_i, u_i(t)\big), \quad 0 < t \le t_{\text{end}}, \\ u_i(0) = a(x_i), \quad u'_i(0) = 0, \quad i = 1, 2, \ldots, N-1, \end{cases}$$

where $\Delta x = 1/N$ is the spatial mesh step, $x_i = i\Delta x$ and $u_i(t) \approx u(x_i, t)$. This system takes the form

$$\begin{cases} \dfrac{d^2 U(t)}{dt^2} + MU(t) = F\big(t, U(t)\big), \quad 0 < t \le t_{\text{end}}, \\ U(0) = \big(a(x_1), \ldots, a(x_{N-1})\big)^{\mathsf{T}}, \quad U'(0) = \mathbf{0}, \end{cases} \tag{2.25}$$

where $U(t) = \big(u_1(t), \ldots, u_{N-1}(t)\big)^{\mathsf{T}}$,

$$M = \frac{1}{\Delta x^2} \begin{pmatrix} 2a(x_1) & -a(x_1) \\ -a(x_2) & 2a(x_2) & -a(x_2) \\ & \ddots & \ddots & \ddots \\ & & -a(x_{N-2}) & 2a(x_{N-2}) & -a(x_{N-2}) \\ & & & -a(x_{N-1}) & 2a(x_{N-1}) \end{pmatrix} + 92 I_{N-1},$$

and

$$F(t, U(t)) = \Big(f\big(t, x_1, u_1(t)\big), \ldots, f\big(t, x_{N-1}, u_{N-1}(t)\big)\Big)^{\mathsf{T}}.$$

Note that the matrix M is not symmetric and the system (2.25) is not a Hamiltonian system. However, (2.25) is an oscillatory system. The system is integrated on the interval $[0, 100]$ with $N = 20$ and the integration stepsizes $h = 1/(40 \times 2^j)$, $j = 0, 1, 2, 3$. Figure 2.4 shows the errors in the positions at $t_{\text{end}} = 100$ versus the computational effort.

It can be observed from Fig. 2.4 that the two novel improved Störmer–Verlet formulae outperform the classical Störmer–Verlet formula and the others in terms of efficiency.

Fig. 2.4 Results for Problem 2.1. The logarithm of the global error (*GE*) over the integration interval against the logarithm of the number of function evaluations

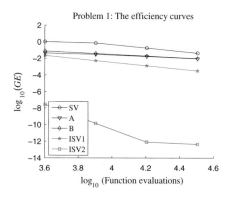

2.4.3 Application 3: Orbital Problems

In this section, three orbital problems are solved using the two improved formulae ISV1 and ISV2, and the two Gautschi-type methods A in [13] and B in [16], as well as the Störmer–Verlet formula. The numerical results are shown.

Problem 2.2 Consider the Hamiltonian equation which governs the motion of an artificial satellite (see [41]) with the Hamiltonian function

$$H(q, p) = \frac{1}{2} p^\mathsf{T} p + \frac{\hbar}{4} q^\mathsf{T} q + \lambda \left(\frac{(q_1 q_3 + q_2 q_4)^2}{r^4} - \frac{1}{12 r^2} \right),$$

where $q = (q_1, q_2, q_3, q_4)^\mathsf{T}$, $p = (p_1, p_2, p_3, p_4)^\mathsf{T}$, $r = q^\mathsf{T} q$, and $\hbar$ is the total energy of the elliptic motion which is defined by $\hbar = \frac{K^2 - 2|p_0|^2}{r_0} - V_0$ with $V_0 = -\frac{\lambda}{12 r_0^3}$. The initial conditions are considered on an elliptic equatorial orbit as

$$q_0 = \sqrt{\frac{r_0}{2}} \left(-1, -\frac{\sqrt{3}}{2}, -\frac{1}{2}, 0 \right)^\mathsf{T}, \quad p_0 = \frac{1}{2} \sqrt{K^2 \frac{1+e}{2}} \left(1, \frac{\sqrt{3}}{2}, \frac{1}{2}, 0 \right)^\mathsf{T}.$$

The parameters of this problem are chosen as $K^2 = 3.98601 \times 10^5$, $r_0 = 6.8 \times 10^3$, $e = 0.1$, $\lambda = \frac{3}{2} K^2 J_2 R^2$, $J_2 = 1.08625 \times 10^{-3}$, $R = 6.37122 \times 10^3$.

The problem is solved on the interval $[0, 100]$. The integration stepsizes are taken as $h = 2^j/300$ with $j = 2, \ldots, 5$. The numerical results are shown in Fig. 2.5a. For the global errors of Hamiltonian, the stepsize used here is $h = 1/100$ on the intervals $[0, 10^i]$, $i = 0, 1, 2, 3$. See Fig. 2.5b for the results.

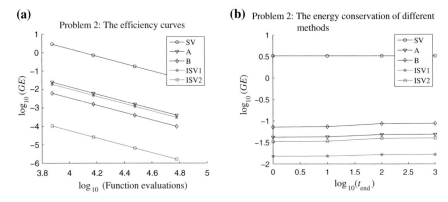

Fig. 2.5 Results for Problem 2.2. **a** The logarithm of the global error (GE) over the integration interval against the logarithm of the number of function evaluations. **b** The logarithm of the maximum global error of Hamiltonian $GEH = \max |H_n - H_0|$ against $\log_{10}(t_{\text{end}})$

Problem 2.3 Consider the orbital problem with perturbation

$$
\begin{cases}
q_1''(t) + q_1(t) = -\dfrac{2\varepsilon + \varepsilon^2}{r^5} q_1(t), & q_1(0) = 1, \ q_1'(0) = 0, \\[2mm]
q_2''(t) + q_2(t) = -\dfrac{2\varepsilon + \varepsilon^2}{r^5} q_2(t), & q_2(0) = 0, \ q_2'(0) = 1 + \varepsilon,
\end{cases}
$$

where $r = \sqrt{q_1^2(t) + q_2^2(t)}$. Its analytic solution is given by

$$
q_1(t) = \cos(t + \varepsilon t), \quad q_2(t) = \sin(t + \varepsilon t).
$$

The problem has been solved on the interval [0, 1000] with $\varepsilon = 10^{-3}$. The stepsizes are taken as $h = 1/(2^j)$ with $j = 3, \ldots, 6$. The numerical results are shown in Fig. 2.6a.

This system is a Hamiltonian system with the Hamiltonian

$$
H = \frac{p_1^2 + p_2^2}{2} + \frac{q_1^2 + q_2^2}{2} - \frac{2\varepsilon + \varepsilon^2}{3(q_1^2 + q_2^2)^{\frac{3}{2}}}.
$$

Accordingly, we integrate this problem with the stepsize $h = 1/8$ on the intervals $[0, 10^i]$, $i = 1, \ldots, 4$. The energy conservation of different methods is shown in Fig. 2.6b.

Problem 2.4 Consider the model for stellar orbits in a galaxy (see [25, 26])

$$
\begin{cases}
q_1''(t) + a^2 q_1(t) = \varepsilon q_2^2(t), & q_1(0) = 1, \ q_1'(0) = 0, \\[1mm]
q_2''(t) + b^2 q_2(t) = 2\varepsilon q_1(t) q_2(t), & q_2(0) = 1, \ q_2'(0) = 0,
\end{cases}
$$

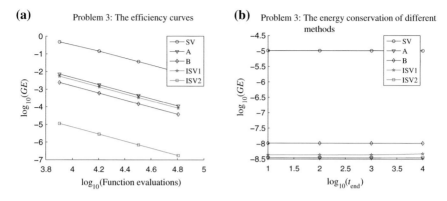

Fig. 2.6 Results for Problem 2.3. **a** The logarithm of the global error (GE) over the integration interval against the logarithm of the number of function evaluations. **b** The logarithm of the maximum global error of Hamiltonian $GEH = \max |H_n - H_0|$ against $\log_{10}(t_{\text{end}})$

where q_1 stands for the radial displacement of the orbit of a star from a reference circular orbit, and q_2 stands for the deviation of the orbit from the galactic plane. The time variable t actually denotes the angle of the planets in a reference coordinate system. Following [1], we choose $a = 2$, $b = 1$. The problem has been solved on the interval [0, 1000] with $\varepsilon = 10^{-3}$. The stepsizes are taken as $h = 1/(2^j)$ with $j = 3, \ldots, 6$. The numerical results are shown in Fig. 2.7a.

It is easy to see that this system is a Hamiltonian system with the Hamiltonian

$$H = \frac{1}{2}(p_1^2 + p_2^2) + \frac{1}{2}(4q_1^2 + q_2^2) - \varepsilon q_1 q_2^2.$$

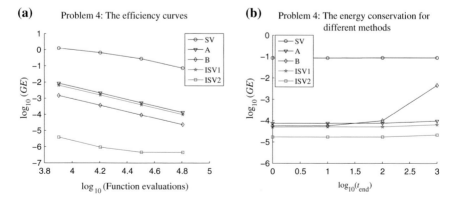

Fig. 2.7 Results for Problem 2.4. **a** The logarithm of the global error (GE) over the integration interval against the logarithm of the number of function evaluations. **b** The logarithm of the maximum global error of Hamiltonian $GEH = \max |H_n - H_0|$ against $\log_{10}(t_{\text{end}})$

Accordingly, we integrate this problem with the stepsize $h = 1/5$ on the interval $[0, 10^i]$, $i = 0, 1, 2, 3$. The energy conservation for different methods is shown in Fig. 2.7b.

2.4.4 Application 4: Fermi–Pasta–Ulam Problem

The Fermi–Pasta–Ulam problem is an important model of non-linear classical and quantum systems of interacting particles in the physics of non-linear phenomena.

Problem 2.5 The Fermi–Pasta–Ulam problem discussed by Hairer et al. [16, 19]. The motion is described by a Hamiltonian system with the total energy

$$H(p, q) = \frac{1}{2}p^\mathsf{T} p + \frac{1}{2}q^\mathsf{T} M q + U(q),$$

where

$$M = \begin{pmatrix} \mathbf{0}_{m \times m} & \mathbf{0}_{m \times m} \\ \mathbf{0}_{m \times m} & \omega^2 I_{m \times m} \end{pmatrix},$$

$$U(q) = \frac{1}{4}\left((q_1 - q_{m+1})^4 + \sum_{i=1}^{m-1}(q_{i+1} - q_{m+i+1} - q_i - q_{m+i})^4 + (q_m + q_{2m})^4\right).$$

Here, q_i represents a scaled displacement of the ith stiff spring, q_{m+i} is a scaled expansion (or compression) of the ith stiff spring, and p_i and p_{m+i} are their velocities (or momenta).

The corresponding Hamiltonian system now becomes

$$\begin{cases} p'(t) = -H_q\big(p(t), q(t)\big), \\ q'(t) = H_p\big(p(t), q(t)\big), \end{cases}$$

which is identical to $q''(t) = -H_q\big(p(t), q(t)\big)$ with $p(t) = q'(t)$. Then we have

$$q''(t) + M q(t) = -\nabla U\big(q(t)\big), \qquad t \in [t_0, t_{\text{end}}],$$

Following Hairer et al. [19], we consider $m = 3$, and choose

$$q_1(0) = 1, \; p_1(0) = 1, \; q_4(0) = \frac{1}{\omega}, \; p_4(0) = 1,$$

with zero for the remaining initial values. The system is integrated on the interval $[0, 10]$ with the stepsizes $h = 0.01/2^j$, $j = 2, \ldots, 5$ for $\omega = 100$ and 200. The results are shown in Fig. 2.8. In Fig. 2.8b, the error $\log_{10}(GE)$ for SV method is very large when $\omega = 200$ and $h = 0.01$, hence the point is not plotted in the graph. For

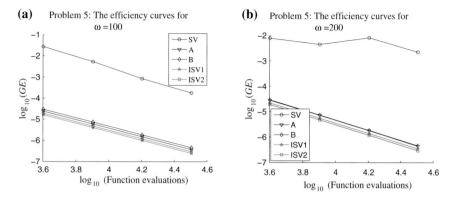

Fig. 2.8 Results for Problem 2.5 with different ω. The logarithm of the global error (GE) over the integration interval against the logarithm of the number of function evaluations

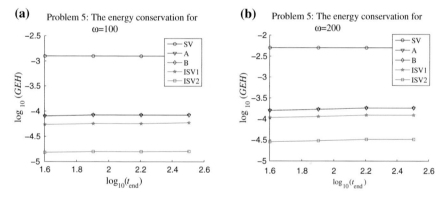

Fig. 2.9 Results for Problem 2.5 with different ω. The logarithm of the maximum global error of Hamiltonian $GEH = \max |H_n - H_0|$ against $\log_{10}(t_{end})$

global errors of Hamiltonian, this problem is integrated with the stepsize $h = 0.001$ on the intervals $[0, 2^i \times 10]$, $i = 2, \ldots, 5$. The results for different $\omega = 100$ and 200 are shown in Fig. 2.9.

For this Hamiltonian problem, long time-step methods may lead to the problem of numerically induced resonance instabilities. Here, in order to discuss the resonance instabilities for this Hamiltonian problem, we compute the maximum global error of the total energy H as a function of the scaled frequency $h\omega$ (stepsize $h = 0.02$). We consider the long-time interval $[0, 1000]$. Figures 2.10 and 2.11 present the results of the different methods. Method SV shows a poor energy conservation when $h\omega$ is more than about $\pi/2$, while methods ISV1 and ISV2 do poorly only near integral multiples of π. Method A shows poor energy conservation near even multiples of π and only the method B shows a uniformly good behaviour for all frequencies.

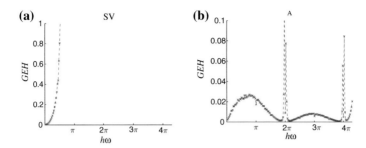

Fig. 2.10 Results for Problem 2.5. The maximum global error (GEH) of the total energy on the interval [0, 1000] for methods SV (*left*), and A (*right*) as a function of $h\omega$ (step size $h = 0.02$)

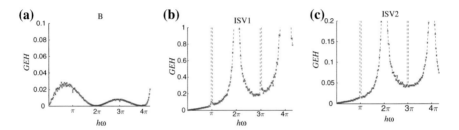

Fig. 2.11 Results for Problem 2.5. The maximum global error (GEH) of the total energy on the interval [0, 1000] for methods B (*left*), ISV1 (*middle*) and ISV2 (*right*) as a function of $h\omega$ (step size $h = 0.02$)

From the numerical results of the four types of problems, it follows that the two improved multidimensional Störmer–Verlet formulae show better accuracy and preserve the Hamiltonian approximately better than the classical Störmer–Verlet formula (2.4) and the two Gautschi-type methods A and B. It is noted that the formula ISV2 behaves better than the formula ISV1. The reason for this numerical behaviour is that the formula ISV2 uses the special structure brought by the linear term My from the problem (2.1) not only in the updates but also in the internal stages, whereas the formula ISV1 uses the special structure only in the updates. In other words, the numerical behaviour of the two improved formulae shows the importance and the superiority of using special structure to obtain structure-preserving algorithms.

2.5 Coupled Conditions for Explicit Symplectic and Symmetric Multi-frequency ERKN Integrators for Multi-frequency Oscillatory Hamiltonian Systems

It can be verified that the symplectic improved Störmer–Verlet formulae ISV2 is also symmetric. Therefore, this section pays attention to the coupled conditions for symplectic and symmetric multi-frequency extended Runge-Kutta-Nyström integrators (SSMERKN) for multi-frequency and multidimensional oscillatory Hamiltonian systems with the Hamiltonian $H(p,q) = p^\mathsf{T} p/2 + q^\mathsf{T} Aq/2 + U(q)$, where A is a symmetric and positive semi-definite matrix implicitly preserving the dominant frequencies of the oscillatory problem, and $U(q)$ is a real-valued function with continuous second derivatives. The solution of the system is a non-linear multi-frequency oscillator. The coupled conditions are presented for SSMERKN integrators. When $A = \mathbf{0} \in \mathbb{R}^{d\times d}$, the explicit SSMERKN integrators reduce to classical symplectic and symmetric RKN methods.

2.5.1 Towards Coupled Conditions for Explicit Symplectic and Symmetric Multi-frequency ERKN Integrators

This section turns to the coupled conditions for symplectic and symmetric multi-frequency ERKN integrators (SSMERKN) for multi-frequency and multidimensional oscillatory Hamiltonian systems

$$\begin{cases} p' = -\nabla_q H(p,q), & p(t_0) = p_0, \\ q' = \nabla_p H(p,q), & q(t_0) = q_0, \end{cases} \tag{2.26}$$

with the Hamiltonian $H(p,q) = p^\mathsf{T} p/2 + q^\mathsf{T} Aq/2 + U(q)$, where A is a symmetric and positive semi-definite matrix implicitly preserving the dominant frequencies of the oscillatory problem, and $U(q)$ is a real-valued function with continuous second derivatives. Clearly, if $f(q) = -\nabla U(q)$, then (2.26) is identical to the multi-frequency and multidimensional oscillatory system of second-order differential equations

$$\begin{cases} q''(t) + Aq(t) = f\big(q(t)\big), \\ q(t_0) = q_0, \quad q'(t_0) = q_0'. \end{cases} \tag{2.27}$$

It is important to observe that the system (2.27) is a *multi-frequency and multidimensional oscillatory problem with multiple time scales*.

A significant amount of early work on the numerical integration of (2.27) focused mainly on the *single-frequency problem, the so-called perturbed oscillator*

$$\begin{cases} q''(t) + \omega^2 q(t) = f\big(q(t)\big), & t \in [t_0, T], \\ q(t_0) = q_0, \quad q'(t_0) = q_0', \end{cases} \tag{2.28}$$

where $\omega > 0$ is the *main frequency of the single-frequency problem* and ω may be known or accurately estimated in advance, and $f(q)$ is a small perturbing force. It is clear that (2.28) is a special case of (2.27). Over the past ten years (and earlier), some novel approaches to modifying classical RKN methods have been proposed for the single-frequency problem (2.28). Franco [10] took advantage of the special structure of (2.28) introduced by the linear term $\omega^2 q$ and modified the updates of classical RKN methods to obtain ARKN methods. For further research following [10], readers are referred to [60, 63]. Also note that Tocino and Vigo-Aguiar [45] presented the symplecticity conditions of modified RKN methods, and Chen et al. [4] gave symmetric and symplectic conditions of ERKN methods for the *single-frequency problem* (2.28). All the methods designed specially for the single-frequency problem (2.28) have coefficients which are analytic functions of v^2 ($v = h\omega$). It is very important to note that, in general, *these methods for the single-frequency problem (2.28) cannot be applied to the multi-frequency oscillatory system (2.27)*. This fact can be shown by the following Hamiltonian system with the Hamiltonian [25, 26]

$$ H = \frac{1}{2} p^{\mathsf{T}} p + \frac{1}{2} q^{\mathsf{T}} \begin{pmatrix} 4 & 0 \\ 0 & 1 \end{pmatrix} q - \varepsilon q_1 q_2^2. \tag{2.29} $$

The problem (2.29) has two coupled oscillators with two different frequencies. However, it is clear that the coefficients of the methods for single-frequency problems (2.28) are functions of $v = h\omega$. Hence the analysis for single-frequency problems (2.28) given in previous publications is not directly applicable to coupled oscillators. Moreover, it has been shown in [59] that a symplectic multi-frequency method requires more coupled conditions than a symplectic single-frequency method, which demonstrates the difference of symplecticity conditions for these two different kinds of methods. Therefore, this section is devoted to developing the coupled conditions of multi-frequency ERKN methods for multi-frequency and multidimensional oscillatory Hamiltonian systems (2.27).

2.5.2 The Analysis of Combined Conditions for SSMERKN Integrators for Multi-frequency and Multidimensional Oscillatory Hamiltonian Systems

Many physical problems have time reversibility and this structure of the original continuous system is preserved by symmetric integrators (for a rigorous definition of reversibility readers are referred to Chap. 5 of Hairer et al. [19]). On the other hand, some real-physical processes with negligible dissipation in applications could be modelled by the Hamiltonian system (2.27). The solution of the system preserves the Hamiltonian $H(p, q)$. Furthermore, the corresponding flow is symplectic in the sense that the differential 2-form $dp \wedge dq$ is invariant. The important property of symplecticity is preserved by symplectic integrators.

Pioneering work on symplectic integrators is due to [6, 7, 35]. The symplecticity conditions for Runge-Kutta methods are obtained in [37], the symplecticity conditions for RKN methods are derived in [43], and the symplecticity conditions for multi-frequency ERKN methods are given in [59]. This section is devoted to deriving the coupled conditions of SSMERKN integrators for the multi-frequency and multidimensional oscillatory Hamiltonian system (2.27).

It is noted that explicit integrators do not require the solution of large and complicated systems of non-linear algebraic or transcendental equations when solving the multi-frequency and multidimensional oscillatory problem (2.27). Therefore, this section pays attention only to explicit SSMERKN integrators for (2.27).

The next theorem gives the coupled conditions of explicit SSMERKN integrators for (2.27).

Theorem 2.3 *An s-stage explicit multi-frequency ERKN integrator for integrating (2.27) is symplectic and symmetric if its coefficients are:*

$$
\begin{cases}
c_i = 1 - c_{s+1-i}, \quad d_i = d_{s+1-i} \neq 0, & i = 1, 2, \ldots, s, \\
b_i(V) = d_i \phi_0(c_{s+1-i}^2 V), & i = 1, 2, \ldots, s, \\[2mm]
\bar{b}_i(V) = d_i c_{s+1-i} \phi_1(c_{s+1-i}^2 V), & i = 1, 2, \ldots, s, \\
\bar{a}_{ij}(V) = \dfrac{1}{d_i} \big(b_i(V) \bar{b}_j(V) - \bar{b}_i(V) b_j(V) \big), & i > j, \ i, j = 1, 2, \ldots, s.
\end{cases}
$$
(2.30)

Proof Consider first the symplecticity.

It follows from the symplecticity conditions in [58, 59] that an s-stage explicit multi-frequency ERKN integrator for (2.27) is symplectic if its coefficients satisfy

$$
\begin{cases}
b_i(V)\phi_0(V) + \bar{b}_i(V)V\phi_1(V) = d_i\phi_0(c_i^2 V), & i = 1, 2, \ldots, s, \quad d_i \in \mathbb{R}, \\
\bar{b}_i(V)\big(\phi_0(V) + c_i V\phi_1(V)\phi_0^{-1}(c_i^2 V)\phi_1(c_i^2 V)\big) \\
\quad = b_i(V)\big(\phi_1(V) - c_i \phi_0(V)\phi_0^{-1}(c_i^2 V)\phi_1(c_i^2 V)\big), & i = 1, 2, \ldots, s, \\
b_i(V)\big(\bar{b}_j(V) - \phi_0(V)\phi_0^{-1}(c_i^2 V)\bar{a}_{ij}(V)\big) - \bar{b}_i(V)V\phi_1(V)\phi_0^{-1}(c_i^2 V)\bar{a}_{ij}(V) \\
\quad = b_j(V)\bar{b}_i(V), & i > j, \ i, j = 1, 2, \ldots, s.
\end{cases}
$$
(2.31)

With the definition (1.7), a careful calculation shows that

$$
\begin{cases}
\phi_0(V)\phi_0(c_i^2 V) + c_i V\phi_1(V)\phi_1(c_i^2 V) = \phi_0\big((1 - c_i)^2 V\big) = \phi_0(c_{s+1-i}^2 V), \\
\phi_1(V)\phi_0(c_i^2 V) - c_i \phi_0(V)\phi_1(c_i^2 V) = (1 - c_i)\phi_1\big((1 - c_i)^2 V\big) = c_{s+1-i}\phi_1(c_{s+1-i}^2 V).
\end{cases}
$$
(2.32)

Solving all the symplecticity conditions (2.31) with $d_i \neq 0$ yields

$$
\begin{cases}
b_i(V) = d_i\big(\phi_0(V)\phi_0(c_i^2 V) + c_i V\phi_1(V)\phi_1(c_i^2 V)\big), & i = 1, 2, \ldots, s, \\
\bar{b}_i(V) = d_i\big(\phi_1(V)\phi_0(c_i^2 V) - c_i \phi_0(V)\phi_1(c_i^2 V)\big), & i = 1, 2, \ldots, s, \\
\bar{a}_{ij}(V) = \dfrac{1}{d_i}\big(b_i(V)\bar{b}_j(V) - \bar{b}_i(V)b_j(V)\big), & i > j, \ i, j = 1, 2, \ldots, s.
\end{cases}
$$
(2.33)

The formulae (2.32) and (2.33) immediately show that the coefficients $b_i(V)$, $\bar{b}_i(V)$, $\bar{a}_{ij}(V)$ in (2.30) satisfy the symplecticity conditions (2.31) for s-stage explicit multi-frequency ERKN integrators.

The purpose of the next step is to verify that the explicit multi-frequency ERKN integrator given by (2.30) is symmetric.

It is known that an integrator $y_{n+1} = \Phi_h(y_n)$ is symmetric if exchanging $y_n \leftrightarrow y_{n+1}$ and $h \leftrightarrow -h$ does not change the integrator. Hence, exchanging $(q_n, p_n) \leftrightarrow (q_{n+1}, p_{n+1})$ and replacing h by $-h$ for an s-stage explicit multi-frequency ERKN method gives

$$
\left\{
\begin{aligned}
&\hat{Q}_i = \phi_0(c_i^2 V)q_{n+1} - hc_i\phi_1(c_i^2 V)p_{n+1} + h^2\sum_{j=1}^{i-1}\bar{a}_{ij}(V)\big(-\nabla U(\hat{Q}_j)\big), \quad i = 1, 2, \ldots, s, \\
&q_n = \phi_0(V)q_{n+1} - h\phi_1(V)p_{n+1} + h^2\sum_{i=1}^{s}\bar{b}_i(V)\big(-\nabla U(\hat{Q}_i)\big), \\
&p_n = hA\phi_1(V)q_{n+1} + \phi_0(V)p_{n+1} - h\sum_{i=1}^{s}b_i(V)\big(-\nabla U(\hat{Q}_i)\big).
\end{aligned}
\right.
$$

$$(2.34)$$

It follows from (2.34) that

$$
\left\{
\begin{aligned}
&\hat{Q}_i = \phi_0((1-c_i)^2 V)q_n + h(1-c_i)\phi_1((1-c_i)^2 V)p_n + h^2\sum_{j=1}^{i-1}\big[\phi_0(c_i^2 V)(\phi_1(V)b_j(V) \\
&\qquad - \phi_0(V)\bar{b}_j(V)) - c_i\phi_1(c_i^2 V)(V\phi_1(V)\bar{b}_j(V) + \phi_0(V)b_j(V)) + \bar{a}_{ij}(V)\big]\big(-\nabla U(\hat{Q}_j)\big), \\
&q_{n+1} = \phi_0(V)q_n + h\phi_1(V)p_n + h^2\sum_{i=1}^{s}\big[\phi_1(V)b_i(V) - \phi_0(V)\bar{b}_i(V)\big]\big(-\nabla U(\hat{Q}_i)\big), \\
&p_{n+1} = -hA\phi_1(V)q_n + \phi_0(V)p_n + h\sum_{i=1}^{s}\big[V\phi_1(V)\bar{b}_i(V) + \phi_0(V)b_i(V)\big]\big(-\nabla U(\hat{Q}_i)\big).
\end{aligned}
\right.
$$

$$(2.35)$$

Replacing all indices i and j in (2.35) by $s+1-i$ and $s+1-j$, respectively, it is easy to see that an s-stage explicit multi-frequency ERKN method is symmetric if the following conditions are true:

$$
\left\{
\begin{aligned}
c_i &= 1 - c_{s+1-i}, & i &= 1, 2, \ldots, s, \\
\bar{b}_i(V) &= \phi_1(V)b_{s+1-i}(V) - \phi_0(V)\bar{b}_{s+1-i}(V), & i &= 1, 2, \ldots, s, \\
b_i(V) &= V\phi_1(V)\bar{b}_{s+1-i}(V) + \phi_0(V)b_{s+1-i}(V), & i &= 1, 2, \ldots, s, \\
\bar{a}_{ij}(V) &= \phi_0(c_{s+1-i}^2 V)\bar{b}_j(V) - c_{s+1-i}\phi_1(c_{s+1-i}^2 V)b_j(V), & i &> j, \; i, \; j = 1, 2, \ldots, s.
\end{aligned}
\right.
$$

$$(2.36)$$

It follows from (2.30) and (1.7) that

$$
\begin{aligned}
\phi_1(V)b_{s+1-i}(V) - \phi_0(V)\bar{b}_{s+1-i}(V) &= d_{s+1-i}c_{s+1-i}\phi_1(c_{s+1-i}^2 V) \\
&= d_i c_{s+1-i}\phi_1(c_{s+1-i}^2 V) = \bar{b}_i(V), \; i = 1, 2, \ldots, s, \\
V\phi_1(V)\bar{b}_{s+1-i}(V) + \phi_0(V)b_{s+1-i}(V) &= d_{s+1-i}\phi_0(c_{s+1-i}^2 V) \\
&= d_i\phi_0(c_{s+1-i}^2 V) = b_i(V), \; i = 1, 2, \ldots, s,
\end{aligned}
$$

and

$$\bar{a}_{ij}(V) = \frac{1}{d_i}\left(b_i(V)\bar{b}_j(V) - \bar{b}_i(V)b_j(V)\right)$$
$$= \phi_0(c_{s+1-i}^2 V)\bar{b}_j(V) - c_{s+1-i}\phi_1(c_{s+1-i}^2 V)b_j(V), \quad i > j, \quad i, j = 1, 2, \ldots, s.$$

This means that an s-stage explicit multi-frequency ERKN integrator with the coefficients (2.30) is symplectic and symmetric. This completes the proof. □

Remark 2.1 It follows from Theorem 2.3 that the coefficients (2.30) are expressed by c_i, d_i, $\phi_0(V)$ and $\phi_1(V)$. In order to compute matrix-valued functions $\phi_0(V)$ and $\phi_1(V)$, recall that a *generalized hypergeometric function* (see [32, 39]) is

$$_mF_n\left[\begin{matrix}\alpha_1, \alpha_2, \ldots, \alpha_m; \\ \beta_1, \beta_2, \ldots, \beta_n;\end{matrix}x\right] = \sum_{l=0}^{\infty} \frac{\prod_{i=1}^{m}(\alpha_i)_l}{\prod_{i=1}^{n}(\beta_i)_l} \frac{x^l}{l!}, \tag{2.37}$$

where the *Pochhammer symbol* $(z)_l$ is defined as $(z)_0 = 1$ and $(z)_l = z(z+1)\cdots(z+l-1)$, $l \in \mathbb{N}$. The parameters α_i and β_i are arbitrary complex numbers, except that β_i can be neither zero nor a negative integer. It is noted that $\phi_0(V)$ and $\phi_1(V)$ can be expressed by the generalized hypergeometric function $_0F_1$:

$$\phi_0(V) = {}_0F_1\left[\begin{matrix}-; \\ \frac{1}{2};\end{matrix} -\frac{V}{4}\right], \quad \phi_1(V) = {}_0F_1\left[\begin{matrix}-; \\ \frac{3}{2};\end{matrix} -\frac{V}{4}\right]. \tag{2.38}$$

Most modern software, e.g. Maple, Mathematica, or Matlab, is well equipped to calculate generalized hypergeometric functions, and there is also much work concerning the evaluation of generalized hypergeometric functions of a matrix argument (see, e.g. [2, 15, 22, 27, 33]). Moreover, there have been some efficient ways to compute the matrix cosine and sine, and readers are referred to [21, 22, 31, 36] for example. All these publications provide different ways to deal with the computation of $\phi_0(V)$ and $\phi_1(V)$.

It is noted that when $V \to \mathbf{0}_{d \times d}$, the multi-frequency ERKN methods reduce to classical RKN methods for solving Hamiltonian systems with the Hamiltonian $H(p, q) = \frac{1}{2}p^\mathsf{T}p + U(q)$. The following result can be obtained from Theorem 2.3.

Theorem 2.4 *An s-stage explicit RKN method with the coefficients*

$$\begin{cases} c_i = 1 - c_{s+1-i}, \quad d_i = d_{s+1-i} \neq 0, & i = 1, 2, \ldots, s, \\ b_i = d_i, \quad \bar{b}_i = d_i c_{s+1-i}, & i = 1, 2, \ldots, s, \\ \bar{a}_{ij} = \frac{1}{d_i}\left(b_i\bar{b}_j - \bar{b}_i b_j\right), & i > j, \ i, j = 1, 2, \ldots, s, \end{cases} \tag{2.39}$$

is symplectic and symmetric. In (2.39), d_i, $i = 1, 2, \ldots, \lfloor \frac{s+1}{2} \rfloor$, are real numbers and can be chosen according to the order conditions of RKN methods or other ways, where $\lfloor \frac{s+1}{2} \rfloor$ denotes the integral part of $\frac{s+1}{2}$.

Proof It follows from (2.39) that

$$
\begin{aligned}
\bar{b}_i &= d_i c_{s+1-i} = b_i (1 - c_i), & i &= 1, 2, \ldots, s, \\
b_i (\bar{b}_j - \bar{a}_{ij}) &= b_i \left(\bar{b}_j - \frac{1}{d_i} (b_i \bar{b}_j - \bar{b}_i b_j) \right) \\
&= b_i (\bar{b}_j - \bar{b}_j) + \bar{b}_i b_j = \bar{b}_i b_j, & i &> j, \ i, \ j = 1, 2, \ldots, s,
\end{aligned}
$$

which are exactly the symplecticity conditions for explicit RKN methods.

Moreover, based on (2.39), we have

$$
\begin{aligned}
c_i &= 1 - c_{s+1-i}, \quad b_i = d_i = d_{s+1-i} = b_{s+1-i}, \quad i = 1, 2, \ldots, s, \\
\bar{a}_{ij} &= \frac{1}{d_i} (b_i \bar{b}_j - \bar{b}_i b_j) = \frac{1}{d_i} (d_i \bar{b}_j - d_i c_{s+1-i} b_j) \\
&= \bar{b}_j - c_{s+1-i} b_j, & i > j, \ i, \ j = 1, 2, \ldots, s,
\end{aligned}
$$

and these are the symmetry conditions for explicit RKN methods. The proof is complete.   □

2.6 Conclusions and Discussions

The well-known Störmer–Verlet formula can be implemented easily, and it has become by far the most widely used numerical scheme. This chapter develops two improved Störmer–Verlet formulae ISV1 and ISV2 by exploiting the special structure of the problem (2.1). The two improved formulae are shown to be symplectic schemes of order two and are generalizations of the classical Störmer–Verlet scheme since when $M \to \mathbf{0}_{d \times d}$, they reduce to the classical Störmer–Verlet formula (2.4). It is known that the advantage of symplectic algorithms is that they show a sort of global stability. Each improved method is a blend of existing trigonometric integrators and symplectic integrators, fully combining the favourable properties of both. Based on the linear test Eq. (2.21), the stability and phase properties for the two improved Störmer–Verlet formulae are also analysed. In order to exhibit the two formulae quantitatively, four different applications, including time-independent Schrödinger equations, non-linear wave equations, orbital problems and the Fermi-Pasta-Ulam problem, are presented. The two improved formulae are compared with the classical Störmer–Verlet formula and the two other improved Störmer–Verlet methods that have appeared already in the literature. The numerical experiments recorded in this chapter demonstrate that the improved Störmer–Verlet formulae are more efficient than the classical Störmer–Verlet formula, and the two other methods that have already appeared in the literature. In particular, it can be concluded that, when applied to a Hamiltonian system, the two improved Störmer–Verlet formulae preserve well the Hamiltonian in the sense of numerical approximation and have better accuracy than those of the classical Störmer–Verlet formula and the two Gautschi-type methods A and B with the same computational cost.

Since the improved Störmer–Verlet formula ISV2 is an explicit symplectic and symmetric multi-frequency ERKN integrator for multi-frequency oscillatory Hamiltonian systems, the coupled conditions for explicit symplectic and symmetric multi-frequency ERKN integrators for multi-frequency oscillatory Hamiltonian systems are investigated in detail.

Last but not least, like the classical Störmer–Verlet formula (2.4), it can be observed clearly that the two improved Störmer–Verlet formulae ISV1 given by (2.15) and ISV2 given by (2.20) are conceptually simple, versatile and easy to code when applied to (2.1) or (2.2). It seems promising that, for the two improved Störmer–Verlet formulae ISV1 and ISV2, other applications may be found in science and engineering.

This chapter is based on the work by Wang et al. [54], and a seminar report by Wang and Wu [52].

References

1. Ariel G, Engquist B, Kim S, Lee Y, Tsai R (2013) A multiscale method for highly oscillatory dynamical systems using a Poincaré map type technique. J Sci Comput 54:247–268
2. Butler RW, Wood ATA (2002) Laplace approximations for hypergeometric functions with matrix argument. Ann Statist 30:1155–1177
3. Candy J, Rozmus W (1991) A symplectic integration algorithm for separable Hamiltonian functions. J Comput Phys 92:230–256
4. Chen Z, You X, Shi W, Liu Z (2012) Symmetric and symplectic ERNK methods for oscillatory Hamiltonian systems. Comput Phys Commun 183:86–98
5. Cohen D, Jahnke T, Lorenz K, Lubich C (2006) Numerical integrators for highly oscillatory Hamiltonian systems: a review. In: Mielke A (ed) Analysis, modeling and simulation of multiscale problems. Springer, Berlin, pp 553–576
6. De Vogelaere R (1956) Methods of integration which preserve the contact transformation property of the Hamiltonian equations. Report No. 4, Department of Mathematics, University of Notre Dame, Notre Dame, Indiana
7. Feng K (1985) On difference schemes and symplectic geometry. In: Proceedings of the 5th international symposium on differential geometry and differential equations, Beijing, 42–58 Aug 1984
8. Feng K, Qin M (2010) Symplectic geometric algorithms for hamiltonian systems. Springer, Berlin
9. Fermi E, Pasta J, Ulam S (1955) Studies of the Nonlinear Problems, I. Los Alamos Report No. LA- 1940, later published in E. Fermi: Collected Papers (Chicago 1965), and Lect Appl Math 15:143 (1974)
10. Franco JM (2002) Runge-Kutta-Nyström methods adapted to the numerical integration of perturbed oscillators. Comput Phys Commun 147:770–787
11. Franco JM (2006) New methods for oscillatory systems based on ARKN methods. Appl Numer Math 56:1040–1053
12. García A, Martín P, González AB (2002) New methods for oscillatory problems based on classical codes. Appl Numer Math 42:141–157

13. García-Archilla B, Sanz-Serna JM, Skeel RD (1999) Long-time-step methods for oscillatory differential equations. SIAM J Sci Comput 20:930–963
14. González AB, Martín P, Farto JM (1999) A new family of Runge-Kutta type methods for the numerical integration of perturbed oscillators. Numer Math 82:635–646
15. Gutiérrez R, Rodriguez J, Sáez AJ (2000) Approximation of hypergeometric functions with matricial argument through their development in series of zonal polynomials. Electron Trans Numer Anal 11:121–130
16. Hairer E, Lubich C (2000) Long-time energy conservation of numerical methods for oscillatory differential equations. SIAM J Numer Anal 38:414–441
17. Hairer E, Lubich C, Wanner G (2003) Geometric numerical integration illustrated by the Störmer-Verlet method. Acta Numerica 12:399–450
18. Hairer E, Lubich C (2000) Energy conservation by Störmer-type numerical integrators. In: Griffiths GF, Watson GA (eds) Numerical analysis 1999. CRC Press LLC, pp 169–190
19. Hairer E, Lubich C, Wanner G (2006) Geometric numerical integration: structure-preserving algorithms for ordinary differential equations, 2nd edn. Springer, Berlin
20. Hairer E, Nørsett SP, Wanner G (1993) Solving ordinary differential equations I: nonstiff problems. Springer, Berlin
21. Hargreaves GI, Higham NJ (2005) Efficient algorithms for the matrix cosine and sine. Numer Algo 40:383–400
22. Higham NJ, Smith MI (2003) Computing the matrix cosine. Numer Algo 34:13–26
23. Hochbruck M, Lubich C (1999) A Gautschi-type method for oscillatory second-order differential equations. Numer Math 83:403–426
24. Kalogiratou Z, Monovasilis Th, Simos TE (2003) Symplectic integrators for the numerical solution of the Schrödinger equation. J Comput Appl Math 158:83–92
25. Kevorkian J, Cole JD (1981) Perturbation methods in applied mathematics. Applied mathematical sciences, vol 34, Springer, New York
26. Kevorkian J, Cole JD (1996) Multiple scale and singular perturbation methods. Applied mathematical sciences, vol 114, Springer, New York
27. Koev P, Edelman A (2006) The efficient evaluation of the hypergeometric function of a matrix argument. Math Comput 75:833–846
28. Li J, Wang B, You X, Wu X (2011) Two-step extended RKN methods for oscillatory systems. Comput Phys Commun 182:2486–2507
29. Li J, Wu X (2013) Adapted Falkner-type methods solving oscillatory second-order differential equations. Numer Algo 62:355–381
30. McLachlan RI, Quispel GRW (2002) Splitting methods. Acta Numerica 11:341–434
31. Püschel M, Moura JMF (2003) The algebraic approach to the discrete cosine and sine transforms and their fast algorithms. SIAM J Comput 32:1280–1316
32. Rainville ED (1960) Special functions. Macmillan, New York
33. Richards DSP (2011) High-dimensional random matrices from the classical matrix groups, and generalized hypergeometric functions of matrix argument. Symmetry 3:600–610
34. Rowlands G (1991) A numerical algorithm for Hamiltonian systems. J Comput Phys 97:235–239
35. Ruth RD (1983) A canonical integration technique. IEEE Trans Nucl Sci 30:2669–2671
36. Serbin SM, Blalock SA (1980) An algorithm for computing the matrix cosine. SIAM J Sci Stat Comput 1:198–204
37. Sanz-Serna JM (1988) Runge-Kutta schems for Hamiltonian systems. BIT Numer Math 28:877–883
38. Simos TE (2010) Exponentially and trigonometrically fitted methods for the solution of the Schrödinger equation. Acta Appl Math 110:1331–1352
39. Slater LJ (1966) Generalized hypergeometric functions. Cambridge University Press, Cambridge
40. Stavroyiannis S, Simos TE (2009) Optimization as a function of the phase-lag order of two-step P-stable method for linear periodic IVPs. Appl Numer Math 59:2467–2474
41. Stiefel EL, Scheifele G (1971) Linear and regular celestial mechanics. Springer, New York

42. Störmer C (1907) Sur les trajectories des corpuscules électrisés. Arch Sci Phys Nat 24:5–18, 113–158, 221–247
43. Suris YB (1989) The canonicity of mapping generated by Runge-Kutta type methods when integrating the systems $\ddot{x} = -\frac{\partial U}{\partial x}$. Zh. Vychisl Mat i Mat Fiz 29:202–211. (In Russian) Translation, U.S.S.S. Comput Maths Math Phys 29:138–144 (1989)
44. Tan X (2005) Almost symplectic Runge-Kutta schemes for Hamiltonian systems. J Comput Phys 203:250–273
45. Tocino A, Vigo-Aguiar J (2005) Symplectic conditions for exponential fitting Runge-Kutta-Nyström methods. Math Comput Modell 42:873–876
46. Berghe VG, Van Daele M (2006) Exponentially-fitted Störmer/Verlet methods. JNAIAM J Numer Anal Ind Appl Math 1:241–255
47. Van der Houwen PJ, Sommeijer BP (1987) Explicit Runge-Kutta(-Nyström) methods with reduced phase errors for computing oscillating solution. SIAM J Numer Anal 24:595–617
48. Van de Vyver H (2006) An embedded phase-fitted modified Runge-Kutta method for the numerical integration of the radial Schrödinger equation. Phys Lett A 352:278–285
49. Verlet L (1967) Computer "experiments" on classical fluids. I. Thermodynamical properties of Lennard-Jones molecules. Phys Rev 159:98–103
50. Vigo-Aguiar J, Simos TE (2002) Family of twelve steps exponential fitting symmetric multistep methods for the numerical solution of the Schrödinger equation. J Math Chem 32:257–270
51. Vigo-Aguiar J, Simos TE, Ferrándiz JM (2004) Controlling the error growth in long-term numerical integration of perturbed oscillations in one or more frequencies. Proc Roy Soc London Ser A 460:561–567
52. Wang B, Wu X (2013) Coupled conditions for explicit symplectic and symmetric multi-frequency ERKN integrators. A seminar report of Nanjing University [preprint]
53. Wang B, Wu X (2012) A new high precision energy-preserving integrator for system of oscillatory second-order differential equations. Phys Lett A 376:1185–1190
54. Wang B, Wu X, Zhao H (2013) Novel improved multidimensional Stormer-Verlet formulas with applications to four aspects in scientific computation. Math Comput Model 57:857–872
55. Wu X (2012) A note on stability of multidimensional adapted Runge-Kutta-Nyström methods for oscillatory systems. Appl Math Modell 36:6331–6337
56. Wu X, Wang B (2010) Multidimensional adapted Runge-Kutta-Nyström methods for oscillatory systems. Comput Phys Commun 181:1955–1962
57. Wu X, Wang B, Shi W (2013) Efficient energy-preserving integrators for oscillatory Hamiltonian systems. J Comput Phys 235:587–605
58. Wu X, Wang B, Xia J (2010) ESRKN methods for Hamiltonian Systems. In: Vigo Aguiar J (ed) Proceedings of the 2010 international conference on computational and mathematical methods in science and engineering, Vol III, Spain, pp 1016–1020
59. Wu X, Wang B, Xia J (2012) Explicit symplectic multidimensional exponential fitting modified Runge-Kutta-Nyström methods. BIT Num Math 52:773–795
60. Wu X, You X, Li J (2009) Note on derivation of order conditions for ARKN methods for perturbed oscillators. Comput Phys Commun 180:1545–1549
61. Wu X, You X, Shi W, Wang B (2010) ERKN integrators for systems of oscillatory second-order differential equations. Comput Phys Commun 181:1873–1887
62. Wu X, You X, Wang B (2013) Structure-preserving algorithms for oscillatory differential equations. Springer, Berlin
63. Wu X, You X, Xia J (2009) Order conditions for ARKN methods solving oscillatory systems. Comput Phys Commun 180:2250–2257

Chapter 3
Improved Filon-Type Asymptotic Methods for Highly Oscillatory Differential Equations

This chapter presents an effective improvement on the existing Filon-type asymptotic methods, so that the improved methods can numerically solve a class of multi-frequency highly oscillatory second-order differential equations with a positive semi-definite singular matrix which implicitly contains and preserves the oscillatory frequencies of the underlying problem. The idea is to combine the existing Filon-type methods with the asymptotic methods, based on the matrix-variation-of-constants formula.

3.1 Motivation

In this chapter, we focus our attention on the multi-frequency highly oscillatory problem

$$x''(t) + Mx(t) = f\big(t, x(t), x'(t)\big), \quad t \in [t_0, t_{\text{end}}], \quad x(t_0) = x_0, \quad x'(t_0) = x_0',$$
$$(3.1)$$

where $f : \mathbb{R} \times \mathbb{R}^d \times \mathbb{R}^d \rightarrow \mathbb{R}^d$ is continuous and $M \in \mathbb{R}^{d \times d}$ is positive semi-definite (not necessarily nonsingular), diagonalizable, and $\|M\| \gg 1$, which implicitly contains and preserves the oscillatory frequencies of the problem. Throughout this chapter, $\|\cdot\|$ presents the spectral norm.

This class of multi-frequency highly oscillatory problems occurs in a wide variety of applications, such as quantum physics, circuit simulations, flexible body dynamics, mechanics and readers are referred to [2, 4, 8, 9, 12, 21, 30] for examples. In the past few years, much research effort has been expended to design and analyse effective numerical methods for highly oscillatory systems (see, e.g. [1, 6, 7, 10, 11, 13–16, 18, 20, 22, 24, 25, 27, 28, 31, 32]). When the matrix M in (3.1) is nonsingular, the highly oscillatory system can be dealt with by using two different approaches. First, we can transform (3.1) into a first-order highly oscillatory system and then apply the approach presented in [18] to dealing with the first-order system. Second, a Filon-type asymptotic approach has been formulated and

© Springer-Verlag Berlin Heidelberg and Science Press, Beijing, China 2015
X. Wu et al., *Structure-Preserving Algorithms for Oscillatory
Differential Equations II*, DOI 10.1007/978-3-662-48156-1_3

analysed in [24], and this is designed specially for solving (3.1) with a nonsingular matrix M. However, neither of them can be applied to the multi-frequency highly oscillatory problem (3.1) when M is singular, such as for the Fermi–Pasta–Ulam problem with multiple time scales, and then the applications of the existing Filon-type asymptotic methods in the literature are limited. In order to effectively solve the multi-frequency highly oscillatory system (3.1) with a positive semi-definite matrix M (not necessarily nonsingular), this chapter is devoted to the formulation and analysis of an effective improvement on Filon-type asymptotic methods.

It is natural to assume that M is allowed to be singular throughout this chapter. As a first step towards an understanding of the improvement on the Filon-type asymptotic approach to solving the highly oscillatory system (3.1), we now consider the following multi-frequency highly oscillatory problem of the form

$$x''(t) + \Omega^2 x(t) = f(t, x(t), x'(t)), \qquad t \in [t_0, t_{\text{end}}], \quad x(t_0) = x_0, \quad x'(t_0) = x_0', \qquad (3.2)$$

where

$$\Omega = \begin{pmatrix} \mathbf{0}_{m_0 \times m_0} & \mathbf{0} & \mathbf{0} & \cdots & \mathbf{0} \\ \mathbf{0} & \omega_1 I_{m_1 \times m_1} & \mathbf{0} & \cdots & \mathbf{0} \\ \mathbf{0} & \mathbf{0} & \omega_2 I_{m_2 \times m_2} & \cdots & \mathbf{0} \\ \vdots & \vdots & \vdots & \ddots & \vdots \\ \mathbf{0} & \mathbf{0} & \mathbf{0} & \cdots & \omega_k I_{m_k \times m_k} \end{pmatrix}, \quad \sum_{i=0}^{k} m_i = d, \qquad (3.3)$$

and $f(t, x(t), x'(t))$ as well as its derivatives have a constant bounded independently of ω_i. Without loss of generality, it is assumed that ω_i $(i = 1, 2, \ldots, k)$ are distinct positive real numbers and $\omega_i \gg 1$ for some i.

A classical example for such kind of problems is the celebrated Fermi–Pasta–Ulam model with multiple time scales, which we will present numerical experiments in Sect. 3.3. Notice that we have assumed that the linear part of the model problem (3.2) brings high frequencies, i.e. the linear part introduces different high frequencies oscillations. However, as pointed in [7, 9], the diagonal form of Ω is not essential since for a general positive semi-definite matrix, we can perform diagonalization. Therefore, following [7, 9], we derive and analyse in detail an improved Filon-type asymptotic approach to solving the multi-frequency highly oscillatory model (3.2) in this chapter.

3.2   Improved Filon-Type Asymptotic Methods

With regard to the exact solution of the multi-frequency highly oscillatory system (3.1) and its derivative, from the matrix-variation-of-constants formula (1.6) in Chap. 1, we have the following result.

Theorem 3.1 *If $f : \mathbb{R} \times \mathbb{R}^d \times \mathbb{R}^d \to \mathbb{R}^d$ is continuous in (3.1), then the solution of (3.1) and its derivative satisfy*

$$
\begin{cases}
x(t_n + h) = \phi_0(V)x(t_n) + h\phi_1(V)x'(t_n) + \displaystyle\int_0^h (h - z)\phi_1\big((h - z)^2 M\big)\hat{f}(t_n + z)\mathrm{d}z, \\[3mm]
x'(t_n + h) = -hM\phi_1(V)x(t_n) + \phi_0(V)x'(t_n) + \displaystyle\int_0^h \phi_0\big((h - z)^2 M\big)\hat{f}(t_n + z)\mathrm{d}z,
\end{cases}
$$

(3.4)

where h is the stepsize, $V = h^2 M$, $\hat{f}(\xi) = f\big(\xi, x(\xi), x'(\xi)\big)$, and the unconditionally convergent matrix-valued functions $\phi_j(V)$, $j = 0, 1$, are defined by (1.7) in Chap. 1.

According to Theorem 3.1, the solution of (3.2) and its derivative satisfy

$$
\begin{cases}
x(t_n + h) = \cos(h\Omega)x(t_n) + \Omega^{-1} \sin(h\Omega)x'(t_n) \\[2mm]
\qquad\qquad + \displaystyle\int_0^h \Omega^{-1} \sin\big((h - z)\Omega\big)\hat{f}(t_n + z)\mathrm{d}z, \\[3mm]
x'(t_n + h) = -\Omega \sin(h\Omega)x(t_n) + \cos(h\Omega)x'(t_n) \\[2mm]
\qquad\qquad + \displaystyle\int_0^h \cos\big((h - z)\Omega\big)\hat{f}(t_n + z)\mathrm{d}z,
\end{cases}
$$

(3.5)

which is exactly the variation-of-constants formula for system (3.2). It can be observed that $\Omega^{-1} \sin(h\Omega)$ is well-defined also for singular Ω.

As an interesting aside, we note that the authors in [3] described a local version of a Filon method, which is built upon an idea first advanced in a special case by Louis Napoleon George Filon (1928) and phrased in more modern terminology by Flinn (1960). For the highly oscillatory integral

$$
I[\tilde{f}] = \int_0^1 \tilde{f}(x)\mathrm{e}^{i\omega g(x)}\mathrm{d}x
$$

with suitably smooth one-dimensional functions $\tilde{f}$, g and $\omega \gg 1$, the Filon method is carried out by interpolating the function $\tilde{f}(x)$ by a polynomial function. In [15], the authors used a Hermite polynomial approximation to $\tilde{f}(x)$ and presented a generalized Filon method. The Filon-type method is an effective method for highly oscillatory integrals and many studies have employed it in solving various highly oscillatory problems (see, e.g. [11, 13, 16, 18, 20, 24]). Most of these studies were devoted to solving one-dimensional highly oscillatory integrals by the Filon method. The authors in [18, 24] considered using the Filon method to solve matrix-valued highly oscillatory integrals. However, the technique developed in [18, 24] requires the considered matrix to be nonsingular and hence it cannot deal with the case of (3.2), where M is permitted to be singular. In this chapter, we design improved Filon-type asymptotic methods for solving the multi-frequency highly oscillatory system

(3.2) with a singular matrix Ω. In what follows, combining the idea of Filon method given in [15] with the matrix-variation-of-constants formula (3.5), we derive and analyse the improved asymptotic methods.

3.2.1   Oscillatory Linear Systems

In this section, we derive an improved Filon-type asymptotic method to solve the particular multi-frequency highly oscillatory second-order linear systems

$$x''(t) + \Omega^2 x(t) = g(t), \qquad t \in [t_0, t_{\text{end}}], \quad x(t_0) = x_0, \quad x'(t_0) = x_0'. \tag{3.6}$$

Following [15], we interpolate the vector-valued function g by a vector-valued polynomial p

$$p(t) = \sum_{l=1}^{v} \sum_{j=0}^{\theta_l - 1} \alpha_{l,j}(t) g^{(j)}(t_n + c_l h), \tag{3.7}$$

such that the following conditions

$$p^{(j)}(t_n + c_l h) = g^{(j)}(t_n + c_l h) \tag{3.8}$$

are held for $l = 1, 2, \ldots, v$, and $j = 0, 1, \ldots, \theta_l - 1$ for each fixed l. It is noted that $c_1, \ldots, c_v$ can be chosen as any values which satisfy $0 = c_1 < c_2 < \cdots < c_v = 1$ and $\theta_1, \ldots, \theta_v$ can be any positive integers. The superscript (j) appearing in this chapter denotes the jth-derivative with respect to t.

Based on the matrix-variation-of-constants formula (3.5), we consider the following Filon-type method for integrating the multi-frequency highly oscillatory linear system (3.6)

$$\begin{cases} x_{n+1} = \cos(h\Omega)x_n + \Omega^{-1}\sin(h\Omega)x_n' + \sum_{l=1}^{v}\sum_{j=0}^{\theta_l-1} I_1[\alpha_{l,j}](t_n)g^{(j)}(t_n + c_l h), \\[2mm] x_{n+1}' = -\Omega\sin(h\Omega)x_n + \cos(h\Omega)x_n' + \sum_{l=1}^{v}\sum_{j=0}^{\theta_l-1} I_2[\alpha_{l,j}](t_n)g^{(j)}(t_n + c_l h), \end{cases} \tag{3.9}$$

where h is the stepsize and I_1, I_2 are defined by

$$\begin{cases} I_1[\alpha_{l,j}](t_n) := \int_0^h (h-z)\alpha_{l,j}(t_n + z)\big((h-z)\Omega\big)^{-1}\sin\big((h-z)\Omega\big)\mathrm{d}z, \\[2mm] I_2[\alpha_{l,j}](t_n) := \int_0^h \alpha_{l,j}(t_n + z)\cos\big((h-z)\Omega\big)\mathrm{d}z, \end{cases} \tag{3.10}$$

respectively. These two highly oscillatory integrals $I_1[\alpha_{l,j}]$, $I_2[\alpha_{l,j}]$ can be approximated by asymptotic methods as follows.

From (3.10), we have

$$I_1[\alpha_{l,j}](t_n) = \int_0^h (h-z)\alpha_{l,j}(t_n+z)\big((h-z)\Omega\big)^{-1}\sin\big((h-z)\Omega\big)dz$$

$$= \begin{pmatrix} A & 0 & 0 & \cdots & 0 \\ 0 & Q_1[\alpha_{l,j},\omega_1](t_n)I_{m_1\times m_1} & 0 & \cdots & 0 \\ 0 & 0 & Q_1[\alpha_{l,j},\omega_2](t_n)I_{m_2\times m_2} & \cdots & 0 \\ \vdots & \vdots & \vdots & \ddots & \vdots \\ 0 & 0 & 0 & \cdots & Q_1[\alpha_{l,j},\omega_k](t_n)I_{m_k\times m_k} \end{pmatrix},$$

$$I_2[\alpha_{l,j}](t_n) = \int_0^h \alpha_{l,j}(t_n+z)\cos\big((h-z)\Omega\big)dz$$

$$= \begin{pmatrix} B & 0 & 0 & \cdots & 0 \\ 0 & Q_2[\alpha_{l,j},\omega_1](t_n)I_{m_1\times m_1} & 0 & \cdots & 0 \\ 0 & 0 & Q_2[\alpha_{l,j},\omega_2](t_n)I_{m_2\times m_2} & \cdots & 0 \\ \vdots & \vdots & \vdots & \ddots & \vdots \\ 0 & 0 & 0 & \cdots & Q_2[\alpha_{l,j},\omega_k](t_n)I_{m_k\times m_k} \end{pmatrix},$$

$$\tag{3.11}$$

where

$$\begin{cases} A := \int_0^h (h-z)\alpha_{l,j}(t_n+z)dzI_{m_0\times m_0}, \\[2mm] B := \int_0^h \alpha_{l,j}(t_n+z)dzI_{m_0\times m_0}, \\[2mm] Q_1[\alpha_{l,j},\omega_i](t_n) := 1/\omega_i \int_0^h \alpha_{l,j}(t_n+z)\sin\big((h-z)\omega_i\big)dz, \\[2mm] Q_2[\alpha_{l,j},,\omega_i](t_n) := \int_0^h \alpha_{l,j}(t_n+z)\cos\big((h-z)\omega_i\big)dz, \qquad i=1,2,\ldots,k. \end{cases}$$

$$\tag{3.12}$$

It is noted that A and B in (3.12) can be evaluated exactly since $\alpha_{l,j}(t)$ is a polynomial function.

In what follows, we turn to the asymptotic approach to dealing with $Q_1[\alpha_{l,j},\omega_i](t_n)$ and $Q_2[\alpha_{l,j},\omega_i](t_n)$.

Applying integration by parts, we arrive at

$$Q_1[\alpha_{l,j},\omega_i](t_n)$$

$$= 1/\omega_i \int_0^h \alpha_{l,j}(t_n+z)\sin\big((h-z)\omega_i\big)dz = 1/\omega_i^2 \int_0^h \alpha_{l,j}(t_n+z)d\cos\big((h-z)\omega_i\big)$$

$$= 1/\omega_i^2\alpha_{l,j}(t_{n+1}) - 1/\omega_i^2\alpha_{l,j}(t_n)\cos(h\omega_i) - 1/\omega_i^2 \int_0^h \alpha_{l,j}'(t_n+z)\cos\big((h-z)\omega_i\big)dz$$

$$= 1/\omega_i^2 \alpha_{l,j}(t_{n+1}) - 1/\omega_i^2 \alpha_{l,j}(t_n) \cos(h\omega_i) + 1/\omega_i^3 \int_0^h \alpha'_{l,j}(t_n + z)\mathrm{d}\sin\big((h-z)\omega_i\big)$$

$$= 1/\omega_i^2 \alpha_{l,j}(t_{n+1}) - 1/\omega_i^2 \alpha_{l,j}(t_n) \cos(h\omega_i) - 1/\omega_i^3 \alpha'_{l,j}(t_n) \sin(h\omega_i)$$

$$\quad - 1/\omega_i^3 \int_0^h \alpha''_{l,j}(t_n + z) \sin\big((h-z)\omega_i\big)\mathrm{d}z$$

$$= 1/\omega_i^2 \alpha_{l,j}(t_{n+1}) - 1/\omega_i^2 \alpha_{l,j}(t_n) \cos(h\omega_i) - 1/\omega_i^3 \alpha'_{l,j}(t_n) \sin(h\omega_i)$$

$$\quad - 1/\omega_i^4 \alpha''_{l,j}(t_{n+1}) + 1/\omega_i^4 \alpha''_{l,j}(t_n) \cos(h\omega_i) + 1/\omega_i^5 \alpha^{(3)}_{l,j}(t_n) \sin(h\omega_i)$$

$$\quad + 1/\omega_i^5 \int_0^h \alpha^{(4)}_{l,j}(t_n + z) \sin\big((h-z)\omega_i\big)\mathrm{d}z$$

$$= \cdots$$

$$= \sum_{k=0}^{\lfloor \deg(\alpha_{l,j})/2 \rfloor} (-1)^k / \omega_i^{2k+2} \Big(\alpha^{(2k)}_{l,j}(t_{n+1}) - \alpha^{(2k)}_{l,j}(t_n) \cos(h\omega_i)$$

$$\quad - 1/\omega_i \alpha^{(2k+1)}_{l,j}(t_n) \sin(h\omega_i) \Big), \quad i = 1, 2, \ldots, k, \tag{3.13}$$

where $\deg(\alpha_{l,j})$ is the degree of $\alpha_{l,j}$ and $\lfloor \deg(\alpha_{l,j})/2 \rfloor$ denotes the integer part of $\deg(\alpha_{l,j})/2$.

Similarly, we obtain

$$Q_2[\alpha_{l,j}, \omega_i](t_n) = \sum_{k=0}^{\lfloor \deg(\alpha_{l,j})/2 \rfloor} (-1)^k / \omega_i^{2k+1} \Big(\alpha^{(2k)}_{l,j}(t_n) \sin(h\omega_i)$$

$$+ 1/\omega_i \alpha^{(2k+1)}_{l,j}(t_{n+1}) - 1/\omega_i^2 \alpha^{(2k+1)}_{l,j}(t_n) \cos(h\omega_i) \Big), \quad i = 1, 2, \ldots, k. \tag{3.14}$$

Therefore, the improved Filon-type asymptotic method for the multi-frequency highly oscillatory linear system (3.6) can be formulated as follows:

Definition 3.1 The improved Filon-type asymptotic method for integrating the multi-frequency highly oscillatory linear system (3.6) is defined as (3.9), where h is the stepsize, $I_1[\alpha_{l,j}](t_n)$ and $I_2[\alpha_{l,j}](t_n)$ are given by (3.11) and are calculated by (3.13) and (3.14), respectively. It is noted that $\Omega^{-1}\sin(h\Omega)$ is well-defined also for singular Ω and thus this method does not require Ω to be nonsingular.

The error bounds of the above method (3.9) are stated by the following theorem.

Theorem 3.2 *Let* $r = \sum_{l=1}^{v} \theta_l$.

$$\begin{cases} \|x_1 - x(h)\| \le \dfrac{Ch^{r+2}}{(r+1)(r+2)}, \\[2mm] \|x_1' - x'(h)\| \le \dfrac{Ch^{r+1}}{r+1}, \end{cases} \tag{3.15}$$

where C depends on $g^{(r)}(t)$ but is independent of Ω.

Proof It follows from (3.5) and (3.9) that

$$x_1 - x(h) = \int_0^h (h-z)\big((h-z)\Omega\big)^{-1} \sin\big((h-z)\Omega\big)\big(p(z) - g(z)\big)dz.$$

Since the vector-valued function g is interpolated by the vector-valued polynomial p in (3.7) with the conditions (3.8), we obtain

$$p(t) - g(t) = Ct^r, \tag{3.16}$$

where C depends on $g^{(r)}(t)$ but is independent of Ω. Then, we arrive at

$$\begin{aligned}
\|x_1 - x(h)\| &= \left\| \int_0^h (h-z)\big((h-z)\Omega\big)^{-1} \sin\big((h-z)\Omega\big)Cz^r dz \right\| \\
&\leq C \int_0^h |(h-z)z^r| \, \left\| \big((h-z)\Omega\big)^{-1} \sin\big((h-z)\Omega\big) \right\| dz \leq C \int_0^h |(h-z)z^r| \, dz \\
&= \frac{Ch^{r+2}}{(r+1)(r+2)}.
\end{aligned}$$

The second inequality in (3.15) can be proved in a similar way.   □

Remark 3.1 Theorem 3.2 shows the convergence of the improved Filon-type asymptotic method and it is noted that C is independent of Ω, which is an important property for solving highly oscillatory systems.

3.2.2 Oscillatory Nonlinear Systems

In what follows, we extend the improved Filon-type asymptotic methods for the multi-frequency highly oscillatory linear system (3.6) in Sect. 3.2.1 to the multi-frequency highly oscillatory nonlinear systems (3.2). Following the Filon-type approach stated above, we interpolate the vector-valued function $\hat{f}$ in (3.5) by a vector-valued polynomial $\hat{p}$:

$$\begin{aligned}
\hat{p}(t, x_n, x_{n+1}, x_n', x_{n+1}') &= \sum_{j=0}^{\theta_1-1} \alpha_{1,j}(t) f^{(j)}(t_n, x_n, x_n') \\
&\quad + \sum_{j=0}^{\theta_2-1} \alpha_{2,j}(t) f^{(j)}(t_{n+1}, x_{n+1}, x_{n+1}'), \tag{3.17}
\end{aligned}$$

such that

$$\begin{cases} \hat{p}^{(j)}(t_n, x_n, x_{n+1}, x_n', x_{n+1}') = f^{(j)}(t_n, x_n, x_n'), & j = 0, 1, \ldots, \theta_1 - 1, \\ \hat{p}^{(j)}(t_{n+1}, x_n, x_{n+1}, x_n', x_{n+1}') = f^{(j)}(t_{n+1}, x_{n+1}, x_{n+1}'), & j = 0, 1, \ldots, \theta_2 - 1, \end{cases} \tag{3.18}$$

where θ_1, θ_2 can be chosen as any positive integers.

Then, replacing the function $\hat{f}$ in the matrix-variation-of-constants formula (3.5) by $\hat{p}$, we obtain the following improved Filon-type asymptotic methods for (3.2).

Definition 3.2 An improved Filon-type asymptotic method for integrating the multi-frequency highly oscillatory nonlinear system (3.2) is defined by

$$\begin{cases} x_{n+1} = \cos(h\Omega)x_n + \Omega^{-1}\sin(h\Omega)x_n' + \sum_{j=0}^{\theta_1-1} I_1[\alpha_{1,j}](t_n)f^{(j)}(t_n, x_n, x_n') \\ \qquad + \sum_{j=0}^{\theta_2-1} I_1[\alpha_{2,j}](t_n)f^{(j)}(t_{n+1}, x_{n+1}, x_{n+1}'), \\ x_{n+1}' = -\Omega\sin(h\Omega)x_n + \cos(h\Omega)x_n' + \sum_{j=0}^{\theta_1-1} I_2[\alpha_{1,j}](t_n)f^{(j)}(t_n, x_n, x_n') \\ \qquad + \sum_{j=0}^{\theta_2-1} I_2[\alpha_{2,j}](t_n)f^{(j)}(t_{n+1}, x_{n+1}, x_{n+1}'), \end{cases} \tag{3.19}$$

where h is the stepsize, $I_1[\alpha_{k,j}](t_n)$ and $I_2[\alpha_{k,j}](t_n)$, $k = 1, 2$ are determined by (3.11), and are calculated by (3.13) and (3.14), respectively.

The following theorem states the error bounds of the above method.

Theorem 3.3 Let $r = \theta_1 + \theta_2$. Then for x_1 produced by (3.19), we have

$$\begin{cases} \|x_1 - x(h)\| \leq \dfrac{Ch^{r+2}}{(r+1)(r+2)}, \\ \|x_1' - x'(h)\| \leq \dfrac{Ch^{r+1}}{r+1}, \end{cases} \tag{3.20}$$

where C depends on $f^{(r)}$.

The proof of this theorem is similar to that of Theorem 3.2 and we skip it for brevity.

Remark 3.2 It can be observed that generally the method (3.19) is implicit and thus an iterative solution is required. There are various iterative algorithms which can be chosen for (3.19). In this chapter, waveform relaxation (WR) algorithms are used.

WR algorithms are a family of iterative techniques designed for analyzing dynamical systems, and have been studied by a number of authors (see, e.g. [17–19, 23, 29]). The classical waveform Picard algorithm for (3.19) is

$$x_{n+1}^{[0]} = x_n^{[k]}, \quad x_{n+1}^{'[0]} = x_n^{'[k]},$$

$$x_{n+1}^{[1]} = \cos(h\Omega)x_n^{[k]} + \Omega^{-1}\sin(h\Omega)x_n^{'[k]} + \sum_{j=0}^{\theta_1-1} I_1[\alpha_{1,j}](t_n)f^{(j)}(t_n, x_n^{[k]}, x_n^{'[k]})$$

$$+ \sum_{j=0}^{\theta_2-1} I_1[\alpha_{2,j}](t_n)f^{(j)}(t_{n+1}, x_{n+1}^{[0]}, x_{n+1}^{'[0]}),$$

$$x_{n+1}^{'[1]} = -\Omega\sin(h\Omega)x_n^{[k]} + \cos(h\Omega)x_n^{'[k]} + \sum_{j=0}^{\theta_1-1} I_2[\alpha_{1,j}](t_n)f^{(j)}(t_n, x_n^{[k]}, x_n^{'[k]})$$

$$+ \sum_{j=0}^{\theta_2-1} I_2[\alpha_{2,j}](t_n)f^{(j)}(t_{n+1}, x_{n+1}^{[0]}, x_{n+1}^{'[0]}),$$

$$\vdots \tag{3.21}$$

$$x_{n+1}^{[k]} = \cos(h\Omega)x_n^{[k]} + \Omega^{-1}\sin(h\Omega)x_n^{'[k]} + \sum_{j=0}^{\theta_1-1} I_1[\alpha_{1,j}](t_n)f^{(j)}(t_n, x_n^{[k]}, x_n^{'[k]})$$

$$+ \sum_{j=0}^{\theta_2-1} I_1[\alpha_{2,j}](t_n)f^{(j)}(t_{n+1}, x_{n+1}^{[k-1]}, x_{n+1}^{'[k-1]}),$$

$$x_{n+1}^{'[k]} = -\Omega\sin(h\Omega)x_n^{[k]} + \cos(h\Omega)x_n^{'[k]} + \sum_{j=0}^{\theta_1-1} I_2[\alpha_{1,j}](t_n)f^{(j)}(t_n, x_n^{[k]}, x_n^{'[k]})$$

$$+ \sum_{j=0}^{\theta_2-1} I_2[\alpha_{2,j}](t_n)f^{(j)}(t_{n+1}, x_{n+1}^{[k-1]}, x_{n+1}^{'[k-1]}),$$

$$x_{n+2}^{[0]} = x_{n+1}^{[k]}, \quad x_{n+2}^{'[0]} = x_{n+1}^{'[k]}.$$

3.3 Practical Methods and Numerical Experiments

In this section, we present one practical numerical method and illustrate the numerical results of the improved Filon-type asymptotic method by the experiments on the Fermi–Pasta–Ulam problem.

As an illustrative example, we choose $\theta_1 = \theta_2 = 2$ in (3.19), and this gives

$$
\begin{cases}
\alpha_{1,0}(t) = \dfrac{1}{h^2}\left(1 + \dfrac{2}{h}(t - t_n)\right)(t - t_{n+1})^2, \\[2mm]
\alpha_{2,0}(t) = \dfrac{1}{h^2}\left(1 - \dfrac{2}{h}(t - t_{n+1})\right)(t - t_n)^2, \\[2mm]
\alpha_{1,1}(t) = \dfrac{1}{h^2}(t - t_n)(t - t_{n+1})^2, \\[2mm]
\alpha_{2,1}(t) = \dfrac{1}{h^2}(t - t_{n+1})(t - t_n)^2.
\end{cases}
\tag{3.22}
$$

The method determined by (3.19) and (3.22) is denoted by FAM.

Since Gautschi-type methods have been used by several authors for solving the Fermi–Pasta–Ulam problem (see, e.g. [6, 7, 9]), the integrators for comparisons are chosen as the Gautschi-type method (denoted by C) given in [6], the another Gautschi-type method (denoted by E) given in [7] and the improved Störmer–Verlet formula ISV2 given in [28]. For FAM, we use the classical waveform Picard algorithm (3.21) and choose $k = 2$ in (3.21), which means that only one iteration is needed at each step. Therefore, FAM can be implemented at a low cost.

Numerical experiments on the Fermi–Pasta–Ulam problem. The Fermi–Pasta–Ulam problem (has been considered by many authors, such as in [7, 9]) is an important model of nonlinear classical and quantum systems of interacting particles in the physics of nonlinear phenomena. It is a model for simulations in statistical mechanics which reveal highly unexpected dynamical behaviours. Consider a chain of springs, where soft nonlinear springs alternate with stiff harmonic springs (see [5]). Let the variables $q_1, \ldots, q_{2m}$ ($q_0 = q_{2m+1} = 0$) stand for the displacements of end-points of the springs, and $p_i = q_i'$ stand for their velocities. The motion is expressed by a Hamiltonian system with total energy

$$
H(p, q) = \frac{1}{2}\sum_{i=1}^{m}(p_{2i-1}^2 + p_{2i}^2) + \frac{\omega^2}{4}\sum_{i=1}^{m}(q_{2i} - q_{2i-1})^2 + \sum_{i=0}^{m}(q_{2i+1} - q_{2i})^4,
$$

where ω is large. With the change of variables

$$
x_i = \frac{q_{2i} + q_{2i-1}}{\sqrt{2}}, \quad x_{m+i} = \frac{q_{2i} - q_{2i-1}}{\sqrt{2}},
$$
$$
y_i = \frac{p_{2i} + p_{2i-1}}{\sqrt{2}}, \quad y_{m+i} = \frac{p_{2i} - p_{2i-1}}{\sqrt{2}},
$$

the movement can again be expressed by a Hamiltonian system with the Hamiltonian

$$
H(y, x) = \frac{1}{2}\sum_{i=1}^{2m}y_i^2 + \frac{\omega^2}{2}\sum_{i=1}^{m}x_{m+i}^2 + \frac{1}{4}[(x_1 - x_{m+1})^4
$$
$$
+ \sum_{i=1}^{m-1}(x_{i+1} - x_{m+i-1} - x_i - x_{m+i})^4 + (x_m + x_{2m})^4],
\tag{3.23}
$$

where x_i is a scaled displacement of the ith stiff spring, x_{m+i} denotes a scaled expansion or compression of the ith stiff spring and y_i, y_{m+i} represent their velocities or momenta.

This Hamiltonian system can be described by

$$x''(t) + \Omega^2 x(t) = -\nabla U(x), \qquad t \in [t_0, t_{\text{end}}],$$

where

$$x = (x_1, x_2, \ldots, x_{2m})^{\mathsf{T}},$$

$$\Omega = \begin{pmatrix} \mathbf{0}_{m \times m} & \mathbf{0}_{m \times m} \\ \mathbf{0}_{m \times m} & \omega I_{m \times m} \end{pmatrix},$$

$$U(x) = \frac{1}{4}\left[(x_1 - x_{m+1})^4 + \sum_{i=1}^{m-1} (x_{i+1} - x_{m+i-1} - x_i - x_{m+i})^4 + (x_m + x_{2m})^4 \right].$$

This system is exactly of the form (3.23).

We consider the problem with three stiff springs ($m = 3$). The total energy of the problem is given by (3.23) with $m = 3$ and the oscillatory energy is

$$I = I_1 + I_2 + I_3 \text{ with } I_j = \frac{1}{2}\dot{x}_{3+j}^2 + \frac{1}{2}\omega^2 x_{3+j}^2, \quad j = 1, 2, 3,$$

where x_{3+j} represents the elongation of the jth stiff spring. The kinetic energy of the mass centre motion and of the relative motion of masses joined by a stiff spring are respectively

$$T_0 = \frac{1}{2}\|\dot{x}_0\|^2, \quad T_1 = \frac{1}{2}\|\dot{x}_1\|^2.$$

The Fermi–Pasta–Ulam model shows different behaviour on different time scales:

- time scale ω^{-1}: almost-harmonic motion of the stiff springs;
- time scale ω^0: motion of the soft springs;
- time scale ω: energy exchange between stiff springs;
- time scale ω^N, $N \geq 2$: almost preservation of the oscillatory energy over very long time intervals.

Following [9], we choose

$$x_1(0) = 1, \quad y_1(0) = 1, \quad x_4(0) = \frac{1}{\omega}, \quad y_4(0) = 1, \tag{3.24}$$

with zero for the remaining initial values. The system is integrated on the interval $[0, 10]$ with the stepsizes $h = 4/(10 \times 2^j)$ for FAM and $h = 1/(10 \times 2^j)$ for the other methods, where $j = 5, \ldots, 8$. The efficiency curves (accuracy vs. the computational cost measured by the number of function evaluations required by each method) for different $\omega = 50, 100, 150, 200, 300, 400$ are shown in Fig. 3.1. It is noted that the

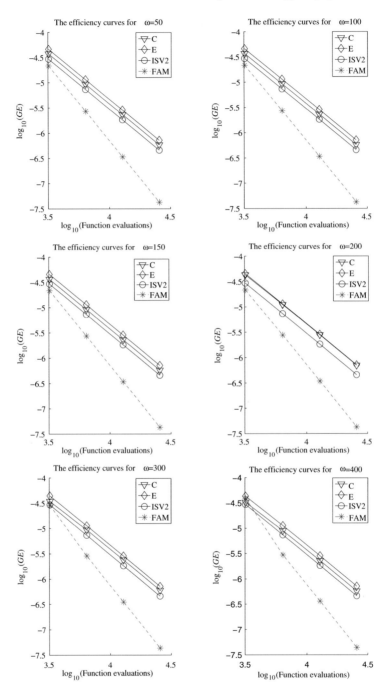

Fig. 3.1 The logarithm of the global error (*GE*) over the integration interval against the logarithm of the number of function evaluations

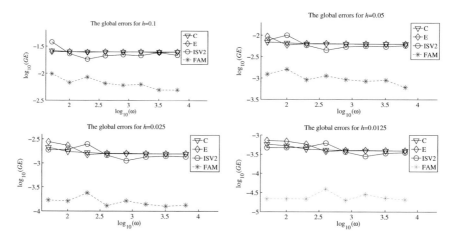

Fig. 3.2 The logarithm of the global error (*GE*) over the integration interval against the logarithm of ω

number of function evaluations means the total for both the number of f evaluations and the number of f' evaluations. Here, it should be remarked for the implicit method FAM that the number of function evaluations also includes those required in the iterative solutions. We use the result of the standard ODE45 method in MATLAB with an absolute and relative tolerance equal to 10^{-13} as the true solution for this problem. We then integrate this system on the interval $[0, 10]$ with the stepsizes $h = 0.1, 0.05, 0.025, 0.0125$ for different $\omega = 50 \times 2^i$, $i = 0, 1, \ldots, 7$. The global errors against ω for the four methods are presented in Fig. 3.2.

It can be observed from Fig. 3.1 that the FAM method derived in this chapter has much better accuracy for the same numbers of the function evaluations than Gautschi-type trigonometric or exponential integrators, and the improved Störmer–Verlet formula. Figure 3.2 illustrates that when ω becomes large, the FAM method still performs much better than the other methods.

We next consider the energy exchange between stiff components. Following [9], we choose $\omega = 50$ and the initial data (3.24) for showing the energy exchange between stiff components. Figure 3.3 shows the oscillatory energies I_1, I_2, I_3, their sum $I = I_1 + I_2 + I_3$ and the total energy $H - 0.8$ as functions of time on the interval $[0, 200]$ with $h = 0.01$ for the four methods. We can see that an exchange of energy takes place, going from the first stiff spring with energy I_1 to the second stiff spring and later to the third one. The numerical results illustrate that all the methods give a good approximation of the energy exchange between the stiff springs. Figure 3.3 also shows that H and I are well preserved over the whole interval.

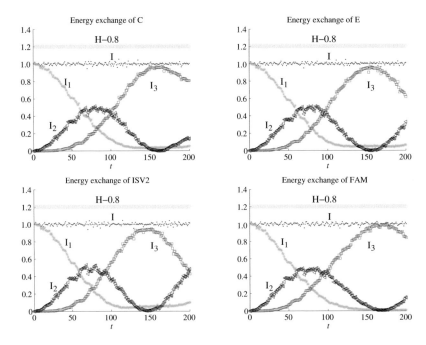

Fig. 3.3 The energy exchange between stiff components as a function of t

When the product of the stepsize and the frequency $h\omega$ is close to a multiple of π, these different methods show widely different behaviour for this problem. Accordingly, we discuss the resonance instabilities for the numerical methods. Figure 3.4 shows the maximum global errors of the Hamiltonian for the four methods as a function of $h\omega$ over the interval $[0, 1000]$ with the stepsize $h = 0.02$.

3.4 Conclusions and Discussions

We note a fact that the existing Filon-type methods in the literature are not applicable to the multi-frequency highly oscillatory problem (3.1) when M is singular. This motivates an effective improvement on existing Filon-type asymptotic methods for multi-frequency highly oscillatory differential equations. Therefore, in this chapter, we have studied further Filon-type asymptotic methods for solving the multi-frequency highly oscillatory second-order differential equations (3.2) with a positive semi-definite matrix (not necessarily nonsingular). These improved methods are a blend of the Filon-type method, asymptotic method and the matrix-variation-of-constants formula. We analysed and presented their error bounds. One practical method FAM was derived, which can be implemented at a low cost. The method FAM

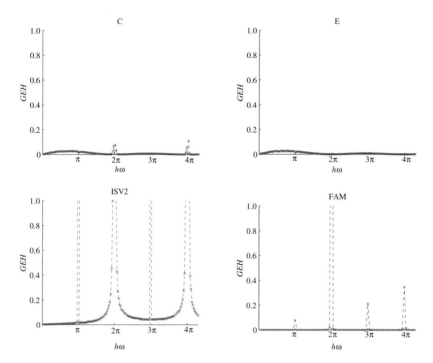

Fig. 3.4 The maximum global error (*GEH*) of the total energy for the four methods as a function of $h\omega$ with the stepsize $h = 0.02$

was compared with Gautschi-type trigonometric or exponential integrators, and with an improved Störmer–Verlet formula by numerical experiments on the Fermi–Pasta–Ulam problem, which showed the remarkable efficiency of the improved method.

This chapter is based on the recent work by Wang and Wu [26].

References

1. Barth E, Schlick T (1998) Overcoming stability limitations in biomolecular dynamics. I. Combining force splitting via extrapolation with Langevin dynamics in LN. J Chem Phys 109:1617
2. Cohen D, Jahnke T, Lorenz K, Lubich C (2006) Numerical integrators for highly oscillatory Hamiltonian systems: a review. In: Mielke A (ed) In analysis, modeling and simulation of multiscale problems. Springer, Berlin, pp 553–576
3. Dahlquist G, Bjorck A (2003) Numerical methods. Dover, Mineola
4. Franco JM (2006) New methods for oscillatory systems based on ARKN methods. Appl Numer Math 56:1040–1053
5. Galgant L, Giorgillt A, Martinoli A, Vanzini S (1992) On the problem of energy equipartition for large systems of the Fermi-Pasta-Ulam type: analytical and numerical estimates. Phys D 59:334–348
6. García-Archilla B, Sanz-Serna JM, Skeel RD (1998) Long-time-step methods for oscillatory differential equations. SIAM J Sci Comput 20:930–963

7. Hairer E, Lubich C (2000) Long-time energy conservation of numerical methods for oscillatory differential equations. SIAM J Numer Anal 38:414–441
8. Hairer E, Lubich C (2009) Oscillations over long times in numerical Hamiltonian systems. In: Engquist B, Fokas A, Hairer E, Iserles A (eds) Highly oscillatory problems. London mathematical society lecture note series, vol 366. Cambridge University Press
9. Hairer E, Lubich C, Wanner G (2006) Geometric numerical integration: structure-preserving algorithms for ordinary differential equations, 2nd edn. Springer, Berlin
10. Hochbruck M, Lubich C (1999) A Gautschi-type method for oscillatory second-order differential equations. Numer Math 83:403–426
11. Huybrechs D, Vandewalle S (2006) On the evaluation of highly oscillatory integrals by analytic continuation. SIAM J Numer Anal 44:1026–1048
12. Iserles A (2002) Think globally, act locally: solving highly-oscillatory ordinary differential equations. Appli Numer Math 43:145–160
13. Iserles A, Levin D (2011) Asymptotic expansion and quadrature of composite highly oscillatory integrals. Math Comp 80:279–296
14. Iserles A, Nøsett SP (2004) On quadrature methods for highly oscillatory integrals and their implementation. BIT Numer Math 44:755–772
15. Iserles A, Nøsett SP (2005) Efficient quadrature of highly oscillatory integrals using derivatives. Proc R Soc Lond Ser A Math Phys Eng Sci 461:1383–1399
16. Iserles A, Nøsett SP (2006) On the computation of highly oscillatory multivariate integrals with critical points. BIT Numer Math 46:549–566
17. Janssen J, Vandewalle S (1997) On SOR waveform relaxation methods. SIAM J Numer Anal 34:2456–2481
18. Khanamiryan M (2008) Quadrature methods for highly oscillatory linear and nonlinear systems of ordinary differential equations: part I. BIT Numer Math 48:743–762
19. Lubich C, Ostermann A (1987) Multigrid dynamic iteration for parabolic equations. BIT Numer Math 27:216–234
20. Olver S (2007) Numerical approximation of vector-valued highly oscillatory integrals. BIT Num Math 47:637–655
21. Petzold LR, Jay LO, Yen J (1997) Numerical solution of highly oscillatory ordinary differential equations. Acta Numer 7:437–483
22. Schlick T, Mandziuk M, Skeel RD, Srinivas K (1998) Nonlinear resonance artifacts in molecular dynamics simulations. J Comput Phys 140:1–29
23. Vandewalle S (1993) Parallel multigrid waveform relaxation for parabolic problems. In: Teubner BG (ed) Teubner scripts on numerical mathematics. Stuttgart (1993)
24. Wang B, Liu K, Wu X (2013) A Filon-type asymptotic approach to solving highly oscillatory second-order initial value problems. J Comput Phys 243:210–223
25. Wang B, Wu X (2012) A new high precision energy-preserving integrator for system of oscillatory second-order differential equations. Phys Lett A 376:1185–1190
26. Wang B, Wu X (2014) Improved Filon-type asymptotic methods for highly oscillatory differential equations with multiple time scales. J Comput Phys 276:62–73
27. Wang B, Wu X, Xia J (2013) Error bounds for explicit ERKN integrators for systems of multi-frequency oscillatory second-order differential equations. Appl Numer Math 74:17–34
28. Wang B, Wu X, Zhao H (2013) Novel improved multidimensional Strömer-Verlet formulas with applications to four aspects in scientific computation. Math Comput Modell 57:857–872
29. White J, Odeh F, Sangiovanni-Vincentelli AL, Ruehli A (1985) Waveform relaxation: theory and practice. EECS Department, University of California, Berkeley
30. Wu X, Wang B, Shi W (2013) Efficient energy-preserving integrators for oscillatory Hamiltonian systems. J Comput Phys 235:587–605
31. Wu X, Wang B, Xia J (2012) Explicit symplectic multidimensional exponential fitting modified Runge-Kutta-Nystrom methods. BIT Numer Math 52:773–795
32. Wu X, You X, Wang B (2013) Structure-preserving algorithms for oscillatory differential equations. Springer, Berlin

Chapter 4
Efficient Energy-Preserving Integrators for Multi-frequency Oscillatory Hamiltonian Systems

This chapter pays special attention to deriving and analysing an efficient energy-preserving formula in detail for the oscillatory or highly oscillatory Hamiltonian system with the Hamiltonian $H(p, q) = p^\mathsf{T} p/2 + q^\mathsf{T} M q/2 + U(q)$. The energy-preserving formula preserves the Hamiltonian exactly. We analyse in detail some important properties of the energy-preserving formula and construct efficient energy-preserving integrators in the sense of numerical implementation. The convergence analysis of the fixed-point iteration is presented for the implicit integrators derived in this chapter and it turns out that the convergence of implicit Average Vector Field methods depends on $\|M\|$, whereas the convergence of the efficient energy-preserving integrators being considered in this chapter is independent of $\|M\|$. Here $\|\cdot\|$ denotes the spectral norm.

4.1 Motivation

The recent growth of geometric numerical integration has led to the development of numerical schemes for differential equations which systematically incorporate qualitative features of the original problem into their structure. It has become a common practice for an oscillatory problem that a numerical method should be designed to preserve as much as possible the physical/geometric properties of the problem. Readers are referred to [10, 18, 30] for a good theoretical foundation of structure-preserving algorithms for ordinary differential equations. The behaviour of first integrals under numerical integration has been discussed for a long time. Examples include automatic preservation of linear integrals by all Runge-Kutta methods, automatic preservation of quadratic integrals by some (the symplectic) Runge-Kutta (-Nyström) methods, some exponential integrators and some linearization-preserving integrators. We refer the reader to [6, 8, 19, 24, 25, 28, 31] for examples on this subject. With this background, we pay attention to numerical methods that preserve energy in Hamiltonian systems. For the Hamiltonian differential equation

© Springer-Verlag Berlin Heidelberg and Science Press, Beijing, China 2015
X. Wu et al., *Structure-Preserving Algorithms for Oscillatory Differential Equations II*, DOI 10.1007/978-3-662-48156-1_4

$$q' = J^{-1}\nabla H(q), \qquad (4.1)$$

with a skew-symmetric constant matrix J and the Hamiltonian $H(q)$, the Average Vector Field (AVF) formula was first written down in [27] and is defined by

$$q_{n+1} = q_n + h \int_0^1 J^{-1}\nabla H\big((1-\tau)q_n + \tau q_{n+1}\big)\mathrm{d}\tau. \qquad (4.2)$$

The authors in [9] showed the existence of energy-preserving B-series methods. The AVF formula is identified with an energy-preserving integrator and as a B-series method in [29]. The AVF formula exactly preserves the energy for an arbitrary Hamiltonian H and only needs evaluations of the vector field. Following [29], we refer the reader to [3–5, 16] for more on the AVF formula. Another interesting class of energy-preserving integrators are "extended collocation methods" and "Hamiltonian boundary value methods", which exactly preserve the energy of polynomial Hamiltonian systems (see, e.g. [1, 2, 21–23]).

In this chapter, we pay attention to efficient energy-preserving integrators for system of nonlinear multi-frequency oscillatory second-order differential equations of the form

$$\begin{cases} q''(t) + Mq(t) = f\big(q(t)\big), & t \in [t_0, T], \\ q(t_0) = q_0, \quad q'(t_0) = q_0', \end{cases} \qquad (4.3)$$

where M is a $d \times d$ symmetric positive semi-definite matrix with $\|M\| \gg 1$ and $f : \mathbb{R}^d \to \mathbb{R}^d$ is the negative gradient of a real-valued function $U(q)$ and is a nonlinear mapping in general. A typical example is the well-known Fermi-Pasta-Ulam problem [11], which is an important model of nonlinear classical and quantum systems of interacting particles in the physics of nonlinear phenomena. Clearly, the system (4.3) is simply the following initial value problem of multi-frequency oscillatory Hamiltonian system

$$\begin{cases} p' = -\nabla_q H(p, q), & p(t_0) = p_0, \\ q' = \nabla_p H(p, q), & q(t_0) = q_0, \end{cases} \qquad (4.4)$$

with the Hamiltonian (see, e.g. [7])

$$H(p, q) = \frac{1}{2}p^{\mathsf{T}}p + \frac{1}{2}q^{\mathsf{T}}Mq + U(q). \qquad (4.5)$$

In recent years, enormous advances in dealing with the oscillatory system (4.3) have appeared and some useful approaches to constructing Runge-Kutta-Nyström (RKN) type methods have been proposed. We refer the reader to [13–15, 17, 20, 26, 32, 36, 37] for examples. The authors in [38] took advantage of the special structure of the system (4.3) brought by the linear term Mq and formulated a standard form of the multi-frequency and multidimensional ERKN methods (extended RKN methods). The ERKN methods perform numerically much better than the classical RKN methods due to incorporating the special structure of the equation, given by the linear term Mq, into the structure of the method. The symplecticity conditions for multi-frequency and multidimensional ERKN methods were investigated and presented in [39]. It is important to note that the symplecticity conditions for ERKN methods reduce to those for the classical RKN methods when $M = \mathbf{0}_{d \times d}$. However, it is known that the symplectic ERKN methods cannot exactly preserve the Hamiltonian of the system (4.4) in general.

If we apply the AVF formula (4.2) to the Hamiltonian system (4.4) or (4.3), we obtain

$$
\begin{cases}
q_{n+1} = q_n + hp_n + \dfrac{h^2}{2} \displaystyle\int_0^1 g\big((1 - \tau)q_n + \tau q_{n+1}\big)\mathrm{d}\tau, \\[2ex]
p_{n+1} = p_n + h \displaystyle\int_0^1 g\big((1 - \tau)q_n + \tau q_{n+1}\big)\mathrm{d}\tau,
\end{cases}
\tag{4.6}
$$

where

$$
g(q) = -Mq - \nabla_q U(q) = -Mq + f(q).
$$

However, it should be noted that the AVF formula (4.6) for the Hamiltonian system (4.4) takes no account of the special structure given by the linear term Mq of the system (4.4), or (4.3), equivalently. The energy-preserving numerical methods taking advantage of special structure deserve to be investigated further. Moreover, we can observe that the linear term Mq now becomes a part of the integrand in the AVF formula (4.6), which makes trouble in applications, since the practical AVF methods are all implicit and iterative computations are required in general. In fact, the linear term Mq in the AVF formula (4.6) is a serious obstacle for the convergence of the fixed-point iteration. The difficulty with this linear term, under the integral in the AVF formula (4.6), is that the convergence of fixed-point iteration depends heavily on $\|M\|$. This motivates an exploration of efficient energy-preserving integrators designed specially for achieving efficient performance on oscillatory or highly oscillatory Hamiltonian systems with the Hamiltonian (4.5). In what follows, we derive and establish an efficient energy-preserving formula for the Hamiltonian system (4.4) or (4.3), which is called an adapted AVF (AAVF) formula. Then the theoretical analysis of the AAVF formula is presented and the corresponding numerical scheme is proposed and verified numerically.

4.2 Preliminaries

This section presents some preliminaries in order to gain an insight into a class of highly efficient energy-preserving integrators for the oscillatory Hamiltonian system (4.4), namely, oscillatory second-order differential equations (4.3).

To begin with, we consider the following unconditionally convergent matrix-valued functions defined first in [37]:

$$\phi_l(M) := \sum_{k=0}^{\infty} \frac{(-1)^k M^k}{(2k+l)!}, \quad l = 0, 1, \ldots \tag{4.7}$$

Remark 4.1 It is noted that for the particular case where M is a symmetric positive semi-definite matrix, we have the decomposition of M as follows

$$M = P^{\mathsf{T}} W^2 P = \Omega_0^2, \text{ with } \Omega_0 = P^{\mathsf{T}} W P,$$

where P is an orthogonal matrix and W is a diagonal matrix with nonnegative diagonal entries which are the square roots of the eigenvalues of M. In this special case, the definition (4.7) reduces to the G-functions used in [15] and the ϕ-functions given in [12]. In particular, an important example are the well-known matrix trigonometric functions corresponding to $l = 0$ and $l = 1$ in (4.7) which are used in the exponential integrators proposed in [18]. However, it is noted that the definition (4.7) depends directly on M and is applicable not only to symmetric matrices but also to nonsymmetric ones. Therefore, we use (4.7).

Some interesting properties of these matrix-valued functions are established in the following proposition.

Proposition 4.1 *For a symmetric positive semi-definite matrix M and $l = 0, 1, \ldots,$ the ϕ-functions defined by (4.7) satisfy:*

(i)

$$\phi_l^{\mathsf{T}}(M) = \phi_l(M). \tag{4.8}$$

(ii)

$$\phi_0^2(M) + M\phi_1^2(M) = I, \tag{4.9}$$

where I is the identity matrix sharing the same order as M.

(iii) For a general matrix M (not necessarily symmetric), it is true that

$$\|\phi_l(M)\| \leq \gamma_l, \ l = 0, 1, 2, \tag{4.10}$$

where γ_l are constants depending on $\|M\|$ in general. However, for the particular and important case, where M is a symmetric positive semi-definite matrix, γ_l can be chosen as $\gamma_l = \dfrac{1}{l!}$.

These results are evident and we omit the details of the proof for brevity.

Following Theorem 1.1 of Chap. 1, the next theorem states a result on the exact solution of the system (4.3) and its derivative.

Theorem 4.1 *If* $f : \mathbb{R}^d \to \mathbb{R}^d$ *is continuous in (4.3), then the solution of (4.3) and its derivative satisfies the following equations*

$$
\begin{cases}
q(t) = \phi_0\big((t - t_0)^2 M\big)q_0 + (t - t_0)\phi_1\big((t - t_0)^2 M\big)p_0 \\
\quad + \displaystyle\int_{t_0}^{t} (t - \xi)\phi_1\big((t - \xi)^2 M\big)\hat{f}(\xi)\mathrm{d}\xi, \\
p(t) = -(t - t_0)M\phi_1\big((t - t_0)^2 M\big)q_0 + \phi_0\big((t - t_0)^2 M\big)p_0 \\
\quad + \displaystyle\int_{t_0}^{t} \phi_0\big((t - \xi)^2 M\big)\hat{f}(\xi)\mathrm{d}\xi,
\end{cases}
\tag{4.11}
$$

where t_0, t *are any real numbers and* $\hat{f}(\xi) = f\big(q(\xi)\big)$.

It can be observed that the variation of constants formula (4.11) is independent of the matrix decomposition of M, which is different from the traditional one appearing in the literature, since the latter is dependent on the matrix decompositions $M = \Omega_0^2$. The class of multi-frequency and multidimensional oscillatory problems which fall within its scope is broader, since in the case, where the matrix M in system (4.3) is not symmetric (see, e.g. Problem 3 in [38]), the decomposition is not always applicable.

It follows immediately from (4.11) that

$$
\begin{cases}
q(t_n + h) = \phi_0(V)q(t_n) + h\phi_1(V)p(t_n) + h^2 \displaystyle\int_0^1 (1 - z)\phi_1\big((1 - z)^2 V\big)\hat{f}(t_n + hz)\mathrm{d}z, \\
p(t_n + h) = -hM\phi_1(V)q(t_n) + \phi_0(V)p(t_n) + h \displaystyle\int_0^1 \phi_0\big((1 - z)^2 V\big)\hat{f}(t_n + hz)\mathrm{d}z,
\end{cases}
\tag{4.12}
$$

where h is the stepsize and $V = h^2 M$.

4.3 The Derivation of the AAVF Formula

In this section, we are concerned with the derivation of the AAVF formula for (4.4) or (4.3), equivalently.

Based on the variation of constants formula (4.12), we consider the following scheme:

$$
\begin{cases}
q_{n+1} = \phi_0(V)q_n + h\phi_1(V)p_n + h^2 I_1, \\
p_{n+1} = -hM\phi_1(V)q_n + \phi_0(V)p_n + h I_2,
\end{cases}
\tag{4.13}
$$

where I_1, I_2 are to be determined such that the energy-preserving condition

$$
H(p_{n+1}, q_{n+1}) = H(p_n, q_n),
$$

is satisfied exactly.

We first compute

$$H(p_{n+1}, q_{n+1}) = \frac{1}{2} p_{n+1}^{\mathsf{T}} p_{n+1} + \frac{1}{2} q_{n+1}^{\mathsf{T}} M q_{n+1} + U(q_{n+1}). \tag{4.14}$$

Keeping the fact in mind that M and all $\phi_l(V)$ are commutative and symmetric, and inserting (4.13) into (4.14), with some manipulation, we obtain

$$\begin{aligned} H(p_{n+1}, q_{n+1}) &= \frac{1}{2} p_n^{\mathsf{T}} \big(\phi_0^2(V) + V\phi_1^2(V)\big) p_n + \frac{1}{2} q_n^{\mathsf{T}} M \big(\phi_0^2(V) + V\phi_1^2(V)\big) q_n \\ &\quad + q_n^{\mathsf{T}} V \big(\phi_0(V) I_1 - \phi_1(V) I_2\big) + h p_n^{\mathsf{T}} \big(\phi_0(V) I_2 + V\phi_1(V) I_1\big) \\ &\quad + \frac{1}{2} h^2 (I_2^{\mathsf{T}} I_2 + I_1^{\mathsf{T}} V I_1) + U(q_{n+1}). \end{aligned} \tag{4.15}$$

It follows from (4.9) that

$$\begin{aligned} H(p_{n+1}, q_{n+1}) &= \frac{1}{2} p_n^{\mathsf{T}} p_n + \frac{1}{2} q_n^{\mathsf{T}} M q_n + q_n^{\mathsf{T}} V \big(\phi_0(V) I_1 - \phi_1(V) I_2\big) \\ &\quad + h p_n^{\mathsf{T}} \big(\phi_0(V) I_2 + V\phi_1(V) I_1\big) + \frac{1}{2} h^2 (I_2^{\mathsf{T}} I_2 + I_1^{\mathsf{T}} V I_1) + U(q_{n+1}). \end{aligned} \tag{4.16}$$

We then calculate

$$\begin{aligned} & U(q_n) - U(q_{n+1}) \\ &= -\int_0^1 \mathrm{d}U\big((1 - \tau)q_n + \tau q_{n+1}\big) = -\int_0^1 (q_{n+1} - q_n)^{\mathsf{T}} \nabla_q U\big((1 - \tau)q_n + \tau q_{n+1}\big) \mathrm{d}\tau \\ &= (q_{n+1} - q_n)^{\mathsf{T}} \int_0^1 f\big((1 - \tau)q_n + \tau q_{n+1}\big) \mathrm{d}\tau. \end{aligned} \tag{4.17}$$

The first equation of (4.13) gives

$$q_{n+1} - q_n = \big(\phi_0(V) - I\big) q_n + \phi_1(V)(h p_n) + h^2 I_1. \tag{4.18}$$

Letting

$$I_f = \int_0^1 f\big((1 - \tau)q_n + \tau q_{n+1}\big) \mathrm{d}\tau,$$

and inserting (4.18) into (4.17), yields

$$U(q_n) - U(q_{n+1}) = q_n^{\mathsf{T}} \big(\phi_0(V) - I\big) I_f + h p_n^{\mathsf{T}} \phi_1(V) I_f + h^2 I_1^{\mathsf{T}} I_f. \tag{4.19}$$

Then the formula (4.16) can be rewritten as

$$
\begin{aligned}
H(p_{n+1}, q_{n+1}) &= \frac{1}{2} p_n^\mathsf{T} p_n + \frac{1}{2} q_n^\mathsf{T} M q_n + U(q_n) \\
&\quad + q_n^\mathsf{T} \left(V \big(\phi_0(V) I_1 - \phi_1(V) I_2 \big) - (\phi_0(V) - I) I_f \right) \\
&\quad + h p_n^\mathsf{T} \left(\big(\phi_0(V) I_2 + V \phi_1(V) I_1 \big) - \phi_1(V) I_f \right) \\
&\quad + h^2 \left(\frac{1}{2} (I_2^\mathsf{T} I_2 + I_1^\mathsf{T} V I_1) - I_1^\mathsf{T} I_f \right) \\
&= H(p_n, q_n) + \triangle_n,
\end{aligned}
\tag{4.20}
$$

where

$$
\begin{aligned}
\triangle_n &= q_n^\mathsf{T} \left(V \big(\phi_0(V) I_1 - \phi_1(V) I_2 \big) - (\phi_0(V) - I) I_f \right) + h p_n^\mathsf{T} \Big(\big(\phi_0(V) I_2 \\
&\quad + V \phi_1(V) I_1 \big) - \phi_1(V) I_f \Big) + h^2 \left(\frac{1}{2} (I_2^\mathsf{T} I_2 + I_1^\mathsf{T} V I_1) - I_1^\mathsf{T} I_f \right).
\end{aligned}
\tag{4.21}
$$

The above analysis gives the following important theorem immediately.

Theorem 4.2 *The formula (4.13) exactly preserves the energy H determined by (4.5), i.e.*

$$
H(p_{n+1}, q_{n+1}) = H(p_n, q_n),
$$

if and only if $\triangle_n = 0$ for $n = 0, 1, \ldots$

Based on Theorem 4.2, an efficient energy-preserving formula can be derived.

Theorem 4.3 *The formula (4.13) preserves H defined by (4.5) exactly if the conditions*

$$
I_1 = \phi_2(V) I_f, \quad I_2 = \phi_1(V) I_f
\tag{4.22}
$$

are satisfied.

Proof It is clear from (4.21) that the requirement of the following three equations

$$
\begin{cases}
V \big(\phi_0(V) I_1 - \phi_1(V) I_2 \big) = (\phi_0(V) - I) I_f, \\
\big(\phi_0(V) I_2 + V \phi_1(V) I_1 \big) = \phi_1(V) I_f, \\
\frac{1}{2} (I_2^\mathsf{T} I_2 + I_1^\mathsf{T} V I_1) = I_1^\mathsf{T} I_f,
\end{cases}
\tag{4.23}
$$

means $\triangle_n = 0$, $n = 0, 1, \ldots$. Thus we have

$$
H(p_{n+1}, q_{n+1}) = H(p_n, q_n).
$$

Solving the first and second equations in (4.23) yields (4.22) and it can be verified that under the condition (4.22), the third equation in (4.23) is satisfied as well. Therefore, (4.22) represents the sufficient conditions for energy preservation of scheme (4.13).   □

We are now in a position to present the following efficient energy-preserving formula, that is, the AAVF formula.

Definition 4.1 The AAVF (adapted AVF) formula for integrating the Hamiltonian system (4.4) or the oscillatory system (4.3) is defined by

$$
\begin{cases}
q_{n+1} = \phi_0(V)q_n + h\phi_1(V)p_n + h^2\phi_2(V)\displaystyle\int_0^1 f\big((1-\tau)q_n + \tau q_{n+1}\big)\mathrm{d}\tau, \\[4mm]
p_{n+1} = -hM\phi_1(V)q_n + \phi_0(V)p_n + h\phi_1(V)\displaystyle\int_0^1 f\big((1-\tau)q_n + \tau q_{n+1}\big)\mathrm{d}\tau,
\end{cases}
$$
$$(4.24)$$

where h is the stepsize, and $\phi_0(V)$, $\phi_1(V)$ and $\phi_2(V)$ are matrix-valued functions of $V = h^2 M$ defined by (4.7).

Remark 4.2 It is noted that the formula (4.24) makes full use of the special structure of the Eq. (4.3) given by the linear term Mq to adapt it to the energy-preserving behaviour of the true solutions (4.12). The AAVF formula is implicit in general, but only the first equation of (4.24) needs iterative computation. Moreover, just as for the AVF formula, formula (4.24) exactly preserves the energy for an arbitrary Hamiltonian (4.5). It can be observed that when $V \to \mathbf{0}_{d\times d}$, (4.24) reduces to the AVF formula (4.6). Therefore, it is natural to call (4.24) an AAVF (adapted AVF) formula, which is an essential generalization of the AVF formula for the Hamiltonian system (4.4) or the oscillatory system (4.3). Moreover, for fixed stepsize h, a numerical method based on the AAVF formula only requires computing the matrix functions $\phi_0(V)$, $\phi_1(V)$ and $\phi_2(V)$ one time, respectively, and then the results of these matrix functions can be used repeatedly in the sequel of computations. Besides, it is noted that in most of practical applications, the matrix M is often real symmetric but sparse, in fact, tridiagonal. Thus, $\phi_0(V)$, $\phi_1(V)$ and $\phi_2(V)$ can be obtained at comparatively small additional computational cost.

Remark 4.3 We note that Hochbruck and Lubich developed Gautschi-type exponential integrators in [20] for the system (4.3) and the scheme of Gautschi-type exponential integrators looks like (4.24) at first sight. However, Gautschi's methods are designed for different purposes and in general they cannot exactly preserve the energy of Hamiltonian systems. For the analysis of energy conservation of Gautschi's methods, we refer the reader to [17, 18]. We also note that the use of filter functions for Gautschi's methods is a good idea, but that is for a different purpose from our key point in this chapter. It is of great importance to observe that our formula (4.24) is derived based on Theorem 4.2. Thus, it combines the existing exponential integrators with the AVF method and then exactly preserves the energy.

Remark 4.4 Another significant point here is that the AAVF formula (4.24) incorporates the special structure of the system (4.3) given by the linear term Mq into the variation of constants formula (4.11), so that the integral in the AAVF formula (4.24) is independent of the matrix M. This property is very important and powerful for an efficient structure-preserving integrator in applications. On the contrary, the AVF formula (4.6) is dependent on the matrix M and this fact leads to a serious disadvantage of the AVF formula in applications to the oscillatory Hamiltonian system (4.4) or (4.3).

4.4 Some Properties of the AAVF Formula

In this section, we further study the AAVF formula and present some important properties in relation to the formula.

4.4.1 Stability and Phase Properties

Taking into account that (4.3) is an oscillatory system, an analysis of the stability and phase properties of the AAVF formula is necessary.

We first consider the stability and phase properties of the AAVF formula (4.24). Following [33, 35], we use a revised second-order scalar homogeneous linear equation as a test problem:

$$q''(t) + \omega^2 q(t) = -\varepsilon q(t), \quad \omega^2 + \varepsilon > 0, \tag{4.25}$$

where ω represents an estimate of the dominant frequency λ and $\varepsilon = \lambda^2 - \omega^2$ is the error of the estimation. Applying the AAVF formula (4.24) to (4.25) yields

$$\begin{pmatrix} q_{n+1} \\ hp_{n+1} \end{pmatrix} = S(V, z) \begin{pmatrix} q_n \\ hp_n \end{pmatrix},$$

where the stability matrix $S(V, z)$ is determined by

$$S(V, z) = \begin{pmatrix} \dfrac{2\phi_0(V) - z\phi_2(V)}{2 + z\phi_2(V)} & \dfrac{2\phi_1(V)}{2 + z\phi_2(V)} \\ -\dfrac{\phi_1(V)(2V + z + z\phi_0(V) + V z\phi_2(V))}{2 + z\phi_2(V)} & \dfrac{2\phi_0(V) + z\phi_0(V)\phi_2(V) - z(\phi_1(V))^2}{2 + z\phi_2(V)} \end{pmatrix} \tag{4.26}$$

with $z = \varepsilon h^2$ and $V = h^2 \omega^2$.

The stability of the AAVF formula (4.24) is characterized by the spectral radius $\rho(S(V, z))$. We give the definition of stability for the AAVF formula (4.24).

Definition 4.2 We call

$$R_s = \{(V, z)|\ V > 0 \text{ and } \rho\big(S(V, z)\big) < 1\}$$

the *stability region* of the AAVF formula (4.24) and

$$R_p = \left\{(V, z)|\ V > 0,\ \rho\big(S(V, z)\big) = 1 \text{ and } \mathrm{tr}\big(S(V, z)\big)^2 < 4\det\big(S(V, z)\big)\right\}$$

the *periodicity region*.

The dispersion error and the dissipation error can be defined similarly to [33].

Definition 4.3 The following quantities

$$\phi(\eta) = \eta - \arccos\left(\frac{\mathrm{tr}\big(S(V, z)\big)}{2\sqrt{\det\big(S(V, z)\big)}}\right),\quad d(\eta) = 1 - \sqrt{\det\big(S(V, z)\big)}$$

are called the dispersion error and the dissipation error of the AAVF formula (4.24), respectively, where $\eta = \sqrt{V + z}$. Thus, the formula is said to be dispersive of order q if $\phi(\eta) = \mathcal{O}(\eta^{q+1})$, and is said to be dissipative of order r if $d(\eta) = \mathcal{O}(\eta^{r+1})$. If $\phi(\eta) = 0$ and $d(\eta) = 0$, the formula is said to be zero dispersive and zero dissipative.

For the AAVF formula (4.24), the stability region is shown in Fig. 4.1. From (4.26), it is easy to verify that $\det\big(S(V, z)\big) = 1$, thus the AAVF formula (4.24) is zero dissipative. The dispersion error of the AAVF formula is given by

$$\phi(\eta) = \frac{\varepsilon}{12(\varepsilon + \omega^2)}\eta^3 + \mathcal{O}(\eta^5).$$

Fig. 4.1 Stability region for the AAVF formula

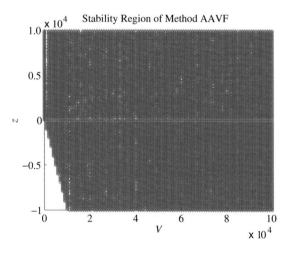

4.4.2 Other Properties

In what follows, we turn to other properties of the AAVF formula (4.24).

Theorem 4.4 *The AAVF formula (4.24) is symmetric.*

Proof Exchanging $q_{n+1} \leftrightarrow q_n$, $p_{n+1} \leftrightarrow p_n$ and replacing h by $-h$ in (4.24), we have

$$
\begin{cases}
q_n = \phi_0(V)q_{n+1} - h\phi_1(V)p_{n+1} + h^2\phi_2(V)\int_0^1 f\big((1-\tau)q_{n+1} + \tau q_n\big)\mathrm{d}\tau, \\[2mm]
p_n = hM\phi_1(V)q_{n+1} + \phi_0(V)p_{n+1} - h\phi_1(V)\int_0^1 f\big((1-\tau)q_{n+1} + \tau q_n\big)\mathrm{d}\tau.
\end{cases}
\tag{4.27}
$$

From formula (4.27), it follows that

$$
\begin{cases}
q_{n+1} = \phi_0(V)q_n + h\phi_1(V)p_n + h^2\big(\phi_1^2(V) \\[1mm]
\qquad - \phi_0(V)\phi_2(V)\big)\int_0^1 f\big((1-\tau)q_{n+1} + \tau q_n\big)\mathrm{d}\tau, \\[2mm]
p_{n+1} = -hM\phi_1(V)q_n + \phi_0(V)p_n + h\big(V\phi_1(V)\phi_2(V) \\[1mm]
\qquad + \phi_0(V)\phi_1(V)\big)\int_0^1 f\big((1-\tau)q_{n+1} + \tau q_n\big)\mathrm{d}\tau.
\end{cases}
\tag{4.28}
$$

Keeping the properties of matrix-valued functions (4.7)

$$
\phi_1^2(V) - \phi_0(V)\phi_2(V) = \phi_2(V), \quad V\phi_1(V)\phi_2(V) + \phi_0(V)\phi_1(V) = \phi_1(V),
$$

in mind, (4.28) becomes

$$
\begin{cases}
q_{n+1} = \phi_0(V)q_n + h\phi_1(V)p_n + h^2\phi_2(V)\int_0^1 f\big((1-\tau)q_{n+1} + \tau q_n\big)\mathrm{d}\tau, \\[2mm]
p_{n+1} = -hM\phi_1(V)q_n + \phi_0(V)p_n + h\phi_1(V)\int_0^1 f\big((1-\tau)q_{n+1} + \tau q_n\big)\mathrm{d}\tau.
\end{cases}
\tag{4.29}
$$

Letting $x = 1 - \tau$ yields

$$
\int_0^1 f\big((1-\tau)q_{n+1} + \tau q_n\big)\mathrm{d}\tau = \int_0^1 f\big(xq_{n+1} + (1-x)q_n\big)\mathrm{d}x
$$

$$
= \int_0^1 f\big((1-\tau)q_n + \tau q_{n+1}\big)\mathrm{d}\tau,
$$

which shows that (4.29) is the same as (4.24). Therefore, the AAVF formula (4.24) is symmetric. $\qquad\square$

It can be observed that the integral in (4.24) can be evaluated exactly if U has a special structure. The following theorem states this point and it can be used to evaluate the integral.

Theorem 4.5 *If $U = U(a^\top q)$, where $a \in \mathbb{R}^d$, then*

$$\int_0^1 f\big((1-\tau)q_n + \tau q_{n+1}\big)\mathrm{d}\tau = \frac{-a}{a^\top q_{n+1} - a^\top q_n}\big(U(a^\top q_{n+1}) - U(a^\top q_n)\big).$$

Proof

$$\int_0^1 f\big((1-\tau)q_n + \tau q_{n+1}\big)\mathrm{d}\tau = -\int_0^1 a U'\Big(a^\top\big((1-\tau)q_n + \tau q_{n+1}\big)\Big)\mathrm{d}\tau$$

$$= \frac{-a}{a^\top q_{n+1} - a^\top q_n}\int_0^1 \frac{\mathrm{d}U\big((1-\tau)q_n + \tau q_{n+1}\big)}{\mathrm{d}\tau}\mathrm{d}\tau$$

$$= \frac{-a}{a^\top q_{n+1} - a^\top q_n}\big(U(a^\top q_{n+1}) - U(a^\top q_n)\big). \qquad \square$$

We next consider and analyse the case that the integral in (4.24) cannot be evaluated exactly. This means that we have to use a numerical integration for the integral appearing in (4.24) which leads to the following theorem.

Theorem 4.6 *Let (b_i, c_i), $i = 1, \ldots, s$ be the weights and nodes of a quadrature rule on $[0, 1]$ that is exact for polynomials of degree $\le n-1$. Let $U(q)$ be a polynomial in q of degree n. Then the s-stage modified AAVF formula*

$$\begin{cases} q_{n+1} = \phi_0(V)q_n + h\phi_1(V)p_n + h^2\phi_2(V)\displaystyle\sum_{i=1}^s b_i f\big(q_n + c_i(q_{n+1} - q_n)\big), \\ p_{n+1} = -hM\phi_1(V)q_n + \phi_0(V)p_n + h\phi_1(V)\displaystyle\sum_{i=1}^s b_i f\big(q_n + c_i(q_{n+1} - q_n)\big), \end{cases}$$

(4.30)

exactly preserves H defined by (4.5).

Proof This conclusion follows immediately from the fact that the quadrature rule is exact for all polynomials of degree $\le n - 1$. $\qquad \square$

We can also apply a quadrature to the integral in the first formula of (4.24) and another different quadrature to the integral in the second formula of (4.24) to obtain various numerical schemes. The following theorem states this point.

Theorem 4.7 *Let $(\bar{b}_i, \bar{c}_i)$, $i = 1, \ldots, s$ be the weights and nodes of a quadrature rule on $[0, 1]$ that is exact for all polynomials of degree $\le m - 1$ and let $(\bar{b}_i, \bar{c}_i)$, $i = 1, \ldots, s$ be the weights and the nodes of a quadrature rule on $[0, 1]$ that is exact for all polynomials of degree $\le n - 1$. Let $U(q)$ be a polynomial in q of degree $z \le \min\{m, n\}$. Then the s-stage modified AAVF formula*

$$
\begin{cases}
q_{n+1} = \phi_0(V)q_n + h\phi_1(V)p_n + h^2\phi_2(V) \sum_{i=1}^{s} \tilde{b}_i f\left(q_n + \tilde{c}_i(q_{n+1} - q_n)\right), \\[2ex]
p_{n+1} = -hM\phi_1(V)q_n + \phi_0(V)p_n + h\phi_1(V) \sum_{i=1}^{s} \bar{b}_i f\left(q_n + \bar{c}_i(q_{n+1} - q_n)\right),
\end{cases}
$$

$$(4.31)$$

exactly preserves H defined by (4.5).

It can be observed that usually the first formula in (4.30) or (4.31) is implicit and requires iterative solution. Here, for implicit methods, we use the well-known fixed-point iteration in practical computation. The key point about the convergence of the fixed-point iteration for the first formula of (4.30) is presented by the following theorem.

Theorem 4.8 *Let f satisfy a Lipschitz condition, i.e. there exists a constant L with the property $\|f(q_1) - f(q_2)\| \leq L \|q_1 - q_2\|$. If*

$$
0 < h \leq \hat{h} < \frac{\sqrt{2}}{\sqrt{L \sum_{i=1}^{s} |b_i c_i|}},
$$

then the fixed-point iteration for the first formula of (4.30) is convergent.

Proof It follows from Proposition 4.1 that $\|\phi_2(V)\| \leq \dfrac{1}{2}$. Let

$$
\varphi(x) = \phi_0(V)q_n + h\phi_1(V)p_n + h^2\phi_2(V) \sum_{i=1}^{s} b_i f\left(q_n + c_i(x - q_n)\right).
$$

Then

$$
\|\varphi(x) - \varphi(y)\| = \left\| h^2\phi_2(V) \sum_{i=1}^{s} b_i \left(f\left(q_n + c_i(x - q_n)\right) - f\left(q_n + c_i(y - q_n)\right) \right) \right\|
$$

$$
\leq h^2 L \|\phi_2(V)\| \sum_{i=1}^{s} |b_i c_i| \, \|x - y\| \leq \frac{1}{2}\hat{h}^2 L \sum_{i=1}^{s} |b_i c_i| \, \|x - y\|
$$

$$
= \rho \|x - y\|,
$$

where

$$
0 < \rho = \frac{1}{2}\hat{h}^2 L \sum_{i=1}^{s} |b_i c_i| < 1. \tag{4.32}
$$

By the assumption and the well-known contraction mapping theorem, the fixed-point iteration for the first formula of (4.30) is convergent. $\qquad\square$

The first equation in the AVF formula (4.6) is also implicit and requires iterative solution. Under the assumption that f satisfies a Lipschitz condition, in order to analyse the convergence for the fixed-point iteration for the AVF formula (4.6), we set the iterative function

$$\psi(x) = q_n + hp_n + \frac{h^2}{2}\sum_{i=1}^{s} b_i g\big(q_n + c_i(x - q_n)\big).$$

Then,

$$
\begin{aligned}
\|\psi(x) - \psi(y)\| &= \left\| \frac{h^2}{2}\sum_{i=1}^{s} b_i \Big(g\big(q_n + c_i(x - q_n)\big) - g\big(q_n + c_i(y - q_n)\big)\Big)\right\| \\
&\le \frac{h^2}{2} \left\| \sum_{i=1}^{s} b_i \Big(f\big(q_n + c_i(x - q_n)\big) - f\big(q_n + c_i(y - q_n)\big)\Big)\right\| \\
&\quad + \frac{h^2}{2} \left\| \sum_{i=1}^{s} b_i \Big(M\big(q_n + c_i(x - q_n)\big) - M\big(q_n + c_i(y - q_n)\big)\Big)\right\| \\
&\le \frac{h^2}{2} L \sum_{i=1}^{s} |b_i c_i|\, \|x - y\| + \frac{h^2}{2} \|M\| \sum_{i=1}^{s} |b_i c_i|\, \|x - y\| \\
&= \frac{h^2}{2}(L + \|M\|) \sum_{i=1}^{s} |b_i c_i|\, \|x - y\|. \\
&\le \frac{1}{2}\tilde{h}^2 (L + \|M\|) \sum_{i=1}^{s} |b_i c_i|\, \|x - y\| = \hat{\rho}\,\|x - y\|,
\end{aligned}
$$

where

$$0 < \hat{\rho} = \frac{1}{2}\tilde{h}^2(L + \|M\|)\sum_{i=1}^{s} |b_i c_i| < 1. \qquad (4.33)$$

Consequently, it is obvious from (4.33) that the convergence of AVF formula depends on $\|M\|$, and *the larger $\|M\|$ becomes, the smaller the required stepsize. Whereas, it is significant to note that the convergence of the AAVF formula is independent of $\|M\|$ from (4.32).* This fact implies that the AAVF formula has better efficiency than the AVF formula, especially when $\|M\|$ is large, such as when the problem (4.3) is highly oscillatory.

4.5 Some Integrators Based on AAVF Formula

Two straightforward examples of Theorem 4.6 are the Simpson's rule,

$$
\begin{cases}
q_{n+1} = \phi_0(V)q_n + h\phi_1(V)p_n + \dfrac{h^2}{6}\phi_2(V)\left(f(q_n) + 4f\left(\dfrac{q_n + q_{n+1}}{2}\right) + f(q_{n+1})\right), \\[2mm]
p_{n+1} = -hM\phi_1(V)q_n + \phi_0(V)p_n + \dfrac{h}{6}\phi_1(V)\left(f(q_n) + 4f\left(\dfrac{q_n + q_{n+1}}{2}\right) + f(q_{n+1})\right),
\end{cases}
\tag{4.34}
$$

and the two-point Gauss-Legendre rule,

$$
\begin{cases}
q_{n+1} = \phi_0(V)q_n + h\phi_1(V)p_n \\[1mm]
\quad + \dfrac{h^2}{2}\phi_2(V)\left(f\left(\dfrac{3+\sqrt{3}}{6}q_n + \dfrac{3-\sqrt{3}}{6}q_{n+1}\right) + f\left(\dfrac{3-\sqrt{3}}{6}q_n + \dfrac{3+\sqrt{3}}{6}q_{n+1}\right)\right), \\[3mm]
p_{n+1} = -hM\phi_1(V)q_n + \phi_0(V)p_n \\[1mm]
\quad + \dfrac{h}{2}\phi_1(V)\left(f\left(\dfrac{3+\sqrt{3}}{6}q_n + \dfrac{3-\sqrt{3}}{6}q_{n+1}\right) + f\left(\dfrac{3-\sqrt{3}}{6}q_n + \dfrac{3+\sqrt{3}}{6}q_{n+1}\right)\right).
\end{cases}
\tag{4.35}
$$

We denote the methods (4.34) and (4.35) by AAVF1 and AAVF2, respectively. We note that these two formulae are both implicit and require iterative solution. In order to obtain an explicit scheme, we apply the left rectangle rule to the integral in the first formula of (4.24) and the trapezoidal rule to the integral in the second formula of (4.24). This yields an explicit scheme

$$
\begin{cases}
q_{n+1} = \phi_0(V)q_n + h\phi_1(V)p_n + h^2\phi_2(V)f(q_n), \\[2mm]
p_{n+1} = -hM\phi_1(V)q_n + \phi_0(V)p_n + \dfrac{h}{2}\phi_1(V)\big(f(q_n) + f(q_{n+1})\big).
\end{cases}
\tag{4.36}
$$

Here, we are not interested in *explicit AAVF schemes since they have only limited accuracy approximating the integral in* (4.24).

Remark 4.5 One can think of the representation (4.30) or (4.31) as an approach to implementing (4.24). It is clear that we can use high-order quadrature to obtain energy-preserving methods that exactly preserve the Hamiltonian with $U(q)$ a polynomial in q of any degree.

Remark 4.6 If $U(q)$ is not a polynomial, in general, (4.30) does not exactly preserve the energy due to the numerical error in approximating the integral

$$
\int_0^1 f\big((1-\tau)q_{n+1} + \tau q_n\big)d\tau = \frac{1}{2}\int_{-1}^1 s\big(\tau(\xi)\big)d\xi,
\tag{4.37}
$$

where

$$
\tau(\xi) = \frac{1}{2}\xi + \frac{1}{2},
$$

and

$$s\big(\tau(\xi)\big) = f\Big(\frac{1+\xi}{2}q_n + \frac{1-\xi}{2}q_{n+1}\Big).$$

Therefore, high-order quadrature, such as Gaussian quadrature is needed in applications, so that the energy is preserved as accurately as possible. An example based on the three-point Gauss–Legendre rule can be found in [34].

In what follows, we analyse and derive higher-order integrators based on the AAVF formula by applying the four-point, and five-point, Gauss–Legendre rules to the integral

$$\int_{-1}^{1} s\big(\tau(\xi)\big)\mathrm{d}\xi.$$

Applying the four-point Gauss–Legendre's quadrature to (4.37) gives

$$
\left\{
\begin{aligned}
q_{n+1} ={}& \phi_0(V)q_n + h\phi_1(V)p_n \\
&+ \frac{h^2}{2}\phi_2(V)\Big(\big(\frac{1}{2}-\frac{1}{6}\sqrt{\frac{5}{6}}\big)\times f\big(\frac{1-\sqrt{\frac{3}{7}+\frac{2\sqrt{\frac{6}{5}}}{7}}}{2}q_n + \frac{1+\sqrt{\frac{3}{7}+\frac{2\sqrt{\frac{6}{5}}}{7}}}{2}q_{n+1}\big) \\
&+ \big(\frac{1}{2}+\frac{1}{6}\sqrt{\frac{5}{6}}\big)\times f\big(\frac{1-\sqrt{\frac{3}{7}-\frac{2\sqrt{\frac{6}{5}}}{7}}}{2}q_n + \frac{1+\sqrt{\frac{3}{7}-\frac{2\sqrt{\frac{6}{5}}}{7}}}{2}q_{n+1}\big) \\
&+ \big(\frac{1}{2}+\frac{1}{6}\sqrt{\frac{5}{6}}\big)\times f\big(\frac{1+\sqrt{\frac{3}{7}-\frac{2\sqrt{\frac{6}{5}}}{7}}}{2}q_n + \frac{1-\sqrt{\frac{3}{7}-\frac{2\sqrt{\frac{6}{5}}}{7}}}{2}q_{n+1}\big) \\
&+ \big(\frac{1}{2}-\frac{1}{6}\sqrt{\frac{5}{6}}\big)\times f\big(\frac{1+\sqrt{\frac{3}{7}+\frac{2\sqrt{\frac{6}{5}}}{7}}}{2}q_n + \frac{1-\sqrt{\frac{3}{7}+\frac{2\sqrt{\frac{6}{5}}}{7}}}{2}q_{n+1}\big)\Big), \\[2mm]
p_{n+1} ={}& -hM\phi_1(V)q_n + \phi_0(V)p_n \\
&+ \frac{h}{2}\phi_1(V)\Big(\big(\frac{1}{2}-\frac{1}{6}\sqrt{\frac{5}{6}}\big)\times f\big(\frac{1-\sqrt{\frac{3}{7}+\frac{2\sqrt{\frac{6}{5}}}{7}}}{2}q_n + \frac{1+\sqrt{\frac{3}{7}+\frac{2\sqrt{\frac{6}{5}}}{7}}}{2}q_{n+1}\big) \\
&+ \big(\frac{1}{2}+\frac{1}{6}\sqrt{\frac{5}{6}}\big)\times f\big(\frac{1-\sqrt{\frac{3}{7}-\frac{2\sqrt{\frac{6}{5}}}{7}}}{2}q_n + \frac{1+\sqrt{\frac{3}{7}-\frac{2\sqrt{\frac{6}{5}}}{7}}}{2}q_{n+1}\big) \\
&+ \big(\frac{1}{2}+\frac{1}{6}\sqrt{\frac{5}{6}}\big)\times f\big(\frac{1+\sqrt{\frac{3}{7}-\frac{2\sqrt{\frac{6}{5}}}{7}}}{2}q_n + \frac{1-\sqrt{\frac{3}{7}-\frac{2\sqrt{\frac{6}{5}}}{7}}}{2}q_{n+1}\big) \\
&+ \big(\frac{1}{2}-\frac{1}{6}\sqrt{\frac{5}{6}}\big)\times f\big(\frac{1+\sqrt{\frac{3}{7}+\frac{2\sqrt{\frac{6}{5}}}{7}}}{2}q_n + \frac{1-\sqrt{\frac{3}{7}+\frac{2\sqrt{\frac{6}{5}}}{7}}}{2}q_{n+1}\big)\Big).
\end{aligned}
\right.
\tag{4.38}
$$

Likewise, applying the five-point Gauss–Legendre's quadrature to (4.37) yields

$$
\begin{cases}
q_{n+1} = \phi_0(V)q_n + h\phi_1(V)p_n \\
\qquad + \dfrac{h^2}{2}\phi_2(V)\Big(\dfrac{21(50+\sqrt{70})}{100(35+2\sqrt{70})}\times f\big(\dfrac{1-\frac{1}{3}\sqrt{5+2\sqrt{\frac{10}{7}}}}{2}q_n + \dfrac{1+\frac{1}{3}\sqrt{5+2\sqrt{\frac{10}{7}}}}{2}q_{n+1}\big) \\
\qquad + \dfrac{322+13\sqrt{70}}{900}\times f\big(\dfrac{1-\frac{1}{3}\sqrt{5-2\sqrt{\frac{10}{7}}}}{2}q_n + \dfrac{1+\frac{1}{3}\sqrt{5-2\sqrt{\frac{10}{7}}}}{2}q_{n+1}\big) \\
\qquad + \dfrac{322+13\sqrt{70}}{900}\times f\big(\dfrac{1+\frac{1}{3}\sqrt{5-2\sqrt{\frac{10}{7}}}}{2}q_n + \dfrac{1-\frac{1}{3}\sqrt{5-2\sqrt{\frac{10}{7}}}}{2}q_{n+1}\big) \\
\qquad + \dfrac{21(50+\sqrt{70})}{100(35+2\sqrt{70})}\times f\big(\dfrac{1+\frac{1}{3}\sqrt{5+2\sqrt{\frac{10}{7}}}}{2}q_n + \dfrac{1-\frac{1}{3}\sqrt{5+2\sqrt{\frac{10}{7}}}}{2}q_{n+1}\big) \\
\qquad + \dfrac{128}{225}\times f\big(\dfrac{1}{2}q_n + \dfrac{1}{2}q_{n+1}\big)\Big), \\[2ex]
p_{n+1} = -hM\phi_1(V)q_n + \phi_0(V)p_n \\
\qquad + \dfrac{h}{2}\phi_1(V)\Big(\dfrac{21(50+\sqrt{70})}{100(35+2\sqrt{70})}\times f\big(\dfrac{1-\frac{1}{3}\sqrt{5+2\sqrt{\frac{10}{7}}}}{2}q_n + \dfrac{1+\frac{1}{3}\sqrt{5+2\sqrt{\frac{10}{7}}}}{2}q_{n+1}\big) \\
\qquad + \dfrac{322+13\sqrt{70}}{900}\times f\big(\dfrac{1-\frac{1}{3}\sqrt{5-2\sqrt{\frac{10}{7}}}}{2}q_n + \dfrac{1+\frac{1}{3}\sqrt{5-2\sqrt{\frac{10}{7}}}}{2}q_{n+1}\big) \\
\qquad + \dfrac{322+13\sqrt{70}}{900}\times f\big(\dfrac{1+\frac{1}{3}\sqrt{5-2\sqrt{\frac{10}{7}}}}{2}q_n + \dfrac{1-\frac{1}{3}\sqrt{5-2\sqrt{\frac{10}{7}}}}{2}q_{n+1}\big) \\
\qquad + \dfrac{21(50+\sqrt{70})}{100(35+2\sqrt{70})}\times f\big(\dfrac{1+\frac{1}{3}\sqrt{5+2\sqrt{\frac{10}{7}}}}{2}q_n + \dfrac{1-\frac{1}{3}\sqrt{5+2\sqrt{\frac{10}{7}}}}{2}q_{n+1}\big) \\
\qquad + \dfrac{128}{225}\times f\big(\dfrac{q_n + q_{n+1}}{2}\big)\Big).
\end{cases}
\tag{4.39}
$$

We denote the methods (4.38) and (4.39) by AAVF3 and AAVF4, respectively.

In order to compare these four new methods with AVF-type methods, we also apply Simpson's rule, the two-point Gauss–Legendre's quadrature, four-point Gauss–Legendre's quadrature and five-point Gauss–Legendre's quadrature to the integral in the AVF formula (4.6), and then obtain the following four AVF-type methods:

$$\begin{cases} q_{n+1} = q_n + hp_n + \dfrac{h^2}{12}\left(g(q_n) + 4g\left(\dfrac{q_n + q_{n+1}}{2}\right) + g(q_{n+1})\right), \\[3mm] p_{n+1} = p_n + \dfrac{h}{6}\left(g(q_n) + 4g\left(\dfrac{q_n + q_{n+1}}{2}\right) + g(q_{n+1})\right), \end{cases} \quad (4.40)$$

$$\begin{cases} q_{n+1} = q_n + hp_n + \dfrac{h^2}{4}\left(g\left(\dfrac{3+\sqrt{3}}{6}q_n + \dfrac{3-\sqrt{3}}{6}q_{n+1}\right) + g\left(\dfrac{3-\sqrt{3}}{6}q_n + \dfrac{3+\sqrt{3}}{6}q_{n+1}\right)\right), \\[3mm] p_{n+1} = p_n + \dfrac{h}{2}\left(g\left(\dfrac{3+\sqrt{3}}{6}q_n + \dfrac{3-\sqrt{3}}{6}q_{n+1}\right) + g\left(\dfrac{3-\sqrt{3}}{6}q_n + \dfrac{3+\sqrt{3}}{6}q_{n+1}\right)\right), \end{cases}$$

$$(4.41)$$

$$\begin{cases} q_{n+1} = q_n + hp_n + \dfrac{h^2}{4}\Bigg(\left(\dfrac{1}{2} - \dfrac{1}{6}\sqrt{\dfrac{5}{6}}\right) \times g\left(\dfrac{1 - \sqrt{\frac{3}{7} + \frac{2\sqrt{\frac{6}{5}}}{7}}}{2}q_n + \dfrac{1 + \sqrt{\frac{3}{7} + \frac{2\sqrt{\frac{6}{5}}}{7}}}{2}q_{n+1}\right) \\[5mm] \qquad + \left(\dfrac{1}{2} + \dfrac{1}{6}\sqrt{\dfrac{5}{6}}\right) \times g\left(\dfrac{1 - \sqrt{\frac{3}{7} - \frac{2\sqrt{\frac{6}{5}}}{7}}}{2}q_n + \dfrac{1 + \sqrt{\frac{3}{7} - \frac{2\sqrt{\frac{6}{5}}}{7}}}{2}q_{n+1}\right) \\[5mm] \qquad + \left(\dfrac{1}{2} + \dfrac{1}{6}\sqrt{\dfrac{5}{6}}\right) \times g\left(\dfrac{1 + \sqrt{\frac{3}{7} - \frac{2\sqrt{\frac{6}{5}}}{7}}}{2}q_n + \dfrac{1 - \sqrt{\frac{3}{7} - \frac{2\sqrt{\frac{6}{5}}}{7}}}{2}q_{n+1}\right) \\[5mm] \qquad + \left(\dfrac{1}{2} - \dfrac{1}{6}\sqrt{\dfrac{5}{6}}\right) \times g\left(\dfrac{1 + \sqrt{\frac{3}{7} + \frac{2\sqrt{\frac{6}{5}}}{7}}}{2}q_n + \dfrac{1 - \sqrt{\frac{3}{7} + \frac{2\sqrt{\frac{6}{5}}}{7}}}{2}q_{n+1}\right)\Bigg), \\[6mm] p_{n+1} = p_n + \dfrac{h}{2}\Bigg(\left(\dfrac{1}{2} - \dfrac{1}{6}\sqrt{\dfrac{5}{6}}\right) \times g\left(\dfrac{1 - \sqrt{\frac{3}{7} + \frac{2\sqrt{\frac{6}{5}}}{7}}}{2}q_n + \dfrac{1 + \sqrt{\frac{3}{7} + \frac{2\sqrt{\frac{6}{5}}}{7}}}{2}q_{n+1}\right) \\[5mm] \qquad + \left(\dfrac{1}{2} + \dfrac{1}{6}\sqrt{\dfrac{5}{6}}\right) \times g\left(\dfrac{1 - \sqrt{\frac{3}{7} - \frac{2\sqrt{\frac{6}{5}}}{7}}}{2}q_n + \dfrac{1 + \sqrt{\frac{3}{7} - \frac{2\sqrt{\frac{6}{5}}}{7}}}{2}q_{n+1}\right) \\[5mm] \qquad + \left(\dfrac{1}{2} + \dfrac{1}{6}\sqrt{\dfrac{5}{6}}\right) \times g\left(\dfrac{1 + \sqrt{\frac{3}{7} - \frac{2\sqrt{\frac{6}{5}}}{7}}}{2}q_n + \dfrac{1 - \sqrt{\frac{3}{7} - \frac{2\sqrt{\frac{6}{5}}}{7}}}{2}q_{n+1}\right) \\[5mm] \qquad + \left(\dfrac{1}{2} - \dfrac{1}{6}\sqrt{\dfrac{5}{6}}\right) \times g\left(\dfrac{1 + \sqrt{\frac{3}{7} + \frac{2\sqrt{\frac{6}{5}}}{7}}}{2}q_n + \dfrac{1 - \sqrt{\frac{3}{7} + \frac{2\sqrt{\frac{6}{5}}}{7}}}{2}q_{n+1}\right)\Bigg), \end{cases}$$

$$(4.42)$$

and

$$
\begin{cases}
q_{n+1} = q_n + h p_n + \dfrac{h^2}{4}\left(\dfrac{21(50+\sqrt{70})}{100(35+2\sqrt{70})} \times g\left(\dfrac{1-\frac{1}{3}\sqrt{5+2\sqrt{\frac{10}{7}}}}{2}q_n + \dfrac{1+\frac{1}{3}\sqrt{5+2\sqrt{\frac{10}{7}}}}{2}q_{n+1}\right)\right. \\[2em]
\quad + \dfrac{322+13\sqrt{70}}{900} \times g\left(\dfrac{1-\frac{1}{3}\sqrt{5-2\sqrt{\frac{10}{7}}}}{2}q_n + \dfrac{1+\frac{1}{3}\sqrt{5-2\sqrt{\frac{10}{7}}}}{2}q_{n+1}\right) \\[2em]
\quad + \dfrac{322+13\sqrt{70}}{900} \times g\left(\dfrac{1+\frac{1}{3}\sqrt{5-2\sqrt{\frac{10}{7}}}}{2}q_n + \dfrac{1-\frac{1}{3}\sqrt{5-2\sqrt{\frac{10}{7}}}}{2}q_{n+1}\right) \\[2em]
\quad + \dfrac{21(50+\sqrt{70})}{100(35+2\sqrt{70})} \times g\left(\dfrac{1+\frac{1}{3}\sqrt{5+2\sqrt{\frac{10}{7}}}}{2}q_n + \dfrac{1-\frac{1}{3}\sqrt{5+2\sqrt{\frac{10}{7}}}}{2}q_{n+1}\right) \\[2em]
\quad \left.+ \dfrac{128}{225} \times g\left(\dfrac{1}{2}q_n + \dfrac{1}{2}q_{n+1}\right)\right), \\[2em]
p_{n+1} = p_n + \dfrac{h}{2}\left(\dfrac{21(50+\sqrt{70})}{100(35+2\sqrt{70})} \times g\left(\dfrac{1-\frac{1}{3}\sqrt{5+2\sqrt{\frac{10}{7}}}}{2}q_n + \dfrac{1+\frac{1}{3}\sqrt{5+2\sqrt{\frac{10}{7}}}}{2}q_{n+1}\right)\right. \\[2em]
\quad + \dfrac{322+13\sqrt{70}}{900} \times g\left(\dfrac{1-\frac{1}{3}\sqrt{5-2\sqrt{\frac{10}{7}}}}{2}q_n + \dfrac{1+\frac{1}{3}\sqrt{5-2\sqrt{\frac{10}{7}}}}{2}q_{n+1}\right) \\[2em]
\quad + \dfrac{322+13\sqrt{70}}{900} \times g\left(\dfrac{1+\frac{1}{3}\sqrt{5-2\sqrt{\frac{10}{7}}}}{2}q_n + \dfrac{1-\frac{1}{3}\sqrt{5-2\sqrt{\frac{10}{7}}}}{2}q_{n+1}\right) \\[2em]
\quad + \dfrac{21(50+\sqrt{70})}{100(35+2\sqrt{70})} \times g\left(\dfrac{1+\frac{1}{3}\sqrt{5+2\sqrt{\frac{10}{7}}}}{2}q_n + \dfrac{1-\frac{1}{3}\sqrt{5+2\sqrt{\frac{10}{7}}}}{2}q_{n+1}\right) \\[2em]
\quad \left.+ \dfrac{128}{225} \times g\left(\dfrac{q_n+q_{n+1}}{2}\right)\right).
\end{cases}
\tag{4.43}
$$

We denote these four methods by AVF1, AVF2, AVF3 and AVF4, respectively.

4.6   Numerical Experiments

In what follows, we report on two numerical experiments with the Fermi–Pasta–Ulam problem and sine-Gordon equation. We use the methods derived in this chapter for comparison.

Problem 4.1   Consider the Fermi–Pasta–Ulam problem (see [17, 18]).
This problem has been considered in Sect. 2.4.4 of Chap. 2.

Following [18], we choose

$$m = 3, \quad q_1(0) = 1, \quad p_1(0) = 1, \quad q_4(0) = \frac{1}{\omega}, \quad p_4(0) = 1,$$

and choose zero for the remaining initial values. Notice that $U(x)$ is a polynomial of degree 4, therefore, the Simpson's rule, and the two-point, four-point and five-point Gauss–Legendre rules are exact for the integral in formulae (4.6) and (4.24). We consider the eight implicit methods and use the same iterative solution algorithm for each method. The system is integrated on the interval $[0, 1000]$ with the stepsize $h = 0.004$ and different $\omega = 250, 300, 350, 400$. We plot the logarithm of the global errors of Hamiltonian $EH = \max |H(p_{5000i}, q_{5000i}) - H_0|$ against $N = 5000i$, where $i = 1, 2, \ldots, 50$. In the fixed-point iteration, we set the error tolerance as 10^{-15} and set the maximum number of iteration as 10. For each step from t_n to t_{n+1}, we choose the values of numerical solution at the previous step as the starting values for the iteration. The results are shown in Figs. 4.2, 4.3, 4.4 and 4.5, respectively. It can be seen from the experiment that the larger ω becomes, the lower the accuracy of the AVF methods becomes in energy preservation for $h = 0.004$, whereas the accuracy of the AAVF methods is almost invariant.

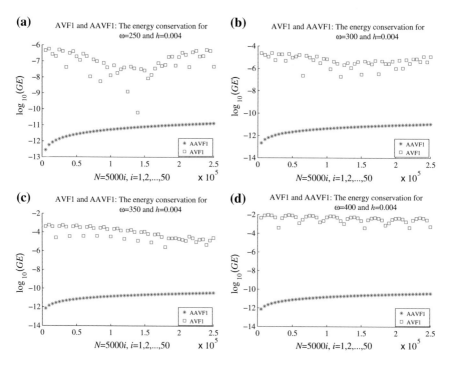

Fig. 4.2 Results of Problem 4.1 for AAVF1 and AVF1 with different ω: The logarithm of the global errors of Hamiltonian $EH = |H(p_{5000i}, q_{5000i}) - H_0|$ against $N = 5000i$, where $i = 1, 2, \ldots, 50$

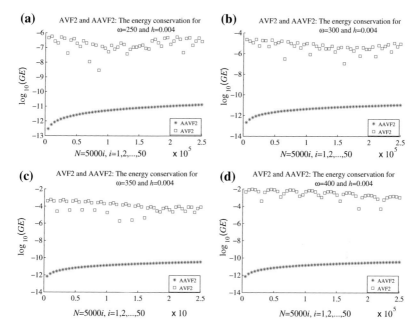

Fig. 4.3 Results of Problem 4.1 for AAVF2 and AVF2 with different ω: The logarithm of the global errors of Hamiltonian $EH = |H(p_{5000i}, q_{5000i}) - H_0|$ against $N = 5000i$, where $i = 1, 2, \ldots, 50$

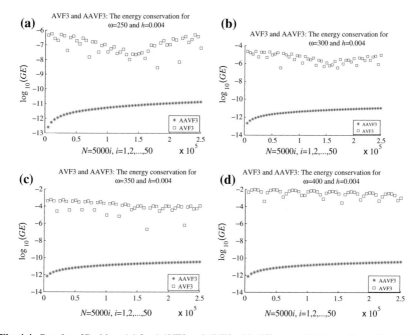

Fig. 4.4 Results of Problem 4.1 for AAVF3 and AVF3 with different ω: The logarithm of the global errors of Hamiltonian $EH = |H(p_{5000i}, q_{5000i}) - H_0|$ against $N = 5000i$, where $i = 1, 2, \ldots, 50$

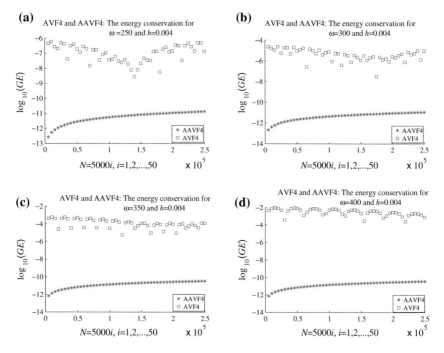

Fig. 4.5 Results of Problem 4.1 for AAVF4 and AVF4 with different ω: The logarithm of the global errors of Hamiltonian $EH = |H(p_{5000i}, q_{5000i}) - H_0|$ against $N = 5000i$, where $i = 1, 2, \ldots, 50$

Problem 4.2 Consider the sine-Gordon equation with periodic boundary conditions (see [20])

$$\begin{cases} \dfrac{\partial^2 u}{\partial t^2} = \dfrac{\partial^2 u}{\partial x^2} - \sin u, & -1 < x < 1, \ t > 0, \\ u(-1, t) = u(1, t). \end{cases}$$

We carry out a semi-discretization of the spatial variable using second-order symmetric differences and obtain the system

$$\frac{d^2 U}{dt^2} + MU = F(t, U), \quad 0 < t \le t_{\text{end}}, \tag{4.44}$$

where $U(t) = (u_1(t), \ldots, u_N(t))^{\mathsf{T}}$ with $u_i(t) \approx u(x_i, t)$, $x_i = -1 + i\Delta x$, $i = 1, \ldots, N$, $\Delta x = 2/N$, and

$$M = \frac{1}{\Delta x^2} \begin{pmatrix} 2 & -1 & & & -1 \\ -1 & 2 & -1 & & \\ & \ddots & \ddots & \ddots & \\ & & -1 & 2 & -1 \\ -1 & & & -1 & 2 \end{pmatrix},$$

$$F(t, U) = -\sin(U) = -\left(\sin u_1, \ldots, \sin u_N\right)^{\mathsf{T}}.$$

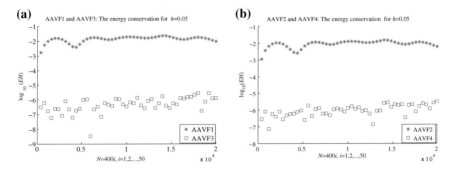

Fig. 4.6 Results of Problem 4.2 for AAVF1-AAVF4: The logarithm of the global errors of Hamiltonian $EH = |H(p_{400i}, q_{400i}) - H_0|$ against $N = 400i$, where $i = 1, 2, \ldots, 50$

The Hamiltonian of (4.44) is given by

$$H(U', U) = \frac{1}{2}U'^{\mathsf{T}}U' + \frac{1}{2}U^{\mathsf{T}}MU - \big(\cos(u_1) + \cdots + \cos(u_N)\big).$$

We take the initial conditions as

$$U(0) = (\pi)_{i=1}^{N}, \quad U_t(0) = \sqrt{N}\left(0.01 + \sin(\frac{2\pi i}{N})\right)_{i=1}^{N}.$$

The problem is integrated on the interval [0, 1000] with the stepsize $h = 0.05$ and $N = 128$ for the eight implicit methods. In the fixed-point iteration, we set the error tolerance as 10^{-15} and the maximum number of iteration as 10. We also choose the values of numerical solution at the previous step as the starting values for the next iteration. For the stepsize $h = 0.05$, the global errors of Hamiltonian of all the AVF methods are too large to plot. The results for all the AAVF methods are given in Fig. 4.6.

4.7 Conclusions

In this chapter, we concentrated our attention on efficient energy-preserving integrators. We derived and analysed the efficient AAVF formula, an energy-preserving formula, for the system of nonlinear oscillatory second-order differential equations (4.3), which is the Hamiltonian system (4.4). This formula exactly preserves the energy H defined by (4.5). Some other useful properties of this new formula, such as symmetry, phase properties and stability, were also analysed. Moreover, we proposed four efficient implicit energy-preserving integrators AAVF1, AAVF2, AAVF3 and AAVF4 in the sense of numerical implementation. The solvability of the implicit schemes was analysed in detail, from which, it turns out that *the convergence of AVF formula depends on $\|M\|$, and the larger $\|M\|$ becomes, the smaller the required* *stepsize. Whereas, the convergence of AAVF formula is independent of* $\|M\|$. This fact shows the superiority of our methods and the importance of taking advantage of the special structure of problems for achieving efficient energy-preserving integrators. These new integrators were applied to the well-known Fermi–Pasta–Ulam problem and the sine-Gordon equation to show their efficiency and robustness. The results of the two numerical experiments demonstrated that for the fixed maximum number of iteration, these new energy-preserving integrators based on (4.24) numerically preserve the Hamiltonian much better than the schemes AVF1, AVF2, AVF3 and AVF4 based on AVF formula.

This chapter is based on the work by Wu et al. [40].

References

1. Betsch P (2006) Energy-consistent numerical integration of mechanical systems with mixed holonomic and nonholonomic constraints. Comput Methods Appl Mech Eng 195:7020–7035
2. Brugnano L, Iavernaro F, Trigiante D (2010) Hamiltonian boundary value methods (energy preserving discrete line integral methods). JNAIAM J Numer Anal Ind Appl Math 5:17–37
3. Celledoni E, McLachlan RI, McLaren DI, Owren B, Quispel GRW, Wright WM (2009) Energy-preserving Runge-Kutta methods. ESAIM: M2AN Math Model Numer Anal 43(4):645–649
4. Celledoni E, McLachlan RI, Owren B, Quispel GRW (2010) On conjugate B-series and their geometric structure. JNAIAM J Numer Anal Ind Appl Math 5:85–94
5. Celledoni E, McLachlan RI, Owren B, Quispel GRW (2010) Energy-preserving integrators and the structure of B-series. Found Comput Math 10:673–693
6. Chartier P, Murua A (2007) Preserving first integrals and volume forms of additively split systems. IMA J Numer Anal 27:381–405
7. Cohen D, Jahnke T, Lorenz K, Lubich C (2006) Numerical integrators for highly oscillatory Hamiltonian systems: a review. In: Mielke A (ed) Analysis, modeling and simulation of multiscale problems. Springer, Berlin, pp 553–576
8. Cooper GJ (1987) Stability of Runge-Kutta methods for trajectory problems. IMA J Numer Anal 7:1–13
9. Faou E, Hairer E, Pham T-L (2004) Energy conservation with non-symplectic methods: examples and counter-examples. BIT Numer Math 44:699–709
10. Feng K, Qin M (2010) Symplectic geometric algorithms for hamiltonian systems. Springer, Berlin
11. Fermi E, Pasta J, Ulam S (1955) Studies of the Nonlinear Problems, I. Los Alamos Report No. LA- 1940, later published in E. Fermi: Collected Papers (Chicago 1965), and Lect Appl Math 15:143 (1974)
12. Franco JM (2002) Runge-Kutta-Nyström methods adapted to the numerical integration of perturbed oscillators. Comput Phys Commun 147:770–787
13. Franco JM (2006) New methods for oscillatory systems based on ARKN methods. Appl Numer Math 56:1040–1053
14. García-Archilla B, Sanz-Serna JM, Skeel RD (1999) Long-time-step methods for oscillatory differential equations. SIAM J Sci Comput 20:930–963
15. González AB, Martín P, Farto JM (1999) A new family of Runge-Kutta type methods for the numerical integration of perturbed oscillators. Numer Math 82:635–646
16. Hairer E (2010) Energy-preserving variant of collocation methods. JNAIAM J Numer Anal Ind Appl Math 5:73–84
17. Hairer E, Lubich C (2000) Long-time energy conservation of numerical methods for oscillatory differential equations. SIAM J Numer Anal 38:414–441

18. Hairer E, Lubich C, Wanner G (2006) Geometric numerical integration: structure-preserving algorithms, 2nd edn. Springer, Berlin
19. Hairer E, McLachlan RI, Skeel RD (2009) On energy conservation of the simplified Takahashi-Imada method. Math Model Numer Anal 43:631–644
20. Hochbruck M, Lubich C (1999) A Gautschi-type method for oscillatory second-order differential equations. Numer Math 83:403–426
21. Iavernaro F, Pace B (2007) S-stage trapezoidal methods for the conservation of Hamiltonian functions of polynomial type. AIP Conf Proc 936:603–606
22. Iavernaro F, Pace B (2008) Conservative Block-Boundary value methods for the solution of polynomial Hamiltonian systems. AIP Conf Proc 1048:888–891
23. Iavernaro F, Trigiante D (2009) High-order symmetric schemes for the energy conservation of polynomial Hamiltonian problems. JNAIAM J Numer Anal Ind Appl Math 4:87–101
24. Iserles A, Zanna A (2000) Preserving algebraic invariants with Runge-Kutta methods. J. Comput. Appl. Math. 125:69–81
25. Iserles A, Quispel GRW, Tse PSP (2007) B-series methods cannot be volume-preserving. BIT Numer Math 47:351–378
26. Li J, Wang B, You X, Wu X (2011) Two-step extended RKN methods for oscillatory systems. Comput Phys Commun 182:2486–2507
27. McLachlan RI, Quispel GRW, Robidoux N (1999) Geometric integration using discrete gradients. Philos Trans R Soc A 357:1021–1046
28. McLachlan RI, Quispel GRW, Tse PSP (2009) Linearization-preserving selfadjoint and symplectic integrators. BIT Numer Math 49:177–197
29. Quispel GRW, McLaren DI (2008) A new class of energy-preserving numerical integration methods. J Phys A Math Theor 41(045206):1–7
30. Sanz-Serna JM (1992) Symplectic integrators for Hamiltonian problems: an overview. Acta Numer 1:243–286
31. Shampine LF (1986) Conservation laws and the numerical solution of ODEs. Comput Math Appl B 12B:1287–1296
32. Stavroyiannis S, Simos TE (2009) Optimization as a function of the phase-lag order of two-step P-stable method for linear periodic IVPs. Appl Numer Math 59:2467–2474
33. Van der Houwen PJ, Sommeijer BP (1987) Explicit Runge-Kutta (-Nyström) methods with reduced phase errors for computing oscillating solution. SIAM J Numer Anal 24:595–617
34. Wang B, Wu X (2012) A new high precision energy-preserving integrator for system of oscillatory second-order differential equations. Phys Lett A 376:1185–1190
35. Wu X (2012) A note on stability of multidimensional adapted Runge-Kutta-Nyström methods for oscillatory systems. Appl Math Model 36:6331–6337
36. Wu X, Wang B (2010) Multidimensional adapted Runge-Kutta-Nyström methods for oscillatory systems. Comput Phys Commun 181:1955–1962
37. Wu X, You X, Xia J (2009) Order conditions for ARKN methods solving oscillatory systems. Comput Phys Commun 180:2250–2257
38. Wu X, You X, Shi W, Wang B (2010) ERKN integrators for systems of oscillatory second-order differential equations. Comput Phys Commun 181:1873–1887
39. Wu X, Wang B, Xia J (2012) Explicit symplectic multidimensional exponential fitting modified Runge-Kutta-Nyström methods. BIT Numer Math 52:773–795
40. Wu X, Wang B, Shi W (2013) Efficient energy-preserving integrators for oscillatory Hamiltonian systems. J Comput Phys 235:587–605

Chapter 5
An Extended Discrete Gradient Formula for Multi-frequency Oscillatory Hamiltonian Systems

The purpose of this chapter is to introduce an extended discrete gradient formula for the multi-frequency oscillatory Hamiltonian system with the Hamiltonian $H(p,q) = \frac{1}{2}p^\mathsf{T}p + \frac{1}{2}q^\mathsf{T}Mq + U(q)$, where $q : \mathbb{R} \to \mathbb{R}^d$ represents generalized positions, $p : \mathbb{R} \to \mathbb{R}^d$ represents generalized momenta, and $M \in \mathbb{R}^{d \times d}$ is a symmetric and positive semi-definite matrix. The solution of this system is a nonlinear oscillator. Basically, many nonlinear oscillatory mechanical systems with a partitioned Hamiltonian function lend themselves to this approach. The extended discrete gradient formula exactly preserves the energy $H(p,q)$. Some properties of the new formula are given. The convergence is analysed for the implicit schemes based on the discrete gradient formula, and it turns out that the convergence is independent of $\|M\|$, which is a significant property for numerically solving the oscillatory Hamiltonian system. This implies that a larger stepsize can be chosen for the extended energy-preserving scheme than for the traditional discrete gradient methods in applications to multi-frequency oscillatory Hamiltonian systems. Illustrative examples show that the new schemes are greatly superior to the traditional discrete gradient methods in the literature.

5.1 Motivation

The design and analysis of numerical integration methods for nonlinear oscillators is an important problem that has received a great deal of attention in the last few years. It is known that the traditional approach to deriving numerical integration methods is based on the natural procedure of discretizing the differential equation in such a way as to make the local truncation errors associated with the discretization as small as possible. A relatively new and increasingly important area in numerical integration methods is geometric integration. A numerical integration method is called geometric if it exactly preserves one or more physical/geometric properties

© Springer-Verlag Berlin Heidelberg and Science Press, Beijing, China 2015
X. Wu et al., *Structure-Preserving Algorithms for Oscillatory Differential Equations II*, DOI 10.1007/978-3-662-48156-1_5

such as first integrals, symplectic structure, symmetries and reversing symmetries, phase-space volume, Lyapunov functions and foliations of the original continuous system. Geometric integration has important applications in many fields, such as fluid dynamics, celestial mechanics, molecular dynamics, quantum physics, plasma physics and meteorology. We refer the reader to [10, 17, 21] for recent surveys of geometric integration.

Consider the Hamiltonian differential equations

$$\dot{y} = J^{-1}\nabla H(y), \tag{5.1}$$

where $y = (p^{\mathsf{T}}, q^{\mathsf{T}})^{\mathsf{T}}$, $q = (q_1, q_2, \ldots, q_d)^{\mathsf{T}}$, $p = (p_1, p_2, \ldots, p_d)^{\mathsf{T}}$, q_i are the position coordinates and p_i the momenta for $i = 1, \ldots, d$, ∇ is the gradient operator

$$\left(\frac{\partial}{\partial p_1}, \ldots, \frac{\partial}{\partial p_d}, \frac{\partial}{\partial q_1}, \ldots, \frac{\partial}{\partial q_d}\right)^{\mathsf{T}},$$

and J is the $2d \times 2d$ skew-symmetric matrix

$$J = \begin{pmatrix} 0 & I \\ -I & 0 \end{pmatrix}.$$

It is known that the Hamiltonian $H(y) = H(p, q)$ is a first integral of the system, which implies that it is a constant along exact solutions of (5.1). In applications, $H(y)$ is the total energy (sum of kinetic and potential energy) so that this property is identical to energy conservation. Thus, it is interesting for a numerical integration to know whether the Hamiltonian remains a constant or nearly a constant along the numerical solution over very long time intervals.

It has been shown that symplectic numerical integrators approximately conserve the total energy over times that are exponentially long in the stepsize [8]. Many energy-preserving methods have been developed for (5.1), such as the discrete gradient methods [3, 4, 14, 16, 18–20] and the Hamiltonian boundary value methods [1, 2, 13]. Numerical integration methods based on the discrete gradient formula were proposed many years ago in order to numerically integrate N-body systems of classical mechanics, with possible applications in molecular dynamics and celestial mechanics [18]. The discrete gradient method for (5.1) is written as [16]

$$\frac{y_{n+1} - y_n}{h} = J^{-1}\overline{\nabla}H(y_n, y_{n+1}). \tag{5.2}$$

For separable Hamiltonian systems with $H(p, q) = T(p) + V(q)$, the method can be expressed by

$$\frac{q_{n+1} - q_n}{h} = \overline{\nabla}T(p_n, p_{n+1}), \quad \frac{p_{n+1} - p_n}{h} = -\overline{\nabla}V(q_n, q_{n+1}), \tag{5.3}$$

where $\overline{\nabla} U$ is the discrete gradient of a scalar function U.

In this chapter, we are concerned with energy-preserving integrators for the multi-frequency oscillatory Hamiltonian system,

$$\begin{cases} \dot{p} = -\nabla_q H(p, q), & p(t_0) = p_0, \\ \dot{q} = \nabla_p H(p, q), & q(t_0) = q_0, \end{cases} \tag{5.4}$$

with the Hamiltonian

$$H(p, q) = \frac{1}{2} p^{\mathsf{T}} p + \frac{1}{2} q^{\mathsf{T}} M q + U(q), \tag{5.5}$$

where $q : \mathbb{R} \to \mathbb{R}^d$ represents generalized positions, $p : \mathbb{R} \to \mathbb{R}^d$ represents generalized momenta, and $M \in \mathbb{R}^{d \times d}$ is a symmetric and positive semi-definite matrix that implicitly contains and preserves the dominant frequencies of the system.

It is easy to see that (5.4) is simply the following multi-frequency oscillatory second-order differential equations

$$\begin{cases} \ddot{q} + M q = f(q), & t \in [t_0, T], \\ q(t_0) = q_0, & \dot{q}(t_0) = p_0, \end{cases} \tag{5.6}$$

where $f : \mathbb{R}^d \to \mathbb{R}^d$ is the negative gradient of $U(q)$, i.e. $f(q) = -\nabla U(q)$.

In recent years, numerical studies of nonlinear effects in physical systems have received much attention and the numerical treatment of multi-frequency oscillatory systems is fundamental for understanding nonlinear phenomena. Accordingly, there has been an enormous advance in dealing with the oscillatory system (5.6). Some useful approaches to constructing Runge-Kutta-Nyström (RKN)-type integrators have been proposed. We refer the reader to [5, 6, 9, 12, 23, 24] for example. Recently, Wu et al. [23] took account of the special structure of the system (5.6) introduced by the linear term Mq, which leads to the high-frequency oscillation in the solution, and formulated a standard form for a multi-frequency ERKN (extended RKN) integrator. ERKN integrators exhibit the correct qualitative behaviour much better than classical RKN methods due to using the special structure of the equation introduced by the linear term Mq. For work on this topic, we refer the reader to [22, 24, 25].

With this background, taking into account both the special structure of the equation introduced by the linear term Mq and integrating the idea of the discrete gradient method for ERKN integrators, we will formulate an extended discrete gradient formula for the multi-frequency oscillatory Hamiltonian system (5.4).

5.2 Preliminaries

This section presents the preliminaries in order to gain an insight into a new class of energy-preserving integrators for the Hamiltonian system (5.4), namely, the second-order differential equation (5.6).

First, we introduce the discrete gradient method. Consider the continuous-time system in linear-gradient form:

$$\dot{y} = L(y)\nabla Q(y), \tag{5.7}$$

where $L(y)$ is a matrix-valued function which is skew-symmetric for all y, and Q is a differentiable function. Note that the terminology 'linear-gradient' has nothing to do with whether the system under consideration is linear or not.

The corresponding discrete gradient method for (5.7) has the following form:

$$\frac{y' - y}{h} = \overline{L}(y, y', h)\overline{\nabla}Q(y, y'), \tag{5.8}$$

where it is required that $\overline{L}(y, y, 0) = L(y)$ and $\overline{\nabla}Q(y, y) = \nabla Q(y)$ for consistency. Here, $\overline{\nabla}Q$ is a discrete gradient of Q, defined as follows (see Gonzalez [7]).

Definition 5.1 Let Q be a differentiable function. Then $\overline{\nabla}Q$ is a discrete gradient of Q provided it is continuous and satisfies

$$\begin{cases} \overline{\nabla}Q(y, y') \cdot (y' - y) = Q(y') - Q(y), \\ \overline{\nabla}Q(y, y) = \nabla Q(y). \end{cases} \tag{5.9}$$

It is obvious that for (5.1), the discrete gradient method (5.8) with $y' = y_{n+1}, y = y_n$ and $\overline{L} = J^{-1}$ reduces to (5.2).

In what follows, we give three examples of discrete gradient of $Q(y)$:

- The *mean value discrete gradient* (Harten et al. [11]):

$$\overline{\nabla}_1 Q(y, y') := \int_0^1 \nabla Q((1 - \tau)y + \tau y')d\tau, \quad y \neq y'. \tag{5.10}$$

- The *midpoint discrete gradient* (Gonzalez [7]):

$$\begin{aligned} \overline{\nabla}_2 Q(y, y') := &\nabla Q\left(\frac{1}{2}(y + y')\right) \\ &+ \frac{Q(y') - Q(y) - \nabla Q(\frac{1}{2}(y + y')) \cdot (y' - y)}{|y' - y|^2}(y' - y), \quad y \neq y'. \end{aligned} \tag{5.11}$$

Both the mean value and the midpoint discrete gradient are second-order approximations to the value of the gradient at the midpoint of the interval $[y, y']$, being exact for linearly varying ∇Q.

- The *coordinate increment discrete gradient* (Itoh and Abe [14]):

$$\overline{\nabla}_3 Q(y, y') := \begin{pmatrix} \dfrac{Q(y'_1, y_2, y_3, \ldots, y_d) - Q(y_1, y_2, y_3, \ldots, y_d)}{y'_1 - y_1} \\ \dfrac{Q(y'_1, y'_2, y_3, \ldots, y_d) - Q(y'_1, y_2, y_3, \ldots, y_d)}{y'_2 - y_2} \\ \vdots \\ \dfrac{Q(y'_1, \ldots, y'_{d-2}, y'_{d-1}, y_d) - Q(y'_1, \ldots, y'_{d-2}, y_{d-1}, y_d)}{y'_{d-1} - y_{d-1}} \\ \dfrac{Q(y'_1, \ldots, y'_{d-2}, y'_{d-1}, y'_d) - Q(y'_1, \ldots, y'_{d-2}, y'_{d-1}, y_d)}{y'_d - y_d} \end{pmatrix}, \tag{5.12}$$

where $0/0$ can be understood as $\partial Q / \partial y_i$.

The coordinate increment discrete gradient is only a first-order approximation of the gradient at the midpoint of the interval $[y, y']$.

It can be observed that for $T(p) = \dfrac{1}{2} p^\mathsf{T} p$, all the three discrete gradients reduce to $\nabla T(p, p') = \dfrac{1}{2}(p + p')$. With this definition, the discrete gradient method for (5.4) with (5.5) becomes

$$\frac{q_{n+1} - q_n}{h} = \frac{1}{2}(p_n + p_{n+1}), \quad \frac{p_{n+1} - p_n}{h} = -\overline{\nabla} V(q_n, q_{n+1}), \tag{5.13}$$

or equivalently,

$$\begin{cases} q_{n+1} = q_n + h p_n - \dfrac{h^2}{2} \overline{\nabla} V(q_n, q_{n+1}), \\ p_{n+1} = p_n - h \overline{\nabla} V(q_n, q_{n+1}), \end{cases} \tag{5.14}$$

where $\overline{\nabla} V(q_n, q_{n+1})$ is the discrete gradient of $V(q) = \dfrac{1}{2} q^\mathsf{T} M q + U(q)$.

In order to describe the extended discrete gradient formula, we need to use the matrix-valued functions given by (4.7) in Chap. 4.

From the matrix-variation-of-constants formula (1.6) in Chap. 1, the solution of (5.6) satisfies

$$\begin{cases} q(t) = \phi_0((t - t_0)^2 M) q_0 + (t - t_0) \phi_1((t - t_0)^2 M) p_0 \\ \quad - \displaystyle\int_{t_0}^t (t - \xi) \phi_1((t - \xi)^2 M) \nabla U(q(\xi)) d\xi, \\ p(t) = -(t - t_0) M \phi_1((t - t_0)^2 M) q_0 + \phi_0((t - t_0)^2 M) p_0 \\ \quad - \displaystyle\int_{t_0}^t \phi_0((t - \xi)^2 M) \nabla U(q(\xi)) d\xi. \end{cases} \tag{5.15}$$

It follows immediately from (5.15) that

$$
\begin{cases}
q(t_n + h) = \phi_0(K)q(t_n) + h\phi_1(K)p(t_n) \\
\quad - h^2 \displaystyle\int_0^1 (1 - z)\phi_1\big((1 - z)^2 K\big)\nabla U\big(q(t_n + hz)\big)\mathrm{d}z, \\
p(t_n + h) = -hM\phi_1(K)q(t_n) + \phi_0(K)p(t_n) \\
\quad - h\displaystyle\int_0^1 \phi_0\big((1 - z)^2 K\big)\nabla U\big(q(t_n + hz)\big)\mathrm{d}z,
\end{cases}
\tag{5.16}
$$

where $K = h^2 M$. In the particular case where $\nabla U(q) \equiv \nabla U_0$ is a constant, (5.16) becomes

$$
\begin{cases}
q(t_n + h) = \phi_0(K)q(t_n) + h\phi_1(K)p(t_n) - h^2\phi_2(K)\nabla U_0, \\
p(t_n + h) = -hM\phi_1(K)q(t_n) + \phi_0(K)p(t_n) - h\phi_1(K)\nabla U_0,
\end{cases}
\tag{5.17}
$$

which gives the exact solution of the Hamiltonian system (5.4).

5.3 An Extended Discrete Gradient Formula Based on ERKN Integrators

In this section, we state and analyse an extended discrete gradient formula based on ERKN integrators for the Hamiltonian system (5.4).

From the formulae (5.16) and (5.17), we consider

$$
\begin{cases}
q_{n+1} = \phi_0(K)q_n + h\phi_1(K)p_n - h^2\phi_2(K)\overline{\nabla} U(q_n, q_{n+1}), \\
p_{n+1} = -hM\phi_1(K)q_n + \phi_0(K)p_n - h\phi_1(K)\overline{\nabla} U(q_n, q_{n+1}),
\end{cases}
\tag{5.18}
$$

where h is the stepsize, $K = h^2 M$ and $\overline{\nabla} U(q_n, q_{n+1})$ is the discrete gradient of $U(q)$. Equation (5.18) is called an extended discrete gradient formula. Each of the three discrete gradients $\overline{\nabla}_1 U(q_n, q_{n+1})$, $\overline{\nabla}_2 U(q_n, q_{n+1})$ and $\overline{\nabla}_3 U(q_n, q_{n+1})$ introduced in Sect. 5.2 can be chosen as $\overline{\nabla} U(q_n, q_{n+1})$.

For the extended discrete gradient formula (5.18), we have the following important theorem.

Theorem 5.1 *The formula (5.18) exactly preserves the Hamiltonian H defined by (5.5).*

Proof We compute

$$
H(p_{n+1}, q_{n+1}) = \frac{1}{2}p_{n+1}^{\mathsf{T}}p_{n+1} + \frac{1}{2}q_{n+1}^{\mathsf{T}}Mq_{n+1} + U(q_{n+1}).
\tag{5.19}
$$

Using the symmetry and commutativity of M and all $\phi_l(K)$ and inserting (5.18) into (5.19), with a tedious and careful computation, we obtain

$$
\begin{aligned}
H(p_{n+1}, q_{n+1}) = {} & \frac{1}{2} p_n^{\mathsf{T}} \big(\phi_0^2(K) + K\phi_1^2(K)\big) p_n + \frac{1}{2} q_n^{\mathsf{T}} M \big(\phi_0^2(K) + K\phi_1^2(K)\big) q_n \\
& + q_n^{\mathsf{T}} K \big(\phi_1(K)^2 - \phi_0(K)\phi_2(K)\big) \overline{\nabla} U(q_n, q_{n+1}) \\
& - h p_n^{\mathsf{T}} \big(\phi_0(K)\phi_1(K) + K\phi_1(K)\phi_2(K)\big) \overline{\nabla} U(q_n, q_{n+1}) \\
& + \frac{1}{2} h^2 \overline{\nabla} U(q_n, q_{n+1})^{\mathsf{T}} \big(\phi_1(K)^2 + K\phi_2(K)^2\big) \overline{\nabla} U(q_n, q_{n+1}) + U(q_{n+1}).
\end{aligned}
$$
(5.20)

From the definition of $\phi_i(K)$, it can be shown that

$$
\begin{cases}
\phi_0^2(K) + K\phi_1^2(K) = I, & K\big(\phi_1(K)^2 - \phi_0(K)\phi_2(K)\big) = I - \phi_0(K), \\
\phi_1(K)^2 + K\phi_2(K)^2 = 2\phi_2(K), & \phi_0(K) + K\phi_2(K) = I,
\end{cases}
$$
(5.21)

where I is the $d \times d$ identity matrix.

Substituting (5.21) into (5.20) gives

$$
\begin{aligned}
H(p_{n+1}, q_{n+1}) = {} & \frac{1}{2} p_n^{\mathsf{T}} p_n + \frac{1}{2} q_n^{\mathsf{T}} M q_n \\
& + q_n^{\mathsf{T}} \big(I - \phi_0(K)\big) \overline{\nabla} U(q_n, q_{n+1}) - h p_n^{\mathsf{T}} \phi_1(K) \overline{\nabla} U(q_n, q_{n+1}) \\
& + h^2 \overline{\nabla} U(q_n, q_{n+1})^{\mathsf{T}} \phi_2(K) \overline{\nabla} U(q_n, q_{n+1}) + U(q_{n+1}).
\end{aligned}
$$
(5.22)

With the definition of discrete gradient and the first equation of (5.18), (5.22) becomes

$$
\begin{aligned}
H(p_{n+1}, q_{n+1}) = {} & \frac{1}{2} p_n^{\mathsf{T}} p_n + \frac{1}{2} q_n^{\mathsf{T}} M q_n + \Big(q_n - \big(\phi_0(K)q_n + h\phi_1(K)p_n \\
& - h^2 \phi_2(K) \overline{\nabla} U(q_n, q_{n+1})\big)\Big)^{\mathsf{T}} \overline{\nabla} U(q_n, q_{n+1}) + U(q_{n+1}) \\
= {} & \frac{1}{2} p_n^{\mathsf{T}} p_n + \frac{1}{2} q_n^{\mathsf{T}} M q_n + \big(q_n - q_{n+1}\big)^{\mathsf{T}} \overline{\nabla} U(q_n, q_{n+1}) + U(q_{n+1}) \\
= {} & \frac{1}{2} p_n^{\mathsf{T}} p_n + \frac{1}{2} q_n^{\mathsf{T}} M q_n + U(q_n) \\
= {} & H(p_n, q_n).
\end{aligned}
$$
(5.23)

The proof is complete.   □

Remark 5.1 It is noted that in the particular case of $M = \mathbf{0}$, the $d \times d$ zero matrix, namely, $H(p, q) = \frac{1}{2} p^{\mathsf{T}} p + U(q)$, and the formula (5.18) reduces to the traditional discrete gradient method (5.13). In fact, the choice of $M = \mathbf{0}$ in (5.18) gives

$$
q_{n+1} = q_n + h p_n - \frac{1}{2} h^2 \overline{\nabla} U(q_n, q_{n+1}), \quad p_{n+1} = p_n - h \overline{\nabla} U(q_n, q_{n+1}), \quad (5.24)
$$

or equivalently,

$$q_{n+1} = q_n + \frac{1}{2}h(p_n + p_{n+1}), \quad p_{n+1} = p_n - h\overline{\nabla}U(q_n, q_{n+1}), \tag{5.25}$$

which is exactly the same as (5.13).

In what follows, we go further in studying the extended discrete gradient formula (5.18) and give some other properties related to the formula.

First, we consider the classical algebraic order of (5.18). An integration formula has order r, if for any smooth problem under consideration, the local truncation errors of the formula satisfy

$$e_{n+1} := q(t_{n+1}) - q_{n+1} = \mathcal{O}(h^{r+1}) \quad \text{and} \quad e'_{n+1} := p(t_{n+1}) - p_{n+1} = \mathcal{O}(h^{r+1}),$$

where $q(t_{n+1})$ and $p(t_{n+1})$ denote the values of the exact solution of the problem and its first derivative at $t_{n+1} = t_n + h$, respectively, and q_{n+1} and p_{n+1} express the one-step numerical results obtained by the formula under the local assumptions $q_n = q(t_n)$ and $p_n = p(t_n)$.

Theorem 5.2 *Assume that $\overline{\nabla}U(q, q')$ is an approximation to the gradient of $U(q)$ of at least first-order at the midpoint of the interval $[q, q']$. Then, the extended discrete gradient formula (5.18) is of order two.*

Proof By (5.16) and (5.18), we have

$$\begin{aligned} q(t_{n+1}) - q_{n+1} = h^2 \int_0^1 -(1-z)\phi_1((1-z)^2 K)\nabla U(q(t_n + hz))\mathrm{d}z \\ + h^2\phi_2(K)\overline{\nabla}U(q_n, q_{n+1}). \end{aligned} \tag{5.26}$$

The first equation of (5.18) gives

$$q_{n+1} - q_n = (\phi_0(K) - I)q_n + h\phi_1(K)p_n - h^2\phi_2(K)\overline{\nabla}U(q_n, q_{n+1}).$$

From $\phi_0(K) - I = \mathcal{O}(h^2)$, it follows that

$$q_{n+1} - q_n = \mathcal{O}(h). \tag{5.27}$$

Under the local assumption $q_n = q(t_n)$, we have

$$q(t_n + hz) - q_n = \mathcal{O}(h), 0 \leqslant z \leqslant 1. \tag{5.28}$$

From (5.27) and (5.28), we obtain

$$q(t_n + hz) - \frac{q_n + q_{n+1}}{2} = \mathcal{O}(h), 0 \leqslant z \leqslant 1. \tag{5.29}$$

Thus, with (5.29) and the assumption of the theorem, (5.26) becomes

$$
q(t_{n+1}) - q_{n+1} = h^2 \int_0^1 -(1-z)\phi_1\big((1-z)^2 K\big)\Big(\nabla U\big(\frac{q_n + q_{n+1}}{2}\big) + \mathcal{O}(h)\Big)
$$
$$
+ \phi_2(K)\Big(\nabla U\big(\frac{q_n + q_{n+1}}{2}\big) + \mathcal{O}(h)\Big)dz
$$
$$
= h^2 \int_0^1 \big(-(1-z)\phi_1((1-z)^2 K) + \phi_2(K)\big)\nabla U\big(\frac{q_n + q_{n+1}}{2}\big) + \mathcal{O}(h)dz
$$
$$
= \mathcal{O}(h^3).
$$

Similarly, we obtain $p(t_{n+1}) - p_{n+1} = \mathcal{O}(h^3)$. □

The symmetry of a method is also very important in long-term integration. The definition of symmetry is given below (see [10]).

Definition 5.2 The adjoint method Φ_h^* of a method Φ_h is defined as the inverse map of the original method with reversed time step $-h$, i.e. $\Phi_h^* := \Phi_{-h}^{-1}$. A method with $\Phi_h^* = \Phi_h$ is called symmetric.

With the definition of symmetry, a method is symmetric if exchanging $n + 1 \leftrightarrow n$, $h \leftrightarrow -h$ leaves the method unaltered.

In what follows, we show a result on symmetry for the extended discrete gradient formula (5.18).

Theorem 5.3 *The extended discrete gradient formula (5.18) is symmetric provided $\overline{\nabla} U$ in (5.18) satisfies the assumption: $\overline{\nabla} U(q, q') = \overline{\nabla} U(q', q)$ for all q and q'.*

Proof Exchanging $q_{n+1} \leftrightarrow q_n$, $p_{n+1} \leftrightarrow p_n$ and replacing h by $-h$ in (5.18) give

$$
\begin{cases}
q_n = \phi_0(K)q_{n+1} - h\phi_1(K)p_{n+1} - h^2\phi_2(K)\overline{\nabla} U(q_{n+1}, q_n), \\
p_n = hM\phi_1(K)q_{n+1} + \phi_0(K)p_{n+1} + h\phi_1(K)\overline{\nabla} U(q_{n+1}, q_n).
\end{cases}
\tag{5.30}
$$

Multiplying both sides of the two equations in (5.30) by $\phi_0(K)$, $\phi_1(K)$, respectively, and with some manipulation, we obtain

$$
\begin{cases}
q_{n+1} = \phi_0(K)q_n + h\phi_1(K)p_n - h^2\big(\phi_1^2(K) - \phi_0(K)\phi_2(K)\big)\overline{\nabla} U(q_{n+1}, q_n), \\
p_{n+1} = -hM\phi_1(K)q_n + \phi_0(K)p_n - h\big(K\phi_1(K)\phi_2(K) + \phi_0(K)\phi_1(K)\big)\overline{\nabla} U(q_{n+1}, q_n).
\end{cases}
\tag{5.31}
$$

Since

$$
\phi_1^2(K) - \phi_0(K)\phi_2(K) = \phi_2(K)
$$

and

$$
K\phi_1(K)\phi_2(K) + \phi_0(K)\phi_1(K) = \phi_1(K),
$$

we obtain

$$\begin{cases} q_{n+1} = \phi_0(K)q_n + h\phi_1(K)p_n - h^2\phi_2(K)\overline{\nabla}U(q_{n+1}, q_n), \\ p_{n+1} = -hM\phi_1(K)q_n + \phi_0(K)p_n - h\phi_1(K)\overline{\nabla}U(q_{n+1}, q_n), \end{cases} \quad (5.32)$$

which shows that the formula (5.18) is symmetric under the stated assumption. □

Remark 5.2 It should be noted that all the three discrete gradients $\overline{\nabla}_1 U$, $\overline{\nabla}_2 U$ and $\overline{\nabla}_3 U$ satisfy the assumption of Theorem 5.3.

5.4   Convergence of the Fixed-Point Iteration for the Implicit Scheme

The previous section derived the extended discrete gradient formula (5.18) and presented some of its properties. In this section, using the discrete gradients given in Sect. 5.2, we propose three practical schemes based on the extended discrete gradient formula (5.18) for the Hamiltonian system (5.4):

- MVDS (mean value discrete gradient):

$$\begin{cases} q_{n+1} = \phi_0(K)q_n + h\phi_1(K)p_n - h^2\phi_2(K)\int_0^1 \nabla U((1-\tau)q_n + \tau q_{n+1})d\tau, \\ p_{n+1} = -hM\phi_1(K)q_n + \phi_0(K)p_n - h\phi_1(K)\int_0^1 \nabla U((1-\tau)q_n + \tau q_{n+1})d\tau. \end{cases} \quad (5.33)$$

- MDS (midpoint discrete gradient):

$$\begin{cases} q_{n+1} = \phi_0(K)q_n + h\phi_1(K)p_n - h^2\phi_2(K)\Big(\nabla U(\tfrac{1}{2}(q_n + q_{n+1})) \\ \qquad + \dfrac{U(q_{n+1}) - U(q_n) - \nabla U(\tfrac{1}{2}(q_n + q_{n+1})) \cdot (q_{n+1} - q_n)}{|q_{n+1} - q_n|^2}(q_{n+1} - q_n)\Big), \\ p_{n+1} = -hM\phi_1(K)q_n + \phi_0(K)p_n - h\phi_1(K)\Big(\nabla U(\tfrac{1}{2}(q_n + q_{n+1})) \\ \qquad + \dfrac{U(q_{n+1}) - U(q_n) - \nabla U(\tfrac{1}{2}(q_n + q_{n+1})) \cdot (q_{n+1} - q_n)}{|q_{n+1} - q_n|^2}(q_{n+1} - q_n)\Big). \end{cases} \quad (5.34)$$

- CIDS (coordinate increment discrete gradient):

$$\begin{cases} q_{n+1} = \phi_0(K)q_n + h\phi_1(K)p_n - h^2\phi_2(K)\overline{\nabla}U(q_n, q_{n+1}), \\ p_{n+1} = -hM\phi_1(K)q_n + \phi_0(K)p_n - h\phi_1(K)\overline{\nabla}U(q_n, q_{n+1}), \end{cases} \quad (5.35)$$

with

$$\overline{\nabla} U(q_n, q_{n+1}) := \begin{pmatrix} \dfrac{U(q_{n+1}^1, q_n^2, q_n^3, \ldots, q_n^d) - U(q_n^1, q_n^2, q_n^3, \ldots, q_n^d)}{q_{n+1}^1 - q_n^1} \\ \dfrac{U(q_{n+1}^1, q_{n+1}^2, q_n^3, \ldots, q_n^d) - U(q_{n+1}^1, q_n^2, q_n^3, \ldots, q_n^d)}{q_{n+1}^2 - q_n^2} \\ \vdots \\ \dfrac{U(q_{n+1}^1, \ldots, q_{n+1}^{d-2}, q_{n+1}^{d-1}, q_n^d) - U(q_{n+1}^1, \ldots, q_{n+1}^{d-2}, q_n^{d-1}, q_n^d)}{q_{n+1}^{d-1} - q_n^{d-1}} \\ \dfrac{U(q_{n+1}^1, \ldots, q_{n+1}^{d-2}, q_{n+1}^{d-1}, q_{n+1}^d) - U(q_{n+1}^1, \ldots, q_{n+1}^{d-2}, q_{n+1}^{d-1}, q_n^d)}{q_{n+1}^d - q_n^d} \end{pmatrix},$$

$$(5.36)$$

where q^i is the ith component of q.

When $M = 0$, the above three methods reduce to

$$\begin{cases} q_{n+1} = q_n + h p_n - \dfrac{h^2}{2} \displaystyle\int_0^1 \nabla V((1 - \tau)q_n + \tau q_{n+1}) d\tau, \\ p_{n+1} = p_n - h \displaystyle\int_0^1 \nabla V((1 - \tau)q_n + \tau q_{n+1}) d\tau, \end{cases} \tag{5.37}$$

$$\begin{cases} q_{n+1} = q_n + h p_n - \dfrac{h^2}{2} \Big(\nabla V(\tfrac{1}{2}(q_n + q_{n+1})) \\ \qquad + \dfrac{V(q_{n+1}) - V(q_n) - \nabla V(\tfrac{1}{2}(q_n + q_{n+1})) \cdot (q_{n+1} - q_n)}{|q_{n+1} - q_n|^2} (q_{n+1} - q_n) \Big), \\ p_{n+1} = p_n - h \Big(\nabla V(\tfrac{1}{2}(q_n + q_{n+1})) \\ \qquad + \dfrac{V(q_{n+1}) - V(q_n) - \nabla V(\tfrac{1}{2}(q_n + q_{n+1})) \cdot (q_{n+1} - q_n)}{|q_{n+1} - q_n|^2} (q_{n+1} - q_n) \Big), \end{cases} \tag{5.38}$$

and

$$\begin{cases} q_{n+1} = q_n + h p_n - \dfrac{h^2}{2} \overline{\nabla} V(q_n, q_{n+1}), \\ p_{n+1} = p_n - h \overline{\nabla} V(q_n, q_{n+1}), \end{cases} \tag{5.39}$$

respectively, where $V(q) = \dfrac{1}{2} q^{\mathsf{T}} M q + U(q)$ and $\overline{\nabla} V(q_n, q_{n+1})$ is the same as (5.36).

Substituting $V(q) = \dfrac{1}{2} q^{\mathsf{T}} M q + U(q)$ into (5.37), (5.38) and (5.39), respectively, yields the following three schemes:

- MVDS0:

$$
\begin{cases}
q_{n+1} = q_n + hp_n - \dfrac{h^2}{2}\left(\dfrac{1}{2}M(q_n + q_{n+1}) + \displaystyle\int_0^1 \nabla U((1-\tau)q_n + \tau q_{n+1})\mathrm{d}\tau\right), \\[4mm]
p_{n+1} = p_n - h\left(\dfrac{1}{2}M(q_n + q_{n+1}) + \displaystyle\int_0^1 \nabla U((1-\tau)q_n + \tau q_{n+1})\mathrm{d}\tau\right).
\end{cases}
$$
(5.40)

- MDS0:

$$
\begin{cases}
q_{n+1} = q_n + hp_n - \dfrac{h^2}{2}\bigg(\dfrac{1}{2}M(q_n + q_{n+1}) + \nabla U\Big(\dfrac{1}{2}(q_n + q_{n+1})\Big) \\[3mm]
\qquad + \dfrac{U(q_{n+1}) - U(q_n) - \nabla U(\frac{1}{2}(q_n + q_{n+1}))\cdot(q_{n+1} - q_n)}{|q_{n+1} - q_n|^2}(q_{n+1} - q_n)\bigg), \\[5mm]
p_{n+1} = p_n - h\bigg(\dfrac{1}{2}M(q_n + q_{n+1}) + \nabla U\Big(\dfrac{1}{2}(q_n + q_{n+1})\Big) \\[3mm]
\qquad + \dfrac{U(q_{n+1}) - U(q_n) - \nabla U(\frac{1}{2}(q_n + q_{n+1}))\cdot(q_{n+1} - q_n)}{|q_{n+1} - q_n|^2}(q_{n+1} - q_n)\bigg).
\end{cases}
$$
(5.41)

- CIDS0:

$$
\begin{cases}
q_{n+1} = q_n + hp_n - \dfrac{h^2}{2}\left(\dfrac{1}{2}M(q_n + q_{n+1}) + \overline{\nabla} U(q_n, q_{n+1})\right), \\[4mm]
p_{n+1} = p_n - h\left(\dfrac{1}{2}M(q_n + q_{n+1}) + \overline{\nabla} U(q_n, q_{n+1})\right),
\end{cases}
$$
(5.42)

where $\overline{\nabla} U(q_n, q_{n+1})$ is defined by (5.36).

The integrals in (5.33) and (5.40) can be approximated by quadrature formulae, in such a way that we can obtain practical numerical schemes. We note that these schemes are all implicit and require iterative solution, in general. In this chapter, we use the well-known fixed-point iteration for these implicit schemes.

In what follows, we analyse the convergence of the fixed-point iteration for these formulae. First, we consider the one-dimensional case.

In the one-dimensional case, the Hamiltonian system (5.4) reduces to

$$
\begin{cases}
\dot{q} = p, \quad \dot{p} = -\omega^2 q - \dfrac{\mathrm{d}U}{\mathrm{d}q}, \quad \omega > 0, \quad t \in [t_0, T], \\[3mm]
q(t_0) = q_0, \quad p(t_0) = p_0,
\end{cases}
$$
(5.43)

and the Hamiltonian reduces to

$$
H(p, q) = \frac{1}{2}p^2 + \frac{1}{2}\omega^2 q^2 + U(q).
$$

The ϕ-functions $\phi_0(K)$, $\phi_1(K)$ and $\phi_2(K)$ in the formulae reduce to $\cos(v)$, $\text{sinc}(v)$ and $\frac{1}{2}\text{sinc}^2(\frac{v}{2})$, respectively, where $v = h\omega$ and $\text{sinc}(\xi) = \frac{\sin(\xi)}{\xi}$. In this case, all the three new energy-preserving schemes and the three corresponding original ones given in this section for (5.4) reduce to the following two schemes:

$$
\begin{cases}
q_{n+1} = \cos(v)q_n + h\,\text{sinc}(v)p_n - \dfrac{h^2}{2}\text{sinc}^2(\dfrac{v}{2})\dfrac{U(q_{n+1}) - U(q_n)}{q_{n+1} - q_n}, \\[2mm]
p_{n+1} = -h\sin(v)q_n + \cos(v)p_n - h\,\text{sinc}(v)\dfrac{U(q_{n+1}) - U(q_n)}{q_{n+1} - q_n},
\end{cases}
\tag{5.44}
$$

$$
\begin{cases}
q_{n+1} = q_n + hp_n - \dfrac{h^2}{2}\left(\dfrac{1}{2}\omega^2(q_{n+1} + q_n) + \dfrac{U(q_{n+1}) - U(q_n)}{q_{n+1} - q_n}\right), \\[2mm]
p_{n+1} = p_n - h\left(\dfrac{1}{2}\omega^2(q_{n+1} + q_n) + \dfrac{U(q_{n+1}) - U(q_n)}{q_{n+1} - q_n}\right).
\end{cases}
\tag{5.45}
$$

We assume that $U(q)$ is twice continuously differentiable. For fixed h, q_n, p_n, the first equation of (5.44) is a nonlinear equation with respect to q_{n+1}. Let

$$
F(q) = \cos(v)q_n + h\,\text{sinc}(v)p_n - \frac{h^2}{2}\text{sinc}^2(\frac{v}{2})\frac{U(q) - U(q_n)}{q - q_n}.
$$

Then q_{n+1} is a fixed-point of $F(q)$ and we have

$$
\begin{aligned}
|F(q) - F(q')| &= \left|\frac{h^2}{2}\text{sinc}^2(\frac{v}{2})\right|\left|\frac{U(q) - U(q_n)}{q - q_n} - \frac{U(q') - U(q_n)}{q' - q_n}\right| \\[2mm]
&= \left|\frac{h^2}{2}\text{sinc}^2(\frac{v}{2})\right|\left|\int_0^1 U'((1 - \tau)q_n + \tau q) - U'((1 - \tau)q_n + \tau q')\,d\tau\right| \\[2mm]
&= \left|\frac{h^2}{2}\text{sinc}^2(\frac{v}{2})\right|\left|\int_0^1\int_0^1 U''((1 - \tau)q_n + (1 - s)\tau q' + s\tau q)\tau(q - q')\,ds\,d\tau\right| \\[2mm]
&\leqslant \frac{h^2}{4}\max_\xi|U''(\xi)|\,|q - q'|.
\end{aligned}
\tag{5.46}
$$

By the fixed-point theorem, if $\frac{h^2}{4}\max_\xi|U''(\xi)| < c < 1$, the fixed-point iteration for the first equation with respect to q_{n+1} in (5.44) is convergent. The important point here is that the convergence of fixed-point iteration for (5.44) is independent of the frequency ω. Similarly, for the first equation in (5.45), let

$$
G(q) = q_n + hp_n - \frac{h^2}{2}\left(\frac{1}{2}\omega^2(q + q_n) + \frac{U(q) - U(q_n)}{q - q_n}\right).
$$

Then

$$
\begin{aligned}
&\bigl|G(q) - G(q')\bigr| \\
&= \frac{h^2}{2}\left|\frac{1}{2}\omega^2(q - q') + \frac{U(q) - U(q_n)}{q - q_n} - \frac{U(q') - U(q_n)}{q' - q_n}\right| \\
&= \frac{h^2}{2}\left|\frac{1}{2}\omega^2(q - q') + \int_0^1 U'((1 - \tau)q_n + \tau q) - U'((1 - \tau)q_n + \tau q')\mathrm{d}\tau\right| \\
&= \frac{h^2}{2}\left|\frac{1}{2}\omega^2(q - q') + \int_0^1\int_0^1 U''((1 - \tau)q_n + (1 - s)\tau q' + s\tau q)\tau(q - q')\mathrm{d}s\mathrm{d}\tau\right| \\
&= \frac{h^2}{2}\left|\frac{1}{2}\omega^2 + \int_0^1\int_0^1 U''((1 - \tau)q_n + (1 - s)\tau q' + s\tau q)\tau\mathrm{d}s\mathrm{d}\tau\right|\left|q - q'\right|. \\
&\leqslant \frac{h^2}{4}\left(\omega^2 + \max_{\xi}\left|U''(\xi)\right|\right)\left|q - q'\right|.
\end{aligned}
\tag{5.47}
$$

Thus, if $\frac{h^2}{4}\left(\omega^2 + \max_{\xi}\left|U''(\xi)\right|\right) < c < 1$, the fixed-point iteration for the first equation with respect to q_{n+1} in the scheme (5.45) is convergent. We note that the convergence of the fixed-point iteration for (5.45) is dependent on the frequency ω.

For the multidimensional case, we only consider (5.33) and (5.40). The analysis for the other cases is similar.

In what follows, we use the Euclidean norm and its induced matrix norm (spectral norm) and denote them by $\|\cdot\|$. Similarly to the one-dimensional case, let the right-hand side of the first equation in (5.33) be

$$
R(q) = \phi_0(K)q_n + h\phi_1(K)p_n - h^2\phi_2(K)\int_0^1 \nabla U\bigl((1 - \tau)q_n + \tau q\bigr)\mathrm{d}\tau,
$$

and in (5.40) let the right-hand side of the first equation be

$$
I(q) = q_n + hp_n - \frac{h^2}{2}\left(\frac{1}{2}M(q_n + q_{n+1}) + \int_0^1 \nabla U\bigl((1 - \tau)q_n + \tau q_{n+1}\bigr)\mathrm{d}\tau\right).
$$

Then,

$$
\begin{aligned}
\bigl\|R(q) - R(q')\bigr\| &= h^2\left\|\phi_2(K)\int_0^1 U'((1 - \tau)q_n + \tau q) - U'((1 - \tau)q_n + \tau q')\mathrm{d}\tau\right\| \\
&= h^2\left\|\phi_2(K)\int_0^1\int_0^1 U''((1 - \tau)q_n + (1 - s)\tau q' + s\tau q)\tau(q - q')\mathrm{d}s\mathrm{d}\tau\right\| \\
&\leqslant \frac{h^2}{2}\|\phi_2(K)\|\max_{\xi}\left\|\nabla^2 U(\xi)\right\|\left\|q - q'\right\| \\
&\leqslant \frac{h^2}{4}\max_{\xi}\left\|\nabla^2 U(\xi)\right\|\left\|q - q'\right\|,
\end{aligned}
\tag{5.48}
$$

where the last inequality is due to the symmetry of M and the definition of $\phi_2(K)$.
Likewise, we have

$$
\|I(q) - I(q')\|
$$

$$
= \frac{h^2}{2} \left\| \frac{1}{2}M(q - q') + \int_0^1 U'\big((1 - \tau)q_n + \tau q\big) - U'\big((1 - \tau)q_n + \tau q'\big)d\tau \right\|
$$

$$
= \frac{h^2}{2} \left\| \frac{1}{2}M(q - q') + \int_0^1 \int_0^1 U''\big((1 - \tau)q_n + (1 - s)\tau q' + s\tau q\big)\tau(q - q')ds d\tau \right\|
$$

$$
\leqslant \frac{h^2}{2} \left\| \frac{1}{2}M + \int_0^1 \int_0^1 U''\big((1 - \tau)q_n + (1 - s)\tau q' + s\tau q\big)\tau ds d\tau \right\| \|q - q'\| .
$$

$$
\leqslant \frac{h^2}{4} \left(\|M\| + \max_\xi \|\nabla^2 U(\xi)\| \right) \|q - q'\| . \tag{5.49}
$$

Compared (5.48) with (5.49), it can be observed that the fixed-point iteration of
extended discrete gradient schemes has a larger convergence domain than that of
traditional discrete gradient schemes, especially for the case $\|\nabla^2 U\| \ll \|M\|$, where
$\nabla^2 U$ is the Hessian matrix of U. Moreover, it is clear from (5.49) that the convergence
of fixed-point iteration for traditional discrete gradient methods depends on $\|M\|$ and
*the larger $\|M\|$ is, the smaller the stepsize is required to be. Whereas, it is important to
see from (5.48) that the convergence of fixed-point iteration for the implicit schemes
based on the extended discrete gradient formula (5.18) is independent of $\|M\|$.*

5.5 Numerical Experiments

Below we apply the methods presented in this chapter to the following three problems
and show their remarkable efficiency.

Problem 5.1 Consider the Duffing equation

$$
\begin{cases} \ddot{q} + \omega^2 q = k^2(2q^3 - q), & t \in [0, t_{\text{end}}], \\ q(0) = 0, & \dot{q}(0) = \omega, \end{cases}
$$

with $0 \leq k < \omega$. The problem is a Hamiltonian system with the Hamiltonian

$$
H(p, q) = \frac{1}{2}p^2 + \frac{1}{2}\omega^2 q^2 + U(q), \quad U(q) = \frac{k^2}{2}(q^2 - q^4).
$$

The analytic solution of this initial value problem is given by

$$
q(t) = sn(\omega t, k/\omega),
$$

and represents a periodic motion in terms of the Jacobian elliptic function *sn*. In this
test, we choose the parameter values $k = 0.03, \ t_{end} = 10000$.

Since it is a one-dimensional problem, only the methods CIDS and CIDS0 are
used. First, for the fixed-point iteration, we set the maximum iteration number as 10
and the error tolerance as 10^{-15}. We plot the logarithm of the errors of the Hamiltonian
$EH = \max |H(p_{5000i}, q_{5000i}) - H(p_0, q_0)|$ against $N = 5000i$ for $\omega = 10$ and $\omega = 20$
with $h = 1/2^4$. The results are shown in Fig. 5.1. We then set the maximum iteration
number as 1000 and the error tolerance as 10^{-15} and plot the logarithm of the error
tolerance of the iteration against the logarithm of the total iteration number for the
two methods with $h = 1/2^4$. The results are shown in Fig. 5.2.

Problem 5.2 Consider the Fermi-Pasta-Ulam problem.

This problem has been considered in Sect. 2.4.4 of Chap. 2.
Following [10], we choose

$$m = 3, \ q_1(0) = 1, \ p_1(0) = 1, \ q_4(0) = \frac{1}{\omega}, \ p_4(0) = 1, \qquad (5.50)$$

and choose zero for the remaining initial values. We apply the methods MDS, MDS0,
CIDS and CIDS0 to the system. The system with the initial values (5.50) is integrated
on the interval [0, 1000] with the stepsize $h = 1/100$ for $\omega = 100$ and $\omega = 200$.
First, for fixed-point iteration, we set the maximum iteration number as 10, the
error tolerance as 10^{-15} and plot the logarithm of the errors of the Hamiltonian
$EH = \max |H(p_{5000i}, q_{5000i}) - H(p_0, q_0)|$ against $N = 5000i, \ i = 1, 2, \ldots$. If the
error is very large, we do not plot the points in the figure. The results are shown in
Fig. 5.3. As is seen from the result, for fixed $h = 1/100$, as ω increases from 100 to

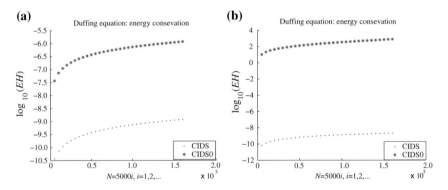

Fig. 5.1 Problem 5.1 (maximum iteration number = 10): the logarithm of the errors of Hamiltonian $EH = |H(p_{5000i}, q_{5000i}) - H_{p_0,q_0}|$ against $N = 5000i, i = 1, 2, \ldots$. **a** $\omega = 10, h = \frac{1}{2^4}$.
b $\omega = 20, h = \frac{1}{2^4}$

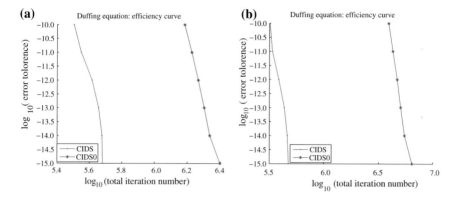

Fig. 5.2 Problem 5.1 (maximum iteration number = 1000): the logarithm of the error tolerance of the iteration against the logarithm of the total iteration number. **a** $\omega = 10, h = \frac{1}{2^4}$. **b** $\omega = 20, h = \frac{1}{2^4}$

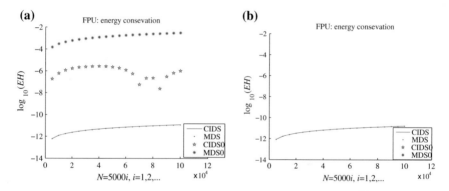

Fig. 5.3 Problem 5.2 (maximum iteration number = 10): the logarithm of the errors of Hamiltonian $EH = |H(p_{5000i}, q_{5000i}) - H_{p_0, q_0}|$ against $N = 5000i, i = 1, 2, \ldots$. **a** $\omega = 100, h = \frac{1}{100}$. **b** $\omega = 200, h = \frac{1}{100}$

200, the traditional discrete gradient methods are not convergent any more. However, the new schemes still converge well.

Next, we set the maximum iteration number as 1000, and choose $h = 1/100$ for $\omega = 100$, $h = 1/200$ for $\omega = 200$. We plot the logarithm of the error tolerance of the iteration against the logarithm of the total iteration number for the four methods. The results are shown in Fig. 5.4.

Problem 5.3 Consider the sine-Gordon equation with periodic boundary conditions (see [12])

$$\begin{cases} \dfrac{\partial^2 u}{\partial t^2} = \dfrac{\partial^2 u}{\partial x^2} - \sin u, & -1 < x < 1, \quad t > 0, \\ u(-1, t) = u(1, t). \end{cases}$$

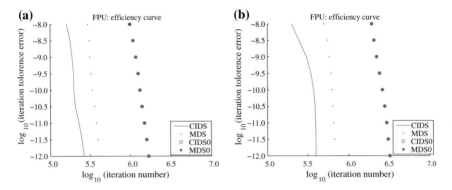

Fig. 5.4 Problem 5.2 (maximum iteration number = 1000): the logarithm of the error tolerance of the iteration against the logarithm of the total iteration number. **a** $\omega = 100, h = \frac{1}{100}$. **b** $\omega = 200, h = \frac{1}{200}$

Using semi-discretization on the spatial variable with second-order symmetric differences, and introducing generalized momenta $p = \dot{q}$, we obtain the Hamiltonian system with the Hamiltonian

$$H(p, q) = \frac{1}{2}p^\mathsf{T}p + \frac{1}{2}q^\mathsf{T}Mq + U(q),$$

where $q(t) = \left(u_1(t), \ldots, u_d(t)\right)^\mathsf{T}$ and $U(q) = -\left(\cos(u_1) + \cdots + \cos(u_d)\right)$ with $u_i(t) \approx u(x_i, t)$, $x_i = -1 + i\Delta x$, $i = 1, \ldots, d$, $\Delta x = 2/d$, and

$$M = \frac{1}{\Delta x^2} \begin{pmatrix} 2 & -1 & & & -1 \\ -1 & 2 & -1 & & \\ & \ddots & \ddots & \ddots & \\ & & -1 & 2 & -1 \\ -1 & & & -1 & 2 \end{pmatrix}. \tag{5.51}$$

We take the initial conditions as

$$q(0) = \left(\pi\right)_{i=1}^d, \quad p(0) = \sqrt{d}\left(0.01 + \sin\left(\frac{2\pi i}{d}\right)\right)_{i=1}^d.$$

The system is integrated on the interval [0, 100] with the methods MDS, MDS0, CIDS and CIDS0. First, for fixed-point iteration, we set the maximum iteration number as 10 and the error tolerance as 10^{-15}. We plot the logarithm of the errors of the Hamiltonian $EH = \max |H(p_{100i}, q_{100i}) - H(p_0, q_0)|$ against $N = 100i$ for $d = 36$ and $d = 144$ with $h = 0.04$. If the error is very large, we do not plot the points in the figure. The results are shown in Fig. 5.5.

We then set the maximum iteration number as 1000 and the error tolerance as 10^{-15}. We apply the four methods to the system on the interval $[0, 10]$ with the stepsize $h = 0.04$ for $d = 36$ and $h = 0.01$ for $d = 144$. We plot the logarithm of the error tolerance of the iteration against the logarithm of the total iteration number for the four methods. The results are shown in Fig. 5.6. It can be observed from the results that compared with traditional discrete gradient methods, the fixed-point iterations of the new methods have much larger convergence domains, and the convergence is faster.

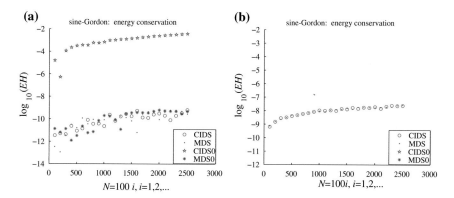

Fig. 5.5 Problem 5.3 (maximum iteration number = 10): the logarithm of the errors of Hamiltonian $EH = |H(p_{100i}, q_{100i}) - H_{p_0, q_0}|$ against $N = 100i$, $i = 1, 2, \ldots$. **a** $d = 36, h = 0.04$. **b** $d = 144, h = 0.04$

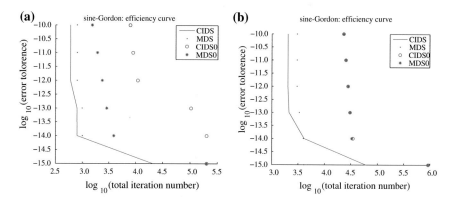

Fig. 5.6 Problem 5.3 (maximum iteration number = 1000): the logarithm of the error tolerance of the iteration against the logarithm of the total iteration number. **a** $d = 36, h = 0.04$. **b** $d = 144, h = 0.01$

5.6 Conclusions

In this chapter, combining the idea of the discrete gradient method with the
ERKN integrator, we discussed an extended discrete gradient formula for the multi-
frequency oscillatory Hamiltonian system. Some properties of the new formula were
analysed. It is the distinguishing feature of the new formula to take advantage of
the special structure of the system introduced by the linear term Mq, so that the
extended discrete gradient methods adapt themselves to the multi-frequency oscillatory system. Since both extended discrete gradient schemes and traditional ones are
implicit, an iterative solution procedure is required. From the convergence analysis
of fixed-point iteration for implicit schemes, it can be seen that a larger stepsize
can be chosen for extended discrete gradient schemes than for traditional discrete
gradient methods, when they are applied to oscillatory Hamiltonian systems. The
convergence rate of extended discrete gradient methods is much higher than that of
traditional methods. The numerical experiments clearly support this point. Another
very important property is that the convergence rate of the fixed-point iteration for
extended discrete gradient schemes is independent of $\|M\|$. Unfortunately, however,
the convergence rate of fixed-point iteration for traditional discrete gradient methods
is dependent on $\|M\|$.

The material of this chapter is based on the work by Liu et al. [15].

References

1. Betsch P (2006) Energy-consistent numerical integration of mechanical systems with mixed holonomic and nonholonomic constraints. Comput Methods Appl Mech Eng 195:7020–7035
2. Brugnano L, Iavernaro F, Trigiante D (2010) Hamiltonian boundary value methods (energy preserving discrete line integral methods). J Numer Anal Ind Appl Math 5:17–37
3. Cieśliński JL, Ratkiewicz B (2011) Energy-preserving numerical schemes of high accuracy for one-dimensional hamiltonian systems. J Phys A: Math Theor 44:155206
4. Dahlby M, Owren B, Yaguchi T (2011) Preserving multiple first integrals by discrete gradients. J Phys A: Math Theor 44:305205
5. García-Archilla B, Sanz-Serna JM, Skeel RD (1999) Long-time-step methods for oscillatory differential equations. SIAM J Sci Comput 20:930–963
6. González AB, Martín P, Farto JM (1999) A new family of Runge-Kutta type methods for the numerical integration of perturbed oscillators. Numer Math 82:635–646
7. Gonzalez O (1996) Time integration and discrete Hamiltonian systems. J Nonlinear Sci 6:449–467
8. Hairer E, Lubich C (1997) The life-span of backward error analysis for numerical integrators. Numer Math 76:441–462
9. Hairer E, Lubich C (2000) Long-time energy conservation of numerical methods for oscillatory differential equations. SIAM J Numer Anal 38:414–441
10. Hairer E, Lubich C, Wanner G (2006) Geometric numerical integration: structure-preserving algorithms, 2nd edn. Springer, Berlin
11. Harten A, Lax PD, van Leer B (1983) On upstream differencing and Godunov-type schemes for hyperbolic conservation laws. SIAM Rev 25:35–61
12. Hochbruck M, Lubich C (1999) A Gautschi-type method for oscillatory second-order differential equations. Numer Math 83:403–426

13. Iavernaro F, Pace B (2007) s-stage trapezoidal methods for the conservation of Hamiltonian functions of polynomial type. AIP Conf Proc 936:603–606

14. Itoh T, Abe K (1988) Hamiltonian conserving discrete canonical equations based on variational difference quotients. J Comput Phys 77:85–102

15. Liu K, Shi W, Wu X (2013) An extended discrete gradient formula for oscillatory Hamiltonian systems. J Phys A: Math Theor 46:165203

16. McLachlan RI, Quispel GRW, Robidoux N (1999) Geometric integration using discrete gradients. Phil Trans R Soc Lond A 357:1021–1045

17. McLachlan RI, Quispel GRW (2002) Splitting methods. Acta Numer 11:341–434

18. Quispel GRW, Capel HW (1996) Solving ODEs numerically while preserving a first integral. Phys Lett A 218:223–228

19. Quispel GRW, McLaren DI (2008) A new class of energy-preserving numerical integration methods. J Phys A: Math Theor 41:045206

20. Quispel GRW, Turner GS (1996) Discrete gradient methods for solving ODEs numerically while preserving a first integral. J Phys A: Math Gen 29:L341–L349

21. Sanz-Serna JM (1992) Symplectic integrators for Hamiltonian problems: an overview. Acta Numer 1:243–286

22. Wu X, You X, Li J (2009) Note on derivation of order conditions for ARKN methods for perturbed oscillators. Comput Phys Commun 180:1545–1549

23. Wu X, You X, Shi W, Wang B (2010) ERKN integrators for systems of oscillatory second-order differential equations. Comput Phys Commun 181:1873–1887

24. Wu X, You X, Xia J (2009) Order conditions for ARKN methods solving oscillatory systems. Comput Phys Commun 180:2250–2257

25. Yang H, Wu X (2008) Trigonometrically-fitted ARKN methods for perturbed oscillators. Appl Numer Math 58:1375–1395

Chapter 6
Trigonometric Fourier Collocation Methods for Multi-frequency Oscillatory Systems

This chapter presents a type of trigonometric Fourier collocation method for solving multi-frequency oscillatory systems $q''(t) + Mq(t) = f(q(t))$ with a principal frequency matrix $M \in \mathbb{R}^{d \times d}$. If M is symmetric and positive semi-definite and $f(q) = -\nabla U(q)$ for a smooth function $U(q)$, then this is a multi-frequency oscillatory Hamiltonian system with the Hamiltonian $H(p, q) = p^{\mathsf{T}}p/2 + q^{\mathsf{T}}Mq/2 + U(q)$. The solution of this system is a non-linear multi-frequency oscillator. The trigonometric Fourier collocation method makes use of the special structure introduced by the linear term Mq, and the construction combines the ideas of collocation methods, the matrix-variation of onstants formula and the local Fourier expansion of the system. The analysis in the chapter shows an important feature, namely that the trigonometric Fourier collocation methods can be of an arbitrarily high order and when $M \to \mathbf{0}$, each trigonometric Fourier collocation method yields a particular Runge-Kutta-Nyström-type Fourier collocation method, which is symplectic under some conditions. This makes it possible to achieve arbitrarily high-order symplectic methods for a special and important class of systems of second-order ordinary differential equations $q''(t) = f(q(t))$ in an efficient way.

6.1 Motivation

This chapter pays attention to systems of multi-frequency oscillatory second-order differential equations of the form

$$q''(t) + Mq(t) = f(q(t)), \qquad q(0) = q_0, \ q'(0) = q'_0, \qquad t \in [0, t_{\text{end}}], \quad (6.1)$$

where M is a $d \times d$ positive semi-definite matrix implicitly containing the dominant frequencies of the oscillatory problem and $f : \mathbb{R}^d \to \mathbb{R}^d$ is an analytic function. The

© Springer-Verlag Berlin Heidelberg and Science Press, Beijing, China 2015
X. Wu et al., *Structure-Preserving Algorithms for Oscillatory*
Differential Equations II, DOI 10.1007/978-3-662-48156-1_6

solution of this system is a multi-frequency non-linear oscillator due to the presence of the linear term Mq. Equation (6.1) occurs in a wide variety of applications such as quantum physics, circuit simulations, flexible body dynamics and mechanics (see, e.g. [11, 12, 15, 20, 21, 57, 61]).

If M is symmetric and f is the negative gradient of a real-valued function $U(q)$, (6.1) is identical to the following initial value Hamiltonian system

$$\begin{cases} \dot{p} = -\nabla_q H(q, p), & p(0) = p_0 \equiv q_0', \\ \dot{q} = \nabla_p H(q, p), & q(0) = q_0, \end{cases} \tag{6.2}$$

with the Hamiltonian

$$H(p, q) = \frac{1}{2} p^{\mathsf{T}} p + \frac{1}{2} q^{\mathsf{T}} M q + U(q). \tag{6.3}$$

Such an important system has been investigated by many authors (see, e.g. [9, 11, 12, 20, 21]). Many mechanical systems with a partitioned Hamiltonian function are modelled by (6.2) with (6.3). As stated in Chap. 1, two fundamental properties of Hamiltonian systems are energy preservation and symplecticity.

An important feature of ordinary differential equations is the invariance of first integrals. There is much work concerning the conservation of invariants (first integrals) by numerical methods, e.g. [7, 13, 17, 21, 22, 31, 32, 38]. Here, we consider the quadratic invariant $Q = q^{\mathsf{T}} D p$ of (6.1). The quadratic form Q is a first integral of (6.1) if and only if $p^{\mathsf{T}} D p + q^{\mathsf{T}} D (f(q) - Mq) = 0$ for all $p, q \in \mathbb{R}^d$. This means that D is a skew-symmetric matrix and that $q^{\mathsf{T}} D(f(q) - Mq) = 0$ for any $q \in \mathbb{R}^d$. It has been pointed out in [17, 21] that symplecticity is a quadratic first integral acting on the vector fields of a differential equation, and every numerical method that preserves quadratic first integrals is a symplectic method for the corresponding Hamiltonian system.

For the Hamiltonian differential equations

$$y' = J^{-1} \nabla H(y) \tag{6.4}$$

with a skew-symmetric constant matrix J and the Hamiltonian $H(y)$, various numerical methods including symplectic methods and energy-preserving methods have been constructed and readers are referred to [1, 3, 4, 6, 8, 10, 19, 22, 28, 29, 35, 38, 40, 42, 43, 45, 48, 50] for some related work. These methods are applicable to (6.2) since (6.2) can be rewritten in the form (6.4). However, it is noted that the system (6.1) or (6.2) is oscillatory because of the special structure introduced by the linear term Mq. In the last few decades, the theory of numerical methods for oscillatory differential equations has reached a certain maturity. Many efficient codes, mainly based on Runge-Kutta-Nyström (RKN) methods and exponentially/trigonometrically fitted methods, have become available. Among them, exponential/trigonometric integrators are a very powerful approach and readers are referred to [26] for a survey work on exponential integrators. Other work related to exponential/trigonometric integrators can be found in [20, 21, 24, 25, 27, 50, 58] for instance. Much effort has been devoted to using the special structure introduced by the term Mq to solve (6.1) effectively and readers are referred to [14–17, 20, 24, 36, 50, 52–54, 56–61] for examples.

An important observation is that the exponential/trigonometric methods taking advantage of the special structure introduced by the term Mq not only produce an improved qualitative behaviour, but also allow for a more accurate long-term integration than with general-purpose methods. This motivates the design and analysis of a kind of trigonometric Fourier collocation method for (6.1).

It is noted that when $M \to \mathbf{0}$, (6.1) reduces to a special and important class of systems of second-order ordinary differential equations expressed in the traditional form

$$q''(t) = f\big(q(t)\big), \qquad q(0) = q_0, \; q'(0) = q_0', \qquad t \in [0, t_{\text{end}}]. \tag{6.5}$$

Its corresponding Hamiltonian system and quadratic invariant can be characterized by the above statements for (6.1) with $M \to \mathbf{0}$. Consequently, it appears promising to design efficient methods for (6.5) based on the methods designed for (6.1) when $M \to \mathbf{0}$.

The main purpose of this chapter is to devise a trigonometric Fourier collocation method of arbitrary order for solving multi-frequency oscillatory systems (6.1) and to obtain arbitrary order symplectic methods for system (6.5) based on the new trigonometric Fourier collocation method. This *trigonometric Fourier collocation (TFC)* method is a kind of collocation method for ordinary differential equations (see, e.g. [17, 19, 21, 30, 55]) with a long history. An interesting feature of collocation methods is that we not only get a discrete set of approximations, but also a continuous approximation to the solution. The TFC methods not only share this feature of collocation methods, but also incorporate the special structure originating from the term Mq. Furthermore, they are derived by truncating the local Fourier expansion of the system and thus it is very simple to get arbitrarily high-order methods for solving (6.1). In addition, the TFC methods are applicable to the system (6.1) with no regard to the symmetry of M. Furthermore, when $M \to \mathbf{0}$, the TFC method also represents a special and efficient arbitrary order symplectic RKN approach to solving the special and important class of second-order ordinary differential equations (6.5). This makes the TFC methods a very powerful and efficient approach in applications.

With this premise, this chapter begins with the local Fourier expansion. Then the TFC methods are formulated and analysed in detail.

6.2   Local Fourier Expansion

Before discussing the Fourier expansion, we begin by recalling the matrix-valued functions:

$$\phi_i(M) := \sum_{l=0}^{\infty} \frac{(-1)^l M^l}{(2l+i)!}, \qquad i = 0, 1, \ldots \tag{6.6}$$

which are unconditionally convergent. These functions reduce to the ϕ-functions used in Gautschi-type trigonometric or exponential integrators in [20, 21, 24] when M is a symmetric and positive semi-definite matrix. We note an important fact, namely that the convergent Taylor expansion in (6.6) depends directly on M and is applicable not only to symmetric matrices but also to nonsymmetric ones. By this definition and the formulation of TCF methods presented in next section, it can be observed that TCF methods are applicable to the system (6.1) with no regard to symmetry of M. This is the reason why the definition (6.6) is used in this chapter.

Concerning the variation of constants formula for (6.1) given in [17], we have the following result on the exact solution of the system (6.1) and its derivative:

$$\begin{cases} q(h) = \phi_0(V)q_0 + h\phi_1(V)p_0 + h^2 \displaystyle\int_0^1 (1-z)\phi_1\big((1-z)^2 V\big) f(q(hz)) \mathrm{d}z, \\[2mm] p(h) = -hM\phi_1(V)q_0 + \phi_0(V)p_0 + h \displaystyle\int_0^1 \phi_0\big((1-z)^2 V\big) f(q(hz)) \mathrm{d}z \end{cases} \tag{6.7}$$

for $h > 0$, where $V = h^2 M$.

We next consider the multi-frequency oscillatory system (6.1) restricted to the interval $[0, h]$:

$$q''(t) + Mq(t) = f(q(t)), \qquad q(0) = q_0, \quad q'(0) = q_0', \qquad t \in [0, h]. \tag{6.8}$$

Choose an orthogonal polynomial basis $\{\widehat{P}_j\}_{j=0}^{\infty}$ on the interval $[0, 1]$: e.g. the shifted Legendre polynomials over the interval $[0, 1]$, scaled in order to be orthonormal. Hence,

$$\int_0^1 \widehat{P}_i(x)\widehat{P}_j(x)\mathrm{d}x = \delta_{ij}, \qquad \deg\big(\widehat{P}_j\big) = j, \qquad i, j \geq 0,$$

where δ_{ij} is the Kronecker symbol. We rewrite the right-hand side of (6.8) as

$$f(q(\xi h)) = \sum_{j=0}^{\infty} \widehat{P}_j(\xi)\gamma_j(q), \quad \xi \in [0, 1]; \qquad \gamma_j(q) := \int_0^1 \widehat{P}_j(\tau)f\big(q(\tau h)\big)\mathrm{d}\tau. \tag{6.9}$$

In what follows, to simplify the notation, we use $\gamma_j(q)$ to denote the coefficients involved in the Fourier expansion, replacing the more complete notation $\gamma_j(h, f(q))$.

Combining (6.7) with (6.9), we have the following result straightforwardly.

Theorem 6.1 *The solution of (6.8) and its derivative satisfy*

$$
\begin{cases}
q(h) = \phi_0(V)q_0 + h\phi_1(V)p_0 + h^2 \displaystyle\sum_{j=0}^{\infty} I_{1,j}\gamma_j(q), \\[3mm]
p(h) = -hM\phi_1(V)q_0 + \phi_0(V)p_0 + h \displaystyle\sum_{j=0}^{\infty} I_{2,j}\gamma_j(q),
\end{cases}
\tag{6.10}
$$

where

$$
I_{1,j} := \int_0^1 \widehat{P}_j(z)(1-z)\phi_1\big((1-z)^2 V\big)\mathrm{d}z, \quad I_{2,j} := \int_0^1 \widehat{P}_j(z)\phi_0\big((1-z)^2 V\big)\mathrm{d}z.
\tag{6.11}
$$

Below we consider the truncated Fourier expansion which appeared first in [3]. The main idea in designing practical schemes to solve (6.1) is truncating the series (6.9) after $r \geq 2$ terms, and this replaces the initial value problem (6.1) by the approximate problem

$$
\begin{cases}
\tilde{q}'(\xi h) = \tilde{p}(\xi h), & \tilde{q}(0) = q_0, \\[2mm]
\tilde{p}'(\xi h) = -M\tilde{q}(\xi h) + \displaystyle\sum_{j=0}^{r-1} \widehat{P}_j(\xi)\gamma_j(\tilde{q}), & \tilde{p}(0) = p_0,
\end{cases}
$$

which is a natural extension of the formulae presented in [2].

The implicit solution of the new problem is given by

$$
\begin{cases}
\tilde{q}(h) = \phi_0(V)q_0 + h\phi_1(V)p_0 + h^2 \displaystyle\sum_{j=0}^{r-1} I_{1,j}\gamma_j(\tilde{q}), \\[3mm]
\tilde{p}(h) = -hM\phi_1(V)q_0 + \phi_0(V)p_0 + h \displaystyle\sum_{j=0}^{r-1} I_{2,j}\gamma_j(\tilde{q}).
\end{cases}
\tag{6.12}
$$

6.3 Formulation of TFC Methods

It is clear that the scheme (6.12) itself falls well short of being a practical method unless the integrals $I_{1,j}$, $I_{2,j}$, $\gamma_j(\tilde{q})$ in (6.12) can be approximated.

6.3.1 The Calculation of $I_{1,j}$, $I_{2,j}$

It follows from the definition of shifted Legendre polynomials on the interval $[0, 1]$,

$$\widehat{P}_j(x) = (-1)^j \sqrt{2j+1} \sum_{k=0}^{j} \binom{j}{k} \binom{j+k}{k}(-x)^k, \qquad j = 0, 1, \ldots, \qquad x \in [0, 1],$$

$$(6.13)$$

that

$$
\begin{aligned}
I_{1,j} &= \int_0^1 \widehat{P}_j(z)(1-z)\phi_1\big((1-z)^2 V\big)\,dz \\
&= \sqrt{2j+1} \sum_{l=0}^{\infty}(-1)^j \sum_{k=0}^{j} \binom{j}{k}\binom{j+k}{k}\int_0^1 (-z)^k (1-z)^{2l+1}\,dz\,\frac{(-1)^l V^l}{(2l+1)!} \\
&= \sqrt{2j+1} \sum_{l=0}^{\infty}\sum_{k=0}^{j}(-1)^{j+k}\binom{j}{k}\binom{j+k}{k}\frac{k!(2l+1)!}{(2l+k+2)!}\frac{(-1)^l V^l}{(2l+1)!} \\
&= \sqrt{2j+1} \sum_{l=0}^{\infty}\sum_{k=0}^{j}\frac{(-1)^{j+k+l}(j+k)!}{k!(j-k)!(2l+k+2)!}V^l, \qquad j = 0, 1, \ldots.
\end{aligned}
$$

Recall that a *generalized hypergeometric function* (see, e.g. [41, 47]) is

$$
{}_m F_n \begin{bmatrix} \alpha_1, \alpha_2, \ldots, \alpha_m; \\ \beta_1, \beta_2, \ldots, \beta_n; \end{bmatrix} = \sum_{l=0}^{\infty} \frac{\prod_{i=1}^{m}(\alpha_i)_l}{\prod_{i=1}^{n}(\beta_i)_l}\frac{x^l}{l!}, \qquad (6.14)
$$

where the *Pochhammer symbol* $(z)_l$ is defined as $(z)_0 = 1$ and $(z)_l = z(z+1)\ldots(z+l-1)$, $l \in \mathbb{N}$. The parameters α_i and β_i are arbitrary complex numbers, except that β_i can be neither zero nor a negative integer. $I_{1,j}$ then can be formulated by

$$
\begin{aligned}
I_{1,j} &= \sqrt{2j+1} \sum_{l=0}^{\infty}\frac{(-1)^{j+l}}{(2l+2)!}{}_2 F_1 \begin{bmatrix} -j, j+1; \\ 2l+3; \end{bmatrix} V^l \\
&= \sqrt{2j+1} \sum_{l=[j/2]}^{\infty}\frac{(-1)^{j+l}(2l+1)!}{(2l+2+j)!(2l+1-j)!}V^l,
\end{aligned}
$$

where we have used the standard formula to sum up ${}_2 F_1$ functions with unit argument (see, e.g. [41]).

For $n \geq 0$, $(2l + n)! = n! 2^{2l} (\frac{n+1}{2})_l (\frac{n+2}{2})_l$ and from this, it follows that

$$
I_{1,2j} = \sqrt{4j+1} \sum_{l=j}^{\infty} \frac{(-1)^{2j+l}(2l+1)!}{(2l+2+2j)!(2l+1-2j)!} V^l
$$

$$
= \sqrt{4j+1}(-1)^j V^j \sum_{l=0}^{\infty} \frac{(-1)^l (2l+2j+1)!}{(2l+4j+2)!(2l+1)!} V^l
$$

$$
= \sqrt{4j+1}(-1)^j V^j \sum_{l=0}^{\infty} \frac{(2j+1)!(j+1)_l(j+\frac{3}{2})_l}{(4j+2)!(2j+\frac{3}{2})_l(2j+2)_l(\frac{3}{2})_l} \frac{(\frac{-V}{4})^l}{l!}
$$

$$
= (-1)^j \sqrt{4j+1} \frac{(2j+1)!}{(4j+2)!} V^j {}_2F_3 \left[\begin{matrix} j+1, j+\frac{3}{2}; \\ 2j+2, \frac{3}{2}, 2j+\frac{3}{2}; \end{matrix} -\frac{V}{4} \right], \qquad j = 0, 1, \ldots.
$$

$$(6.15)$$

$I_{1,2j+1}$ can be dealt with similarly:

$$
I_{1,2j+1} = (-1)^{j+1} \sqrt{4j+3} \frac{(2j+1)!}{(4j+3)!} V^j {}_2F_3 \left[\begin{matrix} j+1, j+\frac{3}{2}; \\ 2j+2, \frac{1}{2}, 2j+\frac{5}{2}; \end{matrix} -\frac{V}{4} \right], \qquad j = 0, 1, \ldots.
$$

$$(6.16)$$

Likewise, we can obtain $I_{2,j}$:

$$
\left\{ \begin{matrix}
I_{2,2j} = (-1)^j \sqrt{4j+1} \dfrac{(2j)!}{(4j+1)!} V^j {}_2F_3 \left[\begin{matrix} j+\frac{1}{2}, j+1; \\ 2j+1, \frac{1}{2}, 2j+\frac{3}{2}; \end{matrix} -\dfrac{V}{4} \right], \\[4mm]
I_{2,2j+1} = (-1)^j \sqrt{4j+3} \dfrac{(2j+2)!}{(4j+4)!} V^{j+1} {}_2F_3 \left[\begin{matrix} j+\frac{3}{2}, j+2; \\ 2j+3, \frac{3}{2}, 2j+\frac{5}{2}; \end{matrix} -\dfrac{V}{4} \right], \qquad j = 0, 1, \ldots.
\end{matrix} \right.
$$

$$(6.17)$$

Based on the formula from MathWorld–A Wolfram Web Resource

$$
{}_2F_3 \left[\begin{matrix} a, a+1/2; \\ 2a, d, 2a-d+1; \end{matrix} z \right] = {}_0F_1 \left[\begin{matrix} -; z \\ d; 4 \end{matrix} \right] {}_0F_1 \left[\begin{matrix} -; & z \\ 2a-d+1; & 4 \end{matrix} \right],
$$

the above results can be simplified as

$$
\left\{ \begin{matrix}
I_{1,2j} = (-1)^j \sqrt{4j+1} \dfrac{(2j+1)!}{(4j+2)!} V^j {}_0F_1 \left[\begin{matrix} -; & V \\ \frac{3}{2}; & -\frac{V}{16} \end{matrix} \right] {}_0F_1 \left[\begin{matrix} -; & V \\ 2j+\frac{3}{2}; & -\frac{V}{16} \end{matrix} \right], \\[4mm]
I_{1,2j+1} = (-1)^{j+1} \sqrt{4j+3} \dfrac{(2j+1)!}{(4j+3)!} V^j {}_0F_1 \left[\begin{matrix} -; & V \\ \frac{1}{2}; & -\frac{V}{16} \end{matrix} \right] {}_0F_1 \left[\begin{matrix} -; & V \\ 2j+\frac{5}{2}; & -\frac{V}{16} \end{matrix} \right], \\[4mm]
I_{2,2j} = (-1)^j \sqrt{4j+1} \dfrac{(2j)!}{(4j+1)!} V^j {}_0F_1 \left[\begin{matrix} -; & V \\ \frac{1}{2}; & -\frac{V}{16} \end{matrix} \right] {}_0F_1 \left[\begin{matrix} -; & V \\ 2j+\frac{3}{2}; & -\frac{V}{16} \end{matrix} \right], \\[4mm]
I_{2,2j+1} = (-1)^j \sqrt{4j+3} \dfrac{(2j+2)!}{(4j+4)!} V^{j+1} {}_0F_1 \left[\begin{matrix} -; & V \\ \frac{3}{2}; & -\frac{V}{16} \end{matrix} \right] {}_0F_1 \left[\begin{matrix} -; & V \\ 2j+\frac{5}{2}; & -\frac{V}{16} \end{matrix} \right], \qquad j = 0, 1, \ldots.
\end{matrix} \right.
$$

$$(6.18)$$

We note that $\phi_0(V)$ and $\phi_1(V)$ can also be expressed by the generalized hypergeometric function $_0F_1$:

$$\phi_0(V) = {}_0F_1\left[\begin{matrix} -; \\ \frac{1}{2}; \end{matrix} -\frac{V}{4}\right], \quad \phi_1(V) = {}_0F_1\left[\begin{matrix} -; \\ \frac{3}{2}; \end{matrix} -\frac{V}{4}\right]. \qquad (6.19)$$

A Bessel function of the first kind can be expressed in terms of function $_0F_1$ by (see, e.g. [39])

$$J_n(x) = \frac{(x/2)^n}{n!} {}_0F_1\left[\begin{matrix} -; \\ n+1; \end{matrix} -\frac{x^2}{4}\right].$$

We also note that $_0F_1\left[\begin{matrix} -; \\ \frac{n}{2}; \end{matrix} -S\right]$ with $n \in \mathbb{N}$ and any symmetric matrix S is also viewed as a Bessel function of matrix argument (see [44]). The importance of this hypergeometric representation is that most modern software, e.g. Maple, Mathematica and Matlab, is well equipped to calculate generalized hypergeometric functions. There is also much work concerning the evaluation of generalized hypergeometric functions of a matrix argument (see, e.g. [5, 18, 33, 44]).

6.3.2 Discretization

The key question now is how to deal with $\gamma_j(\tilde{q})$. This can be achieved by introducing a quadrature formula based at k $(k \geq r)$ abscissae $0 \leq c_1 \leq \cdots \leq c_k \leq 1$, thus obtaining an approximation of the form

$$\gamma_j(\tilde{q}) \approx \sum_{l=1}^{k} b_l \widehat{P}_j(c_l) f\big(\tilde{q}(c_l h)\big), \qquad (6.20)$$

where b_l, $l = 1, 2, \ldots, k$, are the quadrature weights. From (6.12) and (6.20), it is natural to consider the following scheme

$$\begin{cases} v(h) = \phi_0(V)q_0 + h\phi_1(V)p_0 + h^2 \sum_{j=0}^{r-1} I_{1,j} \sum_{l=1}^{k} b_l \widehat{P}_j(c_l) f\big(v(c_l h)\big), \\[2ex] u(h) = -hM\phi_1(V)q_0 + \phi_0(V)p_0 + h \sum_{j=0}^{r-1} I_{2,j} \sum_{l=1}^{k} b_l \widehat{P}_j(c_l) f\big(v(c_l h)\big), \end{cases}$$

which is the exact solution of the oscillatory initial value problem

$$
\begin{cases}
v'(\xi h) = u(\xi h), & v(0) = q_0, \\
u'(\xi h) = -Mv(\xi h) + \displaystyle\sum_{j=0}^{r-1} \widehat{P}_j(\xi) \sum_{l=1}^{k} b_l \widehat{P}_j(c_l) f\big(v(c_l h)\big), & u(0) = p_0.
\end{cases}
\tag{6.21}
$$

It follows from (6.21) that $v(c_i h)$, $i = 1, 2, \ldots, k$, can be obtained by solving the following discrete problems:

$$
v''(c_i h) + Mv(c_i h) = \sum_{j=0}^{r-1} \widehat{P}_j(c_i) \sum_{l=1}^{k} b_l \widehat{P}_j(c_l) f\big(v(c_l h)\big), \quad v(0) = q_0,\ v'(0) = p_0.
\tag{6.22}
$$

By setting, as usual, $v_i = v(c_i h)$, $i = 1, 2, \ldots, k$, (6.22) can be solved by the variation of constants formula in the form:

$$
\begin{aligned}
v_i ={}& \phi_0(c_i^2 V)q_0 + c_i h \phi_1(c_i^2 V)p_0 \\
&+ (c_i h)^2 \sum_{j=0}^{r-1} \int_0^1 \widehat{P}_j(c_i z)(1-z)\phi_1\big((1-z)^2 c_i^2 V\big)\mathrm{d}z \sum_{l=1}^{k} b_l \widehat{P}_j(c_l) f(v_l), \\
& i = 1, 2, \ldots, k.
\end{aligned}
$$

6.3.3 The TFC Methods

We are now in a position to present TFC methods for the multi-frequency oscillatory second-order ordinary differential equations (6.1) or the multi-frequency oscillatory Hamiltonian system (6.2).

Definition 6.1 A trigonometric Fourier collocation (TFC) method for integrating the multi-frequency oscillatory system (6.1) or (6.2) is defined by

$$
\begin{cases}
v_i = \phi_0(c_i^2 V)q_0 + c_i h \phi_1(c_i^2 V)p_0 + (c_i h)^2 \displaystyle\sum_{j=0}^{r-1} I_{1,j,c_i}(V) \sum_{l=1}^{k} b_l \widehat{P}_j(c_l) f(v_l), \\
\quad i = 1, 2, \ldots, k, \\
v(h) = \phi_0(V)q_0 + h\phi_1(V)p_0 + h^2 \displaystyle\sum_{j=0}^{r-1} I_{1,j}(V) \sum_{l=1}^{k} b_l \widehat{P}_j(c_l) f(v_l), \\
u(h) = -hM\phi_1(V)q_0 + \phi_0(V)p_0 + h \displaystyle\sum_{j=0}^{r-1} I_{2,j}(V) \sum_{l=1}^{k} b_l \widehat{P}_j(c_l) f(v_l),
\end{cases}
\tag{6.23}
$$

where h is the stepsize, $\widehat{P}_j$ are defined by (6.13) and c_l, b_l, $l = 1, 2, \ldots, k$ are the node points and the quadrature weights of a quadrature formula, respectively. $I_{1,j}(V)$, $I_{2,j}(V)$ can be calculated by (6.15)–(6.17) and the $I_{1,j,c_i}(V)$ are defined by

$$
\begin{aligned}
I_{1,j,c_i}(V) &:= \int_0^1 \widehat{P}_j(c_i z)(1 - z)\phi_1\big((1 - z)^2 c_i^2 V\big)\mathrm{d}z \\
&= \sqrt{2j+1} \sum_{l=0}^\infty \frac{(-1)^{j+l}}{(2l+2)!} {}_2F_1\left[\begin{array}{c} -j, j+1; \\ 2l+3; \end{array} c_i\right] (c_i^2 V)^l \\
&= (-1)^j \frac{\sqrt{2j+1}(1 - c_i)}{c_i} \sum_{l=0}^\infty P_j^{-2l-2}(1 - 2c_i)\big(c_i(c_i - 1)V\big)^l,
\end{aligned}
$$

where $P_j^{-2l-2}(1 - 2c_i)$ are associated Legendre functions (see, e.g. [46]).

Remark 6.1 The TFC method (6.23) approximates the solution $q(t)$, $p(t)$ of the system (6.2) in the time interval $[0, h]$. Obviously, the values $v(h)$, $u(h)$ can be considered as the initial condition for a new initial value problem approximating $q(t)$, $p(t)$ in the time interval $[h, 2h]$. In general, one can extend the TFC methods in the usual time-stepping manner to the interval $[(i - 1)h, ih]$ for any $i \geq 2$ and finally achieve a TFC method for $q(t)$, $p(t)$ in an arbitrary interval $[0, Nh]$.

Remark 6.2 It can be observed clearly from (6.23) that TFC methods do not require the symmetry of M. Therefore, the method (6.23) can be applied to the system (6.1) with an arbitrary positive semi-definite matrix M. Moreover, the method (6.23) integrates exactly the homogeneous linear system $q'' + Mq = 0$. Thus it has an additional advantage of energy preservation and quadratic invariant preservation for homogeneous linear systems.

Based on (6.23), we use Fourier collocation methods to integrate a special and important class of systems of second-order ordinary differential equations (6.5). When $M \to 0$, $I_{1,j}$ and $I_{2,j}$ in (6.23) become

$$
\begin{cases}
\tilde{I}_{1,j} := \int_0^1 \widehat{P}_j(z)(1 - z)\phi_1(0)\mathrm{d}z = (-1)^j \dfrac{\sqrt{2j+1}}{2} {}_2F_1\left[\begin{array}{c} -j, j+1; \\ 3; \end{array} 1\right], \\[2ex]
\tilde{I}_{2,j} := \int_0^1 \widehat{P}_j(z)\phi_0(0)\mathrm{d}z = (-1)^j \sqrt{2j+1}\, {}_2F_1\left[\begin{array}{c} -j, j+1; \\ 2; \end{array} 1\right].
\end{cases}
$$

In order to simplify the results of $\displaystyle\sum_{j=0}^{r-1} \tilde{I}_{1,j} \sum_{l=1}^k b_l \widehat{P}_j(c_l) f(v_l)$ and $\displaystyle\sum_{j=0}^{r-1} \tilde{I}_{2,j} \sum_{l=1}^k b_l \widehat{P}_j(c_l) f(v_l)$ with $r \geq 2$, we compute

$$\tilde{I}_{1,j} = \begin{cases} \dfrac{1}{2}, & j = 0, \\ \dfrac{-1}{2\sqrt{3}}, & j = 1, \\ 0, & j \geq 2, \end{cases} \qquad \tilde{I}_{2,j} = \begin{cases} 1, & j = 0, \\ 0, & j \geq 1. \end{cases}$$

We then have

$$\sum_{j=0}^{r-1} \tilde{I}_{1,j} \sum_{l=1}^{k} b_l \widehat{P}_j(c_l) f(v_l) = \sum_{l=1}^{k} (1 - c_l) b_l f(v_l),$$

$$\sum_{j=0}^{r-1} \tilde{I}_{2,j} \sum_{l=1}^{k} b_l \widehat{P}_j(c_l) f(v_l) = \sum_{l=1}^{k} b_l f(v_l).$$

All these can be summed up in the following definition.

Definition 6.2 An RKN-type Fourier collocation method for integrating the system (6.5) is defined by

$$\begin{cases} v_i = q_0 + c_i h p_0 + (c_i h)^2 \displaystyle\sum_{j=0}^{r-1} \tilde{I}_{1,j,c_i} \sum_{l=1}^{k} b_l \widehat{P}_j(c_l) f(v_l), & i = 1, 2, \ldots, k, \\[4mm] v(h) = q_0 + h p_0 + h^2 \displaystyle\sum_{l=1}^{k} (1 - c_l) b_l f(v_l), \\[4mm] u(h) = p_0 + h \displaystyle\sum_{l=1}^{k} b_l f(v_l), \end{cases}$$

$$(6.24)$$

where h is the stepsize, c_l, b_l, $l = 1, 2, \ldots, k$, are the node points and the quadrature weights of a quadrature formula, respectively, and $\tilde{I}_{1,j,c_i}$ are defined by

$$\tilde{I}_{1,j,c_i} := (-1)^j \frac{\sqrt{2j+1}}{2} {}_2F_1 \begin{bmatrix} -j, j+1; \\ 3; \end{bmatrix} c_i = (-1)^j \frac{\sqrt{2j+1}(1-c_i)}{c_i} P_j^{-2}(1 - 2c_i).$$

$$(6.25)$$

Remark 6.3 It can be observed that the method (6.24) is the subclass of k-stage RKN methods with the following Butcher tableau:

$$\frac{c \mid \bar{A} = (\bar{a}_{ij})_{k \times k}}{\begin{array}{c} \bar{b}^{\mathsf{T}} \\ b^{\mathsf{T}} \end{array}} = \frac{\begin{array}{ccc} c_1 \left| c_1^2 b_1 \sum_{j=0}^{r-1} \tilde{I}_{1,j,c_1} \widehat{P}_j(c_1) \right. & \cdots & c_1^2 b_k \sum_{j=0}^{r-1} \tilde{I}_{1,j,c_1} \widehat{P}_j(c_k) \\ \vdots & \ddots & \vdots \\ c_k \left| c_k^2 b_1 \sum_{j=0}^{r-1} \tilde{I}_{1,j,c_k} \widehat{P}_j(c_1) \right. & \cdots & c_k^2 b_k \sum_{j=0}^{r-1} \tilde{I}_{1,j,c_k} \widehat{P}_j(c_k) \end{array}}{\begin{array}{ccc} (1-c_1)b_1 & \cdots & (1-c_k)b_k \\ b_1 & \cdots & b_k \end{array}}$$

From the analysis given in next section, it follows that the method (6.24) can attain arbitrary algebraic order and is symplectic when choosing a k-point Gauss–Legendre's quadrature and $r = k$. This feature is significant for solving the traditional second-order ordinary differential equations (6.5). The fact stated above demonstrates the wider applications of the TFC methods (6.23), and it makes TCF methods more efficient and competitive. Problem 6.4 in Sect. 6.5 will demonstrate this point clearly.

6.4 Properties of the TFC Methods

This section first analyses the degree of accuracy of the TFC methods in preserving the solution $p(t)$, $q(t)$, the quadratic invariant $Q = q^{\mathsf{T}} Dp$ of (6.1) and the Hamiltonian H. Other properties including convergence, phase preservation and stability properties are studied as well.

The following result is needed in the underlying analysis and its proof can be found in [3].

Lemma 6.1 Let $g : [0, h] \to \mathbb{R}^d$ have j continuous derivatives on the interval $[0, h]$. Then $\int_0^1 \widehat{P}_j(\tau)g(\tau h)d\tau = \mathcal{O}(h^j)$. As a consequence, $\gamma_j(v) = \mathcal{O}(h^j)$.

Let the quadrature formula in (6.23) be of order $m - 1$, i.e. be exact for polynomials of degree less than or equal to $m - 1$ (we note that $m \geq k$). Then we have

$$\Delta_j(h) := \gamma_j(v) - \sum_{l=1}^k b_l \widehat{P}_j(c_l) f(v_l) = \mathcal{O}(h^{m-j}), \qquad j = 0, 1, \ldots, r - 1.$$

Clearly, since $k \geq r$ is assumed, we obtain $m \geq r$ and this guarantees that the above $\Delta_j(h)$ can have good accuracy for any $j = 0, 1, \ldots, r - 1$. Choosing k large enough, along with a suitable choice of the b_l and c_l, allows us to approximate the given integral $\gamma_j(v)$ to any degree of accuracy.

For the exact solution of (6.2), let $\mathbf{y}(h) = \left(q^{\mathsf{T}}(h), p^{\mathsf{T}}(h) \right)^{\mathsf{T}}$. Then the oscillatory Hamiltonian system (6.2) can be rewritten in the form

$$\mathbf{y}'(\xi h) = F(\mathbf{y}(\xi h)) := \begin{pmatrix} p(\xi h) \\ -Mq(\xi h) + \sum\limits_{j=0}^{\infty} \widehat{P}_j(\xi)\gamma_j(q) \end{pmatrix}, \quad \mathbf{y}_0 = \begin{pmatrix} q_0 \\ p_0 \end{pmatrix}. \quad (6.26)$$

The Hamiltonian is given by

$$H(\mathbf{y}) = \frac{1}{2}p^{\mathsf{T}}p + \frac{1}{2}q^{\mathsf{T}}Mq + U(q). \quad (6.27)$$

On the other hand, for the TFC method (6.23), denoting $\omega(h) = \left(v^{\mathsf{T}}(h), u^{\mathsf{T}}(h) \right)^{\mathsf{T}}$, the numerical solution satisfies

$$\omega'(\xi h) = \begin{pmatrix} u(\xi h) \\ -Mv(\xi h) + \sum\limits_{j=0}^{r-1} \widehat{P}_j(\xi) \sum\limits_{l=1}^{k} b_l \widehat{P}_j(c_l) f\left(v(c_l h) \right) \end{pmatrix}, \quad \omega_0 = \begin{pmatrix} q_0 \\ p_0 \end{pmatrix}.$$
$$(6.28)$$

6.4.1 The Order

To express the dependence of the solutions of $\mathbf{y}'(t) = F(\mathbf{y}(t))$ on the initial values, for any given $\tilde{t} \in [0, h]$, we denote by $\mathbf{y}(\cdot, \tilde{t}, \tilde{\mathbf{y}})$ the solution satisfying the initial condition $\mathbf{y}(\tilde{t}, \tilde{t}, \tilde{\mathbf{y}}) = \tilde{\mathbf{y}}$ and set

$$\Phi(s, \tilde{t}, \tilde{\mathbf{y}}) = \frac{\partial \mathbf{y}(s, \tilde{t}, \tilde{\mathbf{y}})}{\partial \tilde{\mathbf{y}}}. \quad (6.29)$$

Recalling the elementary theory of ordinary differential equations, we have the following standard result (see, e.g. [23])

$$\frac{\partial \mathbf{y}(s, \tilde{t}, \tilde{\mathbf{y}})}{\partial \tilde{t}} = -\Phi(s, \tilde{t}, \tilde{\mathbf{y}}) F(\tilde{\mathbf{y}}). \quad (6.30)$$

The following theorem states the result on the order of TFC methods.

Theorem 6.2 *Let the quadrature formula in (6.23) be of order* $m - 1$. *Then for the TFC method (6.23)*

$$\mathbf{y}(h) - \omega(h) = \mathcal{O}(h^{n+1}) \text{ with } n = \min\{m, 2r\}.$$

The RKN-type Fourier collocation method (6.24) is also of order n.

Proof According to Lemma 6.1, (6.29) and (6.30), we arrive at

$$\mathbf{y}(h) - \omega(h) = \mathbf{y}(h, 0, \mathbf{y}_0) - \mathbf{y}\big(h, h, \omega(h)\big) = -\int_0^h \frac{d\mathbf{y}\big(h, \tau, \omega(\tau)\big)}{d\tau} d\tau$$

$$= -\int_0^h \left[\frac{\partial \mathbf{y}\big(h, \tau, \omega(\tau)\big)}{\partial \tilde{t}} + \frac{\partial \mathbf{y}\big(h, \tau, \omega(\tau)\big)}{\partial \tilde{\mathbf{y}}} \omega'(\tau) \right] d\tau$$

$$= h \int_0^1 \Phi\big(h, \xi h, \omega(\xi h)\big) \Big[F\big(\omega(\xi h)\big) - \omega'(\xi h) \Big] d\xi$$

$$= h \int_0^1 \Phi\big(h, \xi h, \omega(\xi h)\big) \left(\begin{matrix} \mathbf{0} \\ \sum\limits_{j=0}^{r-1} \widehat{P}_j(\xi) \Delta_j(h) + \sum\limits_{j=r}^{\infty} \widehat{P}_j(\xi) \gamma_j(v) \end{matrix} \right) d\xi.$$

We rewrite $\Phi(h, \xi h, \omega(\xi h))$ as a block matrix,

$$\Phi(h, \xi h, \omega(\xi h)) = \begin{pmatrix} \Phi_{11}(\xi h) & \Phi_{12}(\xi h) \\ \Phi_{21}(\xi h) & \Phi_{22}(\xi h) \end{pmatrix},$$

where Φ_{ij} $(i, j = 1, 2)$ are all $d \times d$ matrices. We then obtain

$$\mathbf{y}(h) - \omega(h) = h \left(\begin{matrix} \int_0^1 \Phi_{12}(\xi h)\Big(\sum\limits_{j=0}^{r-1} \widehat{P}_j(\xi) \Delta_j(h) + \sum\limits_{j=r}^{\infty} \widehat{P}_j(\xi) \gamma_j(v) \Big) d\xi \\ \int_0^1 \Phi_{22}(\xi h)\Big(\sum\limits_{j=0}^{r-1} \widehat{P}_j(\xi) \Delta_j(h) + \sum\limits_{j=r}^{\infty} \widehat{P}_j(\xi) \gamma_j(v) \Big) d\xi \end{matrix} \right)$$

$$= h \left(\begin{matrix} \sum\limits_{j=0}^{r-1} \int_0^1 \Phi_{12}(\xi h)\widehat{P}_j(\xi)d\xi \, \Delta_j(h) + \sum\limits_{j=r}^{\infty} \int_0^1 \Phi_{12}(\xi h)\widehat{P}_j(\xi)d\xi \, \gamma_j(v) \\ \sum\limits_{j=0}^{r-1} \int_0^1 \Phi_{22}(\xi h)\widehat{P}_j(\xi)d\xi \, \Delta_j(h) + \sum\limits_{j=r}^{\infty} \int_0^1 \Phi_{22}(\xi h)\widehat{P}_j(\xi)d\xi \, \gamma_j(v) \end{matrix} \right)$$

$$= h \left(\sum_{j=0}^{r-1} \mathcal{O}(h^j \times h^{m-j}) + \sum_{j=r}^{\infty} \mathcal{O}(h^j \times h^j) \right) = \mathcal{O}(h^{m+1}) + \mathcal{O}(h^{2r+1}),$$

which completes the proof.   □

6.4.2 The Order of Energy Preservation and Quadratic Invariant Preservation

In what follows, we are concerned with how accurately the Hamiltonian energy is preserved by the TFC method.

Theorem 6.3 *Under the condition in Theorem 6.2, we have*

$$H(\omega(h)) - H(\mathbf{y}_0) = \mathcal{O}(h^{n+1}) \text{ with } n = \min\{m, 2r\}.$$

The RKN-type Fourier collocation method (6.24) *has the same order in preserving the Hamiltonian energy as the TFC methods.*

Proof By virtue of Lemma 6.1, (6.27) and (6.28), one has

$$H(\omega(h)) - H(\mathbf{y}_0) = h \int_0^1 \nabla H(\omega(\xi h))^{\mathsf{T}} \omega'(\xi h) \mathrm{d}\xi$$

$$= h \int_0^1 \left(\left(Mv(\xi h) - \sum_{j=0}^{\infty} \widehat{P}_j(\xi) \gamma_j(v) \right)^{\mathsf{T}}, \ u(\xi h)^T \right)$$

$$\times \left(\begin{array}{c} u(\xi h) \\ -Mv(\xi h) + \sum_{j=0}^{r-1} \widehat{P}_j(\xi) \sum_{l=1}^{k} b_l \widehat{P}_j(c_l) f\big(v(c_l h)\big) \end{array} \right) \mathrm{d}\xi$$

$$= h \int_0^1 u(\xi h)^{\mathsf{T}} \left(\sum_{j=0}^{r-1} \widehat{P}_j(\xi) \sum_{l=1}^{k} b_l \widehat{P}_j(c_l) f\big(v(c_l h)\big) - \sum_{j=0}^{\infty} \widehat{P}_j(\xi) \gamma_j(v) \right) \mathrm{d}\xi$$

$$= h \int_0^1 u(\xi h)^{\mathsf{T}} \left(-\sum_{j=0}^{r-1} \widehat{P}_j(\xi) \Delta_j(h) - \sum_{j=r}^{\infty} \widehat{P}_j(\xi) \gamma_j(v) \right) \mathrm{d}\xi$$

$$= -h \sum_{j=0}^{r-1} \int_0^1 u(\xi h)^{\mathsf{T}} \widehat{P}_j(\xi) \mathrm{d}\xi \Delta_j(h) - h \sum_{j=r}^{\infty} \int_0^1 u(\xi h)^{\mathsf{T}} \widehat{P}_j(\xi) \mathrm{d}\xi \gamma_j(v)$$

$$= h \sum_{j=0}^{r-1} \mathcal{O}(h^j \times h^{m-j}) + h \sum_{j=r}^{\infty} \mathcal{O}(h^j \times h^j) = \mathcal{O}(h^{m+1}) + \mathcal{O}(h^{2r+1}).$$

This gives the result of the theorem.   □

We then consider the quadratic invariant $Q(\mathbf{y}) = q^{\mathsf{T}} Dp$ of (6.1). The following result states the degree of accuracy of the TFC method (6.23) in its preservation.

Theorem 6.4 *Under the condition in Theorem 6.2, we have*

$$Q(\omega(h)) - Q(\mathbf{y}_0) = \mathcal{O}(h^{n+1}) \text{ with } n = \min\{m, 2r\}.$$

Proof From $Q(\mathbf{y}) = q^\mathsf{T} D p$ and $D^\mathsf{T} = -D$, it follows that

$$Q(\omega(h)) - Q(\mathbf{y}_0) = h \int_0^1 \nabla Q(\omega(\xi h))^\mathsf{T} \omega'(\xi h) \mathrm{d}\xi$$

$$= h \int_0^1 \Big(-u(\xi h)^\mathsf{T} D, \ v(\xi h)^\mathsf{T} D \Big) \begin{pmatrix} u(\xi h) \\ -M v(\xi h) + \sum_{j=0}^{r-1} \widehat{P}_j(\xi) \sum_{l=1}^{k} b_l \widehat{P}_j(c_l) f \big(v(c_l h) \big) \end{pmatrix} \mathrm{d}\xi.$$

Since $q^\mathsf{T} D(f(q) - Mq) = 0$ for any $q \in \mathbb{R}^d$, we obtain

$$Q(\omega(h)) - Q(\mathbf{y}_0) = h \int_0^1 v(\xi h)^\mathsf{T} D\Big(-M v(\xi h) + \sum_{j=0}^{r-1} \widehat{P}_j(\xi) \sum_{l=1}^{k} b_l \widehat{P}_j(c_l) f \big(v(c_l h) \big) \Big) \mathrm{d}\xi$$

$$= h \int_0^1 v(\xi h)^\mathsf{T} D\Big(\sum_{j=0}^{r-1} \widehat{P}_j(\xi) \sum_{l=1}^{k} b_l \widehat{P}_j(c_l) f \big(v(c_l h) \big) - \sum_{j=0}^{\infty} \widehat{P}_j(\xi) \gamma_j(v) \Big) \mathrm{d}\xi$$

$$= h \int_0^1 v(\xi h)^\mathsf{T} D\Big(-\sum_{j=0}^{r-1} \widehat{P}_j(\xi) \Delta_j(h) - \sum_{j=r}^{\infty} \widehat{P}_j(\xi) \gamma_j(v) \Big) \mathrm{d}\xi$$

$$= -h \sum_{j=0}^{r-1} \int_0^1 v(\xi h)^\mathsf{T} D \widehat{P}_j(\xi) \mathrm{d}\xi \, \Delta_j(h) - h \sum_{j=r}^{\infty} \int_0^1 v(\xi h)^\mathsf{T} D \widehat{P}_j(\xi) \mathrm{d}\xi \, \gamma_j(v)$$

$$= h \sum_{j=0}^{r-1} \mathcal{O}(h^j \times h^{m-j}) + h \sum_{j=r}^{\infty} \mathcal{O}(h^j \times h^j) = \mathcal{O}(h^{m+1}) + \mathcal{O}(h^{2r+1}).$$

This gives the result of the theorem.   □

It is well known that nth order numerical methods can preserve the Hamiltonian energy or the quadratic invariant with at least nth degree of accuracy but unfortunately it follows from the proofs of these two theorems that TCF methods preserve the Hamiltonian energy and the quadratic invariant only with nth degree of accuracy. However, it is proved in the next theorem that when $M \to \mathbf{0}$, the Fourier collocation methods can be symplectic, i.e. they exactly preserve the quadratic invariant.

Theorem 6.5 *Under the condition that c_l, $l = 1, 2, \ldots, k$ are chosen as the node points of a k-point Gauss–Legendre's quadrature over the integral $[0, 1]$ and $r = k$, the RKN-type Fourier collocation method (6.24) is symplectic, i.e. it preserves the quadratic invariants of (6.5) exactly.*

Proof For any $m \in (1, 2, \ldots, k)$, let

$$h(x) := x^2 \sum_{j=0}^{k-1} \widehat{P}_j(c_m) \tilde{I}_{1,j,x} - c_m^2 \sum_{j=0}^{k-1} \tilde{I}_{1,j,c_m} \widehat{P}_j(x) - x + c_m.$$

From the formulae (6.13), (6.25) and the fact that c_l are the node points of a Gauss–Legendre's quadrature, it is true that

$$h(x) = (ax + b)\widehat{P}_k(x),$$

where

$$a = \frac{(-1)^k}{\sqrt{2k+1}}\left[c_m^2 \sum_{j=0}^{k-1}(-1)^j\sqrt{2j+1}\,\tilde{I}_{1,j,c_m}(j^2 + j - k^2 - k) + k(k+1)c_m - 1\right],$$

$$b = \frac{(-1)^k}{\sqrt{2k+1}}\left[c_m - c_m^2 \sum_{j=0}^{k-1}(-1)^j\sqrt{2j+1}\,\tilde{I}_{1,j,c_m}\right].$$

By the fact that $\widehat{P}_k(c_n) \equiv 0$, $n = 1, 2, \ldots, k$, we obtain $h(c_n) \equiv 0$ and thus

$$c_m^2 \sum_{j=0}^{k-1}\tilde{I}_{1,j,c_m}\widehat{P}_j(c_n) - c_n^2 \sum_{j=0}^{k-1}\tilde{I}_{1,j,c_n}\widehat{P}_j(c_m) = c_m - c_n.$$

Therefore, we have

$$b_m(\bar{b}_n - \bar{a}_{mn}) - b_n(\bar{b}_m - \bar{a}_{nm}) = b_m\big((1 - c_n)b_n - \bar{a}_{mn}\big) - b_n\big((1 - c_m)b_m - \bar{a}_{nm}\big)$$

$$= b_m b_n \left(c_m - c_n - c_m^2 \sum_{j=0}^{k-1}\tilde{I}_{1,j,c_m}\widehat{P}_j(c_n) + c_n^2 \sum_{j=0}^{k-1}\tilde{I}_{1,j,c_n}\widehat{P}_j(c_m)\right) = b_m b_n 0 = 0.$$

The symplecticity conditions of RKN methods (see [17, 21]) ensure the conclusion immediately. □

Remark 6.4 This result means that choosing a suitable k-point Gauss–Legendre's quadrature formula as well as $r = k$ in (6.24) yields a symplectic method of arbitrary order. This manipulation is very simple and convenient and it opens up the possibility of using high-order symplectic methods to solve the second-order ordinary differential equations (6.5).

6.4.3 Convergence Analysis of the Iteration

It is very clear that usually the TFC method (6.23) constitutes of a system of implicit equations for the determination of v_i and it requires iterative computation. In this chapter, we use the fixed-point iteration in practical computation. The maximum norm for a matrix or a vector is denoted by $\| \cdot \|$. Insofar as the convergence of the fixed-point iteration for the TFC method (6.23) is concerned, we have the following result.

Theorem 6.6 *Assume that M is symmetric and positive semi-definite and that f satisfies a Lipschitz condition in the variable q, i.e. there exists a constant L with the property that*

$$\|f(q_1) - f(q_2)\| \le L \|q_1 - q_2\|.$$

If

$$0 < h < \frac{1}{\sqrt{Lr^2 \max\limits_{i=1,\dots,k} c_i^2 \max\limits_{j=1,\dots,k} |b_j|}}, \tag{6.31}$$

then the fixed-point iteration for the TFC method (6.23) is convergent.

Proof Following Definition 6.1, the first formula of (6.23) can be rewritten by

$$\Lambda = \phi_0(c^2 V)q_0 + c\phi_1(c^2 V)hp_0 + h^2 A(V)f(\Lambda), \tag{6.32}$$

where $c = (c_1, c_2, \dots, c_k)^\mathsf{T}$, $\Lambda = (v_1, v_2, \dots, v_k)^\mathsf{T}$, $A(V) = (a_{ij}(V))_{k \times k}$ and $a_{ij}(V)$ are defined by

$$a_{ij}(V) := c_i^2 b_j \sum_{l=0}^{r-1} I_{1,l,c_i} \widehat{P}_l(c_j).$$

From (6.13), it follows that $|\widehat{P}_j| \le \sqrt{2j+1}$. We then obtain

$$\|a_{ij}(V)\| \le c_i^2 |b_j| \sum_{l=0}^{r-1} \sqrt{2l+1} \int_0^1 |\widehat{P}_l(c_i z)| \, \big\| (1-z)\phi_1\big((1-z)^2 c_i^2 V\big) \big\| \, dz$$

$$\le c_i^2 |b_j| \sum_{l=0}^{r-1} (2l+1) \int_0^1 \big\| (1-z)\phi_1\big((1-z)^2 c_i^2 V\big) \big\| \, dz.$$

By Proposition 2.1 in [37], we know that $\big\| \phi_1\big((1-z)^2 c_i^2 V\big) \big\| \le 1$ and then we get

$$\|a_{ij}(V)\| \le c_i^2 |b_j| \sum_{l=0}^{r-1} (2l+1) = r^2 c_i^2 |b_j|,$$

which yields $\|A(V)\| \le r^2 \max\limits_{i=1,\dots,k} c_i^2 \max\limits_{j=1,\dots,k} |b_j|$. Let

$$\varphi(x) = \phi_0(c^2 V)q_0 + c\phi_1(c^2 V)hp_0 + h^2 A(V)f(x).$$

Then

$$\|\varphi(x) - \varphi(y)\| = \left\|h^2 A(V) f(x) - h^2 A(V) f(y)\right\| \le h^2 L \|A(V)\| \|x - y\|$$
$$\le h^2 L r^2 \max_{i=1,\ldots,k} c_i^2 \max_{j=1,\ldots,k} |b_j| \|x - y\|,$$

which shows that $\varphi(x)$ is a contraction from the assumption (6.31). The well-known contraction mapping Theorem then ensures the convergence of the fixed-point iteration. □

If the matrix M is not symmetric, it can be observed that the restriction on h becomes $0 < h^2 \|A(V)\| < \dfrac{1}{L}$. For the RKN-type Fourier collocation method (6.24), the restriction on h is $0 < h < \dfrac{1}{\sqrt{L \max\limits_{1 \le i, j \le k} \{\bar{a}_{ij}\}}}$. If the assumption on f in Theorem 6.6 is only satisfied in a neighbourhood of the initial value, then further restrictions on h are required in order that the argument of f remains in this neighbourhood.

Remark 6.5 For a quadrature formula, generally speaking, not all of the node points c_i $(i = 1, 2, \ldots, k)$ are equal to zero and this ensures that $\max\limits_{i=1,\ldots,k} c_i^2 \max\limits_{j=1,\ldots,k} |b_j| \ne 0$ in (6.31) of Theorem 6.6.

6.4.4 Stability and Phase Properties

This part pays attention to the stability and phase properties. As stated in [34], the stability of RKN methods is generally analysed by

$$y''(t) = -\lambda^2 y(t) \ \text{with} \ \lambda > 0. \tag{6.33}$$

Applying the RKN-type Fourier collocation method (6.24) to (6.33) yields the recursion

$$\begin{pmatrix} v_1 \\ h u_1 \end{pmatrix} = W(\vartheta) \begin{pmatrix} q_0 \\ h p_0 \end{pmatrix},$$

where

$$W(\vartheta) = \begin{pmatrix} 1 - \vartheta^2 \bar{b}^\mathsf{T} N^{-1} e & 1 - \vartheta^2 \bar{b}^\mathsf{T} N^{-1} c \\ -\vartheta^2 b^\mathsf{T} N^{-1} e & 1 - \vartheta^2 b^\mathsf{T} N^{-1} c \end{pmatrix},$$

with $N = I + \vartheta^2 \bar{A}$, $\vartheta = \lambda h$, $e = (1, \ldots, 1)^\mathsf{T}$.

For an RKN method, we have the following definitions:

- $I_s = \{\vartheta > 0 \mid \rho(W) < 1\}$ is called the *interval of stability*.
- $I_p = \{\vartheta > 0 \mid \rho(W) = 1 \ \text{and} \ \text{tr}(W)^2 < 4 \det(W)\}$ is called the *interval of periodicity*.

• The quantities

$$\phi(\vartheta) = \vartheta - \arccos\left(\frac{\mathrm{tr}(W)}{2\sqrt{\det(W)}}\right), \quad d(\vartheta) = 1 - \sqrt{\det(W)}$$

are called the *dispersion error* and the *dissipation error*, respectively. An RKN method is said to be *dispersive of order q* and *dissipative of order r*, if $\phi(\vartheta) = O(\vartheta^{q+1})$ and $d(\vartheta) = O(\vartheta^{r+1})$, respectively. If $\phi(\vartheta) = 0$ and $d(\vartheta) = 0$, then the method is said to be *zero dispersive* and *zero dissipative*, respectively.

For the TFC method (6.23), following [49], we use the scalar revised test equation:

$$q''(t) + \omega^2 q(t) = -\varepsilon q(t) \text{ with } \omega^2 + \varepsilon > 0, \tag{6.34}$$

where ω represents an estimation of the dominant frequency λ and $\varepsilon = \lambda^2 - \omega^2$ is the error of that estimation. Applying (6.23) to (6.34) produces

$$\begin{cases} \Lambda = \phi_0(c^2 V)q_0 + c\phi_1(c^2 V)hp_0 - zA(V)\Lambda, \quad z = \varepsilon h^2, \quad V = h^2\omega^2, \\ v_1 = \phi_0(V)q_0 + \phi_1(V)hp_0 - z\bar{b}^\mathsf{T}(V)\Lambda, \\ hu_1 = -V\phi_1(V)q_0 + \phi_0(V)hp_0 - zb^\mathsf{T}(V)\Lambda, \end{cases} \tag{6.35}$$

where c, Λ, $A(V)$ are defined in Sect. 6.4.3 and

$$\bar{b}(V) = \left(b_1 \sum_{j=0}^{r-1} I_{1,j}\widehat{P}_j(c_1), \ldots, b_k \sum_{j=0}^{r-1} I_{1,j}\widehat{P}_j(c_k)\right)^\mathsf{T},$$

$$b(V) = \left(b_1 \sum_{j=0}^{r-1} I_{2,j}\widehat{P}_j(c_1), \ldots, b_k \sum_{j=0}^{r-1} I_{2,j}\widehat{P}_j(c_k)\right)^\mathsf{T}.$$

From (6.35), it follows that

$$\begin{pmatrix} v_1 \\ hu_1 \end{pmatrix} = S(V, z)\begin{pmatrix} q_0 \\ hp_0 \end{pmatrix},$$

where the stability matrix $S(V, z)$ is given by

$$S(V, z) = \begin{pmatrix} \phi_0(V) - z\bar{b}^\mathsf{T}(V)N^{-1}\phi_0(c^2 V) & \phi_1(V) - z\bar{b}^\mathsf{T}(V)N^{-1}\left(c \cdot \phi_1(c^2 V)\right) \\ -V\phi_1(V) - zb^\mathsf{T}(V)N^{-1}\phi_0(c^2 V) & \phi_0(V) - zb^\mathsf{T}(V)N^{-1}\left(c \cdot \phi_1(c^2 V)\right) \end{pmatrix}$$

with $N = I + zA(V)$.

Accordingly, we have the following definitions of stability for the TCF method (6.23).

Definition 6.3 $R_s = \{(V, z)|\ V > 0 \text{ and } \rho(S) < 1\}$ is called the *stability region of the method* (6.23). $R_p = \{(V, z)|\ V > 0,\ \rho(S) = 1 \text{ and } \text{tr}(S)^2 < 4 \det(S)\}$ is called the *periodicity region of the method* (6.23).

We also have the following definitions of the phase properties [49] (dispersion order and dissipation order) for the numerical methods.

Definition 6.4 The quantities

$$\phi(\zeta) = \zeta - \arccos\left(\frac{\text{tr}(S)}{2\sqrt{\det(S)}}\right),\quad d(\zeta) = 1 - \sqrt{\det(S)}$$

are called the dispersion error and the dissipation error of the method (6.23), respectively, where $\zeta = \sqrt{V + z}$. Then, a method is said to be dispersive of order q and dissipative of order r, if $\phi(\zeta) = \mathcal{O}(\zeta^{q+1})$ and $d(\zeta) = \mathcal{O}(\zeta^{r+1})$, respectively. If $\phi(\zeta) = 0$ and $d(\zeta) = 0$, then the method is said to be zero dispersive and zero dissipative, respectively.

Remark 6.6 In the previous analysis, the shifted Legendre polynomials are chosen as an example of orthonormal basis. It is noted that a different choice of the an orthonormal basis (e.g. Lagrange basis) can be made, and then the above analysis is accordingly modified. We do not discuss this point further in this chapter.

Remark 6.7 The TFC methods and the whole analysis presented in the chapter are applicable to the non-autonomous problem $q''(t) + Mq(t) = f\big(t, q(t)\big)$. In fact, by appending the equation $t'' = 0$, it can be turned into the autonomous form

$$u''(t) + \widetilde{M}u(t) = g\big(u(t)\big).$$

where $u(t) = \big(t, q^{\mathsf{T}}(t)\big)^{\mathsf{T}}$, $g\big(u(t)\big) = \big(t, f^{\mathsf{T}}(t, q(t))\big)^{\mathsf{T}}$ and

$$\widetilde{M} = \begin{pmatrix} 1 & 0_{1\times d} \\ 0_{d\times 1} & M \end{pmatrix}.$$

6.5 Numerical Experiments

As an example of the TFC methods, we choose a Gauss–Legendre's quadrature that is exact for all polynomials of degree ≤ 5 as the quadrature formula in (6.23), which means that

$$\begin{cases} c_1 = \dfrac{5 - \sqrt{15}}{10},\quad c_2 = \dfrac{1}{2},\quad c_3 = \dfrac{5 + \sqrt{15}}{10}, \\[2mm] b_1 = \dfrac{5}{18},\quad\quad b_2 = \dfrac{4}{9},\quad b_3 = \dfrac{5}{18}. \end{cases} \tag{6.36}$$

We then choose $r = 3$ in (6.23) and denote the corresponding trigonometric Fourier collocation methods by TFC1. Obviously, various examples of TFC methods can be obtained by choosing different values of k and r, here we do not go further on this point for brevity. According to the analysis given in Sect. 6.4, we have the following result for the method TFC1.

Theorem 6.7 *Let $t_{\text{end}} = Nh$ and apply the method TFC1 with the step size h to (6.1) on the interval $[0, t_{\text{end}}]$. We denote the numerical solution at Nh by $\omega_N(h)$ and then we have*

$$\omega_N(h) - \mathbf{y}(Nh) = \mathcal{O}(h^6),$$
$$Q(\omega_N(h)) - Q(\mathbf{y}_0) = \mathcal{O}(h^6),$$
$$H(\omega_N(h)) - H(\mathbf{y}_0) = \mathcal{O}(h^6).$$

Here all the constants implicit in $\mathcal{O}$ are independent of N and h but depend on t_{end}. When $M \to \mathbf{0}$, the method TFC1 is symplectic and it exactly preserves the quadratic invariant of (6.5).

For the method TFC1, the stability region is shown in Fig. 6.1. When $M \to \mathbf{0}$, the interval of periodicity of RNK-type TFC1 is

$$\left(0, \sqrt{54 - 2\sqrt{489}}\,\right] \bigcup \left[\sqrt{10}, 4\sqrt{15/7}\,\right] \bigcup \left[2\sqrt{15}, \sqrt{54 + 2\sqrt{489}}\,\right].$$

Fig. 6.1 Stability region (*shaded region*) of the method TFC1

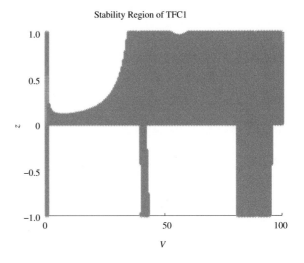

In order to show the efficiency and robustness of the method TFC1, the integrators we select for comparison are:

- TFC1: the TFC method derived in this section;
- AAVF–GL: a high precision energy-preserving integrator AAVF–GL using the Gauss–Legendre's rule (6.36) in [53];
- LIIIA: the Labatto IIIA method of order six in [21];
- HBVM(3,3): the Hamiltonian Boundary Value Method of order six in [3] which coincides with the three-stage Gauss–Legendre collocation method in [21].

It is noted that these five methods are all implicit, and we use a fixed-point iteration in the practical computations. In all the problems, we set 10^{-16} as the error tolerance and 10 as the maximum number of each iteration for showing the efficiency curve as well as energy conservation for a Hamiltonian system. For each problem, we also present the requisite total numbers of iterations for each method when choosing different error tolerances in the fixed-point iteration. The iteration procedure is terminated and the result is accepted once the error tolerance is reached, or the maximum number of iterations is exceeded.

Problem 6.1 Consider the oscillatory non-linear system (see [15])

$$q'' + \begin{pmatrix} 13 & -12 \\ -12 & 13 \end{pmatrix} q = -\frac{\partial U}{\partial q},$$

with $U(q) = q_1 q_2 (q_1 + q_2)^3$. Following [15], the initial conditions are chosen as

$$q(0) = \begin{pmatrix} -1 \\ 1 \end{pmatrix}, \quad q'(0) = \begin{pmatrix} -5 \\ 5 \end{pmatrix}, \tag{6.37}$$

such that the analytic solution of this system is

$$q(t) = \begin{pmatrix} -\cos(5t) - \sin(5t) \\ \cos(5t) + \sin(5t) \end{pmatrix}.$$

The Hamiltonian is given by

$$H(q, q') = \frac{1}{2} q'^\mathsf{T} q' + \frac{1}{2} q^\mathsf{T} \begin{pmatrix} 13 & -12 \\ -12 & 13 \end{pmatrix} q + U(q).$$

We first solve the problem on the interval $[0, 1000]$ with different stepsizes $h = 0.1/2^i$, $i = 0, 1, 2, 3$. The global errors ($GE = \|q(t_{\text{end}}) - v(t_{\text{end}})\|$) are presented in Fig. 6.2a. We then integrate this problem with the stepsize $h = 0.1$ on the interval $[0, 10000]$. See Fig. 6.2b for the energy conservation of different methods. We also solve the problem on the interval $[0, 10]$ with the stepsize $h = 0.01$ and display the total numbers of iterations in Table 6.1 for different error tolerances (tol) chosen in the fixed-point iteration.

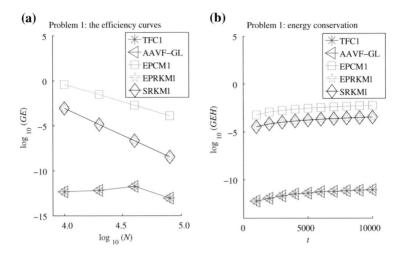

Fig. 6.2 Results for Problem 6.1 with the initial condition (6.37). **a** The logarithm of the global error (GE) over the integration interval against the logarithm of $N = t_{\mathrm{end}}/h$. **b** The logarithm of the global error of Hamiltonian $GEH = |H_n - H_0|$ against t

Table 6.1 Results for Problem 6.1 with the initial condition (6.37): The total numbers of iterations for different error tolerances (tol)

Methods	tol = 1.0e-006	tol = 1.0e-008	tol = 1.0e-010	tol = 1.0e-012
TFC1	1000	1000	1000	1000
AAVF–GL	1000	1000	1000	1000
LIIIA	3793	4787	5722	6603
HBVM(3,3)	3600	4547	5331	6000

Then we change the initial conditions into

$$q(0) = \begin{pmatrix} -1 \\ 1.1 \end{pmatrix}, \quad q'(0) = \begin{pmatrix} -5 \\ 5 \end{pmatrix}, \tag{6.38}$$

and solve the problem in $[0,1000]$ with $h = 0.1/2^i$, $i = 0,1,2,3$. See Fig. 6.3a for the global errors. We also integrate this problem with the stepsize $h = 0.05$ in $[0,5000]$ and present the energy conservation in Fig. 6.3b. Besides, the problem with the initial condition (6.38) is solved in $[0,10]$ with the stepsize $h = 0.01$, and the total numbers of iterations for different error tolerances are listed in Table 6.2.

Problem 6.2 Consider the Fermi–Pasta–Ulam problem of Chap. 2.

We choose $m = 3$, $\omega = 50$, $q_1(0) = 1$, $p_1(0) = 1$, $q_4(0) = \dfrac{1}{\omega}$, $p_4(0) = 1$, with zero for the remaining initial values. The system is integrated on the interval $[0,100]$ with the stepsizes $h = 0.1/2^k$, $k = 0,1,2,3$. The logarithm of the global errors is

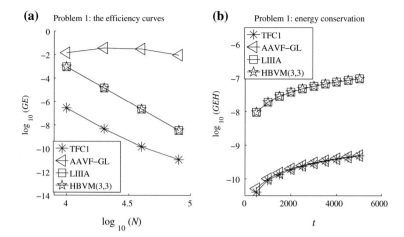

Fig. 6.3 Results for Problem 6.1 with the initial condition (6.38). **a** The logarithm of the global error (GE) over the integration interval against the logarithm of $N = t_{\mathrm{end}}/h$. **b** The logarithm of the global error of Hamiltonian $GEH = |H_n - H_0|$ against t

Table 6.2 Results for Problem 6.1 with the initial condition (6.38): The total numbers of iterations for different error tolerances (tol)

Methods	tol = 1.0e-006	tol = 1.0e-008	tol = 1.0e-010	tol = 1.0e-012
TFC1	1000	1516	1964	2000
AAVF–GL	1000	1667	1978	2283
LIIIA	3799	4795	5731	6614
HBVM(3,3)	3612	4560	5356	6000

plotted in Fig. 6.4a. Here it is noted that for LIIIA and HBVM(3,3), the global errors are too large for some values of h, thus we do not plot the corresponding points in Fig. 6.4a. Then we solve this problem in the interval [0, 10000] with the stepsize $h = 0.005$ and present the energy conservation in Fig. 6.4b. Besides, we solve the problem in the interval [0, 10] with the stepsize $h = 0.01$ to show the convergence rate of iterations for different methods. Table 6.3 lists the total numbers of iterations for different error tolerances.

Problem 6.3 Consider a non-linear wave equation (see [49])

$$
\begin{cases}
\dfrac{\partial^2 u}{\partial t^2} - gd(x)\dfrac{\partial^2 u}{\partial x^2} = \dfrac{1}{4}\lambda^2(x, u)u, & 0 < x \le b, \quad t > 0, \\[2mm]
\dfrac{\partial u}{\partial x}(t, 0) = \dfrac{\partial u}{\partial x}(t, b) = 0, \quad u(0, x) = \sin\left(\dfrac{\pi x}{b}\right), \quad u_t(0, x) = -\dfrac{\pi}{b}\sqrt{gd(x)}\cos\left(\dfrac{\pi x}{b}\right),
\end{cases}
$$

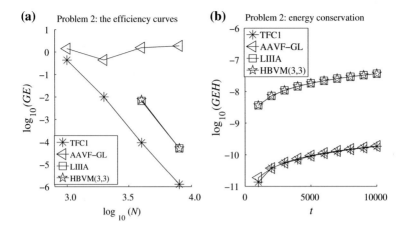

Fig. 6.4 Results for Problem 6.2. **a** The logarithm of the global error (GE) over the integration interval against the logarithm of $N = t_{end}/h$. **b** The logarithm of the global error of Hamiltonian $GEH = |H_n - H_0|$ against t

Table 6.3 Results for Problem 6.2: The total numbers of iterations for different error tolerances (tol)

Methods	tol = 1.0e-006	tol = 1.0e-008	tol = 1.0e-010	tol = 1.0e-012
TFC1	1164	2000	2036	2992
AAVF–GL	1631	2000	2706	3000
LIIIA	6726	8724	10878	12913
HBVM(3,3)	6353	8529	10789	12821

where $d(x)$ is the depth function given by $d(x) = d_0\left[2 + \cos(\dfrac{2\pi x}{b})\right]$, g denotes the acceleration of gravity, and $\lambda(x, u)$ is the coefficient of bottom friction defined by $\lambda(x, u) = \dfrac{g|u|}{C^2 d(x)}$ with Chezy coefficient C.

By using second-order symmetric differences, this problem is converted into a system of ordinary differential equations in time:

$$\begin{cases} \dfrac{d^2 u_i}{dt^2} - g d(x_i)\dfrac{u_{i+1} - 2u_i + u_{i-1}}{\Delta x^2} = \dfrac{1}{4}\lambda^2(x_i, u_i)u_i, \quad 0 < t \le t_{end}, \\[2mm] u_i(0) = \sin\left(\dfrac{\pi x_i}{b}\right), \quad u_i'(0) = -\dfrac{\pi}{b}\sqrt{g d(x_i)}\cos\left(\dfrac{\pi x_i}{b}\right), \quad i = 1, 2\ldots, N, \end{cases}$$

where $\Delta x = \dfrac{1}{N}$ is the spatial mesh step and $x_i = i\Delta x$. This semi-discrete oscillatory system has the form

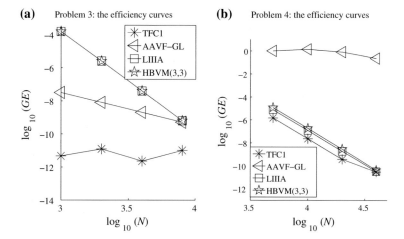

Fig. 6.5 Results for Problems 6.3 and 6.4. The logarithm of the global error (GE) over the integration interval against the logarithm of $N = t_{end}/h$

$$\begin{cases} \dfrac{d^2 U}{dt^2} + MU = F(U), \quad 0 < t \le t_{end}. \\[2mm] U(0) = \left(\sin\left(\dfrac{\pi x_1}{b}\right), \ldots, \sin\left(\dfrac{\pi x_N}{b}\right) \right)^{\mathsf{T}}, \\[2mm] U'(0) = \left(-\dfrac{\pi}{b}\sqrt{gd(x_1)}\cos\left(\dfrac{\pi x_1}{b}\right), \ldots, -\dfrac{\pi}{b}\sqrt{gd(x_N)}\cos\left(\dfrac{\pi x_N}{b}\right) \right)^{\mathsf{T}}, \end{cases} \tag{6.39}$$

where $U(t)$ denotes the N-dimensional vector with entries $u_i(t) \approx u(x_i, t)$,

$$M = \frac{g}{\Delta x^2} \begin{pmatrix} 2d(x_1) & -2d(x_2) \\ -d(x_2) & 2d(x_2) & -d(x_2) \\ & \ddots & \ddots & \ddots \\ & & -d(x_{N-1}) & 2d(x_{N-1}) & -d(x_{N-1}) \\ & & & -2d(x_N) & 2d(x_N) \end{pmatrix}, \tag{6.40}$$

and

$$F(U) = \left(\frac{1}{4}\lambda^2(x_1, u_1)u_1, \ldots, \frac{1}{4}\lambda^2(x_N, u_N)u_N \right)^{\mathsf{T}}.$$

For the parameters in this problem, we choose $b = 100$, $g = 9.81$, $d_0 = 10$, $C = 50$. The system is integrated on the interval $[0, 100]$ with $N = 32$ and the integration stepsizes $h = 0.1/2^j$ for $j = 0, 1, 2, 3$. The global errors are shown in Fig. 6.5a. Table 6.4 gives the total numbers of iterations when applying the different methods to this problem on the interval $[0, 10]$ with the stepsize $h = 0.01$ and different error

Table 6.4 Results for Problem 6.3: The total numbers of iterations for different error tolerances (tol)

Methods	tol = 1.0e-006	tol = 1.0e-008	tol = 1.0e-010	tol = 1.0e-012
TFC1	1000	1000	1000	1000
AAVF–GL	1000	1000	1000	1000
LIIIA	3000	4020	5896	6989
HBVM(3,3)	3000	4000	5000	6217

tolerances in the fixed-point iteration. Since the matrix M in this problem is not symmetric, (6.39) is not equivalent to a Hamiltonian system and thus there is no Hamiltonian energy to conserve. However, this problem can still be used to compare the accuracy of the numerical solution produced by each method. It also provides an example for supporting the assertion that our TFC methods are still applicable to (6.1) with a nonsymmetric matrix M. This is the reason why we apply these numerical methods to this non-Hamiltonian problem.

Now we use the next problem to show that the TFC methods exhibit good performance for a special and important class of second-order ordinary differential equations $q'' = f(q)$ ($M = 0$ in (6.1)).

Problem 6.4 Consider a perturbed Kepler's problem:

$$\begin{cases} q_1'' = -\dfrac{q_1}{(q_1^2 + q_2^2)^{3/2}} - \dfrac{(2\varepsilon + \varepsilon^2)q_1}{(q_1^2 + q_2^2)^{5/2}}, & q_1(0) = 1, \quad q_1'(0) = 0, \\[3mm] q_2'' = -\dfrac{q_2}{(q_1^2 + q_2^2)^{3/2}} - \dfrac{(2\varepsilon + \varepsilon^2)q_2}{(q_1^2 + q_2^2)^{5/2}}, & q_2(0) = 0, \quad q_2'(0) = 1 + \varepsilon. \end{cases}$$

The exact solution of this problem is given by

$$q_1(t) = \cos(t + \varepsilon t), \quad q_2(t) = \sin(t + \varepsilon t),$$

and the Hamiltonian is

$$H = \frac{1}{2}(q_1'^2 + q_2'^2) - \frac{1}{\sqrt{q_1^2 + q_2^2}} - \frac{(2\varepsilon + \varepsilon^2)}{3(q_1^2 + q_2^2)^{3/2}}.$$

The system also has the angular momentum $L = q_1 q_2' - q_2 q_1'$ as a first integral. We take the parameter value $\varepsilon = 10^{-3}$.

We first solve the problem on the interval [0, 1000] with different stepsizes $h = 0.1/2^{i-1}$, $i = 0, 1, 2, 3$ and present the global errors in Fig. 6.5b. We then integrate this problem with the stepsize $h = 0.1$ on the interval [0, 10000] and Fig. 6.6 gives the energy conservation and angular momentum conservation. This problem is

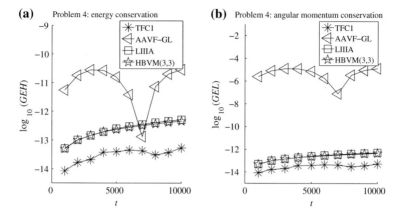

Fig. 6.6 Results for Problem 6.4. **a** The logarithm of the global error of Hamiltonian $GEH = |H_n - H_0|$ against t. **b** The logarithm of the global error of angular momentum $GEL = |L_n - L_0|$ against t

Table 6.5 Results for Problem 6.4: The total numbers of iterations for different error tolerances (tol)

Methods	tol = 1.0e-006	tol = 1.0e-008	tol = 1.0e-010	tol = 1.0e-012
TFC1	1000	2000	2000	2000
AAVF–GL	1000	2000	2000	3000
LIIIA	2000	3000	4000	5000
HBVM(3,3)	2000	3000	4000	5000

also solved on the interval $[0, 10]$ with the stepsize $h = 0.01$ and Table 6.5 lists the total numbers of iterations for different error tolerances. It is noted that the RKN-type Fourier collocation methods (6.24) are closely related to the classical Gauss–Legendre collocation methods adapted to second-order differential equations, and the numerical results suggest that their computational complexity is similar. However, it follows from the results of this problem that the new RKN-type Fourier collocation methods share more favourable error constants than the Gauss-Legendre collocation methods.

From the results of the numerical experiments, it can be observed clearly that the TCF method provides a considerably more accurate numerical solution than other methods and preserves well the Hamiltonian energy. Moreover, the TCF method requires less fixed-point iterations than other methods, which is significant in long-term computations.

6.6 Conclusions and Discussions

This chapter described a framework for the derivation and analysis of a class of TFC methods (6.23) for the multi-frequency oscillatory system (6.1) or (6.2). These trigonometric Fourier collocation methods are based on the matrix-variation of constants formula and a local Fourier expansion of the problem. Based on the TFC methods (6.23), a special RKN-type Fourier collocation method (6.24) was devised for solving a special and important class of second-order ordinary differential equations (6.5). It has been shown that the TFC methods can have arbitrarily high order, and we can obtain symplectic RKN-type Fourier collocation methods (6.24) with an arbitrary order of accuracy for (6.5) in a very convenient and simple way. Numerical experiments carried out in this chapter and the numerical results demonstrate clearly that the novel TFC methods have excellent numerical behaviour in comparison with some existing methods in the scientific literature.

Although we focus our attention on trigonometric Fourier collocation methods for solving (6.1) in this chapter, it seems possible to extend this approach to dealing with other problems, such as differential equations of the form

$$q'(t) + Aq(t) = f\big(q(t)\big).$$

This chapter is based on the recent work by Wang et al. [51].

References

1. Brugnano L, Iavernaro F (2012) Line integral methods which preserve all invariants of conservative problems. J Comput Appl Math 236:3905–3919
2. Brugnano L, Iavernaro F, Trigiante D (2011) A note on the efficient implementation of Hamiltonian BVMs. J Comput Appl Math 236:375–383
3. Brugnano L, Iavernaro F, Trigiante D (2012) A simple framework for the derivation and analysis of effective one-step methods for ODEs. Appl Math Comput 218:8475–8485
4. Brugnano L, Iavernaro F, Trigiante D (2012) Energy and quadratic invariants preserving integrators based upon gauss collocation formulae. SIAM J Numer Anal 50:2897–2916
5. Butler RW, Wood ATA (2002) Laplace approximations for hypergeometric functions with matrix argument. Ann Statist 30:1155–1177
6. Celledoni E, McLachlan RI, Owren B, Quispel GRW (2010) Energy-preserving integrators and the structure of B-series. Found Comput Math 10:673–693
7. Chartier P, Murua A (2007) Preserving first integrals and volume forms of additively split systems. IMA J Numer Anal 27:381–405
8. Cieslinski JL, Ratkiewicz B (2011) Energy-preserving numerical schemes of high accuracy for one-dimensional Hamiltonian systems. J Phys A: Math Theor 44:155206
9. Cohen D (2006) Conservation properties of numerical integrators for highly oscillatory Hamiltonian systems. IMA J Numer Anal 26:34–59
10. Cohen D, Hairer E (2011) Linear energy-preserving integrators for Poisson systems. BIT Numer Math 51:91–101
11. Cohen D, Hairer E, Lubich C (2005) Numerical energy conservation for multi-frequency oscillatory differential equations. BIT Numer Math 45:287–305

12. Cohen D, Jahnke T, Lorenz K, Lubich C (2006) Numerical integrators for highly oscillatory Hamiltonian systems: a review. In: Mielke A (ed.) Analysis, modeling and simulation of multiscale problems. Springer, Berlin, pp. 553–576
13. Dahlby M, Owren B, Yaguchi T (2011) Preserving multiple first integrals by discrete gradients. J Phys A: Math Theor 44:305205
14. Franco JM (2002) Runge-Kutta-Nyström methods adapted to the numerical integration of perturbed oscillators. Comput Phys Commun 147:770–787
15. Franco JM (2006) New methods for oscillatory systems based on ARKN methods. Appl Numer Math 56:1040–1053
16. García A, Martín P, González AB (2002) New methods for oscillatory problems based on classical codes. Appl Numer Math 42:141–157
17. García-Archilla B, Sanz-Serna JM, Skeel RD (1999) Long-time-step methods for oscillatory differential equations. SIAM J Sci Comput 20:930–963
18. Gutiérrez R, Rodriguez J, Sáez AJ (2000) Approximation of hypergeometric functions with matricial argument through their development in series of zonal polynomials. Electron Trans Numer Anal 11:121–130
19. Hairer E (2010) Energy-preserving variant of collocation methods. JNAIAM J Numer Anal Ind. Appl Math 5:73–84
20. Hairer E, Lubich C (2000) Long-time energy conservation of numerical methods for oscillatory differential equations. SIAM J Numer Anal 38:414–441
21. Hairer E, Lubich C, Wanner G (2006) Geometric numerical integration: structure-preserving algorithms for ordinary differential equations, 2nd edn. Springer, Berlin, Heidelberg
22. Hairer E, McLachlan RI, Skeel RD (2009) On energy conservation of the simplified Takahashi-Imada method. Math Model Numer Anal 43:631–644
23. Hale JK (1980) Ordinary differential equations. Roberte E. Krieger Publishing company, Huntington, New York
24. Hochbruck M, Lubich C (1999) A Gautschi-type method for oscillatory second-order differential equations. Numer Math 83:403–426
25. Hochbruck M, Ostermann A (2005) Explicit exponential Runge-Kutta methods for semilineal parabolic problems. SIAM J Numer Anal 43:1069–1090
26. Hochbruck M, Ostermann A (2010) Exponential integrators. Acta Numer 19:209–286
27. Hochbruck M, Ostermann A, Schweitzer J (2009) Exponential rosenbrock-type methods. SIAM J Numer Anal 47:786–803
28. Iavernaro F, Pace B (2008) Conservative block-boundary value methods for the solution of polynomial hamiltonian systems. AIP Conf Proc 1048:888–891
29. Iavernaro F, Trigiante D (2009) High-order symmetric schemes for the energy conservation of polynomial Hamiltonian problems. JNAIAM J. Numer. Anal. Ind. Appl. Math. 4:787–101
30. Iserles A (2008) A first course in the numerical analysis of differential equations, 2nd edn. Cambridge University Press, Cambridge
31. Iserles A, Quispel GRW, Tse PSP (2007) B-series methods cannot be volume-preserving. BIT Numer Math 47:351–378
32. Iserles A, Zanna A (2000) Preserving algebraic invariants with Runge-Kutta methods. J Comput Appl Math 125:69–81
33. Koev P, Edelman A (2006) The efficient evaluation of the hypergeometric function of a matrix argument. Math Comput 75:833–846
34. Lambert JD, Watson IA (1976) Symmetric multistep methods for periodic initialvalue problems. J Inst Math Appl 18:189–202
35. Leok M, Shingel T (2012) Prolongation-collocation variational integrators. IMA J Numer Anal 32:1194–1216
36. Li J, Wang B, You X, Wu X (2011) Two-step extended RKN methods for oscillatory systems. Comput Phys Commun 182:2486–2507
37. Li J, Wu X (2013) Adapted Falkner-type methods solving oscillatory second-order differential equations. Numer Algo 62:355–381

38. McLachlan RI, Quispel GRW, Tse PSP (2009) Linearization-preserving self-adjoint and symplectic integrators. BIT Numer Math 49:177–197
39. Petkovšek M, Wilf HS, Zeilberger D (1996) A=B. AK Peters Ltd, Wellesley, MA
40. Quispel GRW, McLaren DI (2008) A new class of energy-preserving numerical integration methods. J Phys A 41:045206
41. Rainville ED (1960) Special functions. Macmillan, New York
42. Reich S (1996) Symplectic integration of constrained Hamiltonian systems by composition methods. SIAM J Numer Anal 33:475–491
43. Reich S (1997) On higher-order semi-explicit symplectic partitioned Runge-Kutta methods for constrained Hamiltonian systems. Numer Math 76:231–247
44. Richards DSP (2011) High-dimensional random matrices from the classical matrix groups, and generalized hypergeometric functions of matrix argument. Symmetry 3:600–610
45. Sanz-Serna JM (1992) Symplectic integrators for Hamiltonian problems: an overview. Acta Numer 1:243–286
46. Scherr CW, Ivash EV (1963) Associated legendre functions. Am. J. Phys. 31:753
47. Slater LJ (1966) Generalized hypergeometric functions. Cambridge University Press, Cambridge
48. Sun G (1993) Construction of high order symplectic Runge-Kutta methods. J Comput Math 11:250–260
49. Van der Houwen PJ, Sommeijer BP (1987) Explicit Runge-Kutta (-Nyström) methods with reduced phase errors for computing oscillating solutions. SIAM J Numer Anal 24:595–617
50. Van de Vyver H (2006) A fourth-order symplectic exponentially fitted integrator. Comput Phys Commun 174:115–130
51. Wang B, Iserles A, Wu X (2014) Arbitrary order trigonometric Fourier collocation methods for multi-frequency oscillatory systems. Found Comput Math. doi:10.1007/s10208-014-9241-9
52. Wang B, Liu K, Wu X (2013) A Filon-type asymptotic approach to solving highly oscillatory second-order initial value problems. J Comput Phys 243:210–223
53. Wang B, Wu X (2012) A new high precision energy-preserving integrator for system of oscillatory second-order differential equations. Phys Lett A 376:1185–1190
54. Wang B, Wu X, Zhao H (2013) Novel improved multidimensional Strömer-Verlet formulas with applications to four aspects in scientific computation. Math Comput Modell 57:857–872
55. Wright K (1970) Some relationships between implicit Runge-Kutta, collocation and Lanczos τ methods, and their stability properties. BIT Numer Math 10:217–227
56. Wu X, Wang B (2010) Multidimensional adapted Runge-Kutta-Nyström methods for oscillatory systems. Comput Phys Commun 181:1955–1962
57. Wu X, Wang B, Shi W (2013) Efficient energy-preserving integrators for oscillatory Hamiltonian systems. J Comput Phys 235:587–605
58. Wu X, Wang B, Xia J (2012) Explicit symplectic multidimensional exponential fitting modified Runge-Kutta-Nyström methods. BIT Num Math 52:773–795
59. Wu X, You X, Shi W, Wang B (2010) ERKN integrators for systems of oscillatory second-order differential equations. Comput Phys Commun 181:1873–1887
60. Wu X, You X, Xia J (2009) Order conditions for ARKN methods solving oscillatory systems. Comput Phys Commun 180:2250–2257
61. Wu X, You X, Wang B (2013) Structure-preserving algorithms for oscillatory differential equations. Springer, Berlin, Heidelberg

Chapter 7
Error Bounds for Explicit ERKN Integrators for Multi-frequency Oscillatory Systems

A substantial issue in numerical analysis is the investigation and estimation of errors. The extended Runge-Kutta-Nyström (ERKN) integrators proposed by Wu et al. [29] are important generalizations of the classical Runge-Kutta-Nyström (RKN) methods, in the sense that both updates and internal stages have been modified so that the quantitative behaviour of ERKN integrators is adapted to oscillatory properties of the true solution. This chapter addresses the error bounds for the ERKN integrators for systems of multi-frequency oscillatory second-order differential equations.

7.1 Motivation

We are concerned with systems of multi-frequency oscillatory second-order differential equations of the form

$$
\begin{cases}
q''(t) + Mq(t) = f\big(q(t)\big), & t \in [t_0, T], \\
q(t_0) = q_0, \quad q'(t_0) = q'_0,
\end{cases}
\tag{7.1}
$$

where $M \in \mathbb{R}^{d \times d}$ is a constant positive semi-definite matrix implicitly containing and preserving the dominant frequencies of the problem, and $q : [t_0, T] \to X$ is the solution of (7.1). Here $X = \mathscr{D}(q)$ denotes the range of q. We note that the matrix M in (7.1) is not necessarily symmetric. A simple example of the oscillatory system (7.1) is

$$
\begin{cases}
q''(t) + \begin{pmatrix} 1 & 1 \\ 0 & 2 \end{pmatrix} q(t) = 0, & t \in [0, T], \\
q(0) = q_0, \quad q'(0) = q'_0,
\end{cases}
$$

© Springer-Verlag Berlin Heidelberg and Science Press, Beijing, China 2015
X. Wu et al., *Structure-Preserving Algorithms for Oscillatory Differential Equations II*, DOI 10.1007/978-3-662-48156-1_7

where M is not symmetric. The Fermi-Pasta-Ulam problem [10, 13], and the spatial semi-discretization of the wave equation with the method of lines are two classical examples of the form (7.1). Much research effort has been expended on the oscillatory system (7.1) in attempts to obtain effective integrators. We refer the reader to [20] by Petzold et al., [2] by Cohen et al., and [32] by Wu et al. for surveys of existing mathematical and numerical approaches for oscillatory differential equations. For the case where M is a symmetric and positive semi-definite matrix, Hochbruck and Lubich developed Gautschi-type exponential integrators in [15]. The development of these methods can be traced back to [6] by Gautschi in 1961. Recently, some novel revised RKN methods have also been proposed for this case. See, for example, [1, 4, 21, 24, 26, 28, 30, 31] and the references therein. We refer the reader to [19, 22, 23, 27] for some other modified RKN methods for solving (7.1) with the matrix M not necessarily symmetric. Wu et al. [29] formulated a standard form of ERKN method for the multi-frequency oscillatory system (7.1). This form of ERKN integrator makes use of the special structure brought by the linear term Mq in (7.1), and does not require M to be symmetric. The corresponding order conditions for ERKN integrators are also given in [29]. On the other hand, it is very important to present the error analysis for ERKN integrators, which was not discussed in [29]. In particular, it is interesting to show that the error bounds of an ERKN integrator are independent of M if M is symmetric and positive semi-definite. The error analysis of numerical methods has been much studied (see, e.g. [3, 12, 14, 16–18]). For error analyses related to numerical methods for the oscillatory system (7.1) with a symmetric and positive semi-definite matrix M, readers are referred to [5, 7–9, 15]. In this chapter, we present error bounds of explicit ERKN integrators for the multi-frequency oscillatory system (7.1) with a positive semi-definite matrix M which is not required to be symmetric. This means that the error analysis in this chapter has been generalized to the case where the analysis does not depend on matrix decompositions.

We note an important fact that, in comparison with fully implicit or partially implicit methods, the advantage of explicit integrators is that they do not require the solution of large and complicated systems of nonlinear equations when solving (7.1). Consequently, we only consider the error analysis of explicit ERKN integrators in this chapter. We present the error bounds of explicit ERKN integrators up to stiff order three based on a series of useful lemmas. The resulting error bounds from the error analysis developed here provide new insight into ERKN integrators. For the numerical solution of the system (7.1), we design a novel explicit ERKN integrator of order three with minimal dispersion error and dissipation error. Numerical results show that the new ERKN integrator is superior to some existing methods.

7.2   Preliminaries for Explicit ERKN Integrators

In this section, we first reformulate explicit ERKN integrators for the system (7.1) and restate the corresponding order conditions. We then analyse the stability and phase properties of explicit ERKN integrators.

In order to obtain asymptotic expansions for the exact solution of (7.1) and its derivative, we need to use the matrix-valued functions defined by (4.7) in Chap. 4.

Some interesting properties of these functions are established in the following proposition.

Proposition 7.1 *For $l = 0, 1, \ldots,$ the matrix-valued functions defined by (4.7) satisfy:*

(i) $\lim\limits_{M \to 0} \phi_l(M) = \dfrac{1}{l!}I,$ *where I is the identity matrix of the same order as M;*

(ii) *Next,*

$$\|\phi_l(M)\| \le C, \tag{7.2}$$

where C is a constant depending on $\|M\|$ in general. However, for the particular and important case, where M is a symmetric and positive semi-definite matrix, C is independent of $\|M\|$;

(iii) *Also,*

$$\begin{cases} \displaystyle\int_0^1 \frac{(1 - \xi)\phi_1\big(a^2(1 - \xi)^2 M\big)\xi^j}{j!}\,\mathrm{d}\xi = \phi_{j+2}(a^2 M), \\[4mm] \displaystyle\int_0^1 \frac{\phi_0\big(a^2(1 - \xi)^2 M\big)\xi^j}{j!}\,\mathrm{d}\xi = \phi_{j+1}(a^2 M), \quad a \in \mathbb{R}. \end{cases} \tag{7.3}$$

Proof (i) It is evident. We prove (ii) and (iii).

(ii) It is trivial that the series $\sum\limits_{k=0}^{\infty} \dfrac{x^k}{(2k + l)!}$ has radius of convergence $r = +\infty$, and therefore

$$\|\phi_l(M)\| \le \sum_{k=0}^{\infty} \frac{\|M\|^k}{(2k + l)!}.$$

Thus (7.2) is true.

In the important particular case where M is symmetric and positive semi-definite, we have the following decomposition of M:

$$M = P^\mathsf{T} W^2 P = \Omega_0^2 \text{ with } \Omega_0 = P^\mathsf{T} W P,$$

where P is an orthogonal matrix and W is a diagonal matrix with nonnegative diagonal entries which are square roots of the eigenvalues of M. In this case, we have

$$\phi_0(M) = \cos(P^\mathsf{T} W P) = P^\mathsf{T} \cos(W)P,$$
$$\phi_1(M) = (P^\mathsf{T} W P)^{-1} \sin(P^\mathsf{T} W P) = P^\mathsf{T} W^{-1} P P^\mathsf{T} \sin(W)P = P^\mathsf{T} W^{-1} \sin(W)P,$$

which show that the inequalities (7.2) with $l = 0, 1$ are true and that C is independent of $\|M\|$. It is noted here that $(P^\mathsf{T} W P)^{-1} \sin(P^\mathsf{T} W P)$ and $W^{-1} \sin(W)$ are still well-defined even if the matrix W is non-invertible. For $l \geq 2$, (7.2) and the property that C is independent of $\|M\|$ follow in a similar way.

(iii) The first formula in (7.3) can be shown as follows:

$$
\int_0^1 \frac{(1-\xi)\phi_1(a^2(1-\xi)^2M)\xi^j}{j!}\mathrm{d}\xi = \int_0^1 \frac{(1-\xi)\sum_{k=0}^\infty \frac{(-1)^k a^{2k}(1-\xi)^{2k}M^k}{(2k+1)!}\xi^j}{j!}\mathrm{d}\xi
$$

$$
= \sum_{k=0}^\infty \int_0^1 \frac{(1-\xi)^{2k+1}\xi^j}{(2k+1)!j!}\mathrm{d}\xi(-1)^k a^{2k}M^k = \sum_{k=0}^\infty \frac{(-1)^k a^{2k}M^k}{(2k+j+2)!} = \phi_{j+2}(a^2M).
$$

Likewise, the second formula of (7.3) can be obtained in a straightforward way. $\quad\square$

7.2.1   Explicit ERKN Integrators and Order Conditions

For the exact solution of (7.1) and its derivative, we have the following matrix-variation-of-constants formula derived in Chap. 1.

Theorem 7.1 *If $f : \mathbb{R}^d \to \mathbb{R}^d$ is continuous in (7.1), then for any real numbers t_0 and t, the solution of (7.1) and its derivative satisfy*

$$
\begin{cases}
q(t) = \phi_0\big((t-t_0)^2M\big)q_0 + (t-t_0)\phi_1\big((t-t_0)^2M\big)q_0' \\
\qquad + \displaystyle\int_{t_0}^t (t-\zeta)\phi_1\big((t-\zeta)^2M\big)\hat{f}(\zeta)\mathrm{d}\zeta, \\
q'(t) = -(t-t_0)M\phi_1\big((t-t_0)^2M\big)q_0 + \phi_0\big((t-t_0)^2M\big)q_0' \\
\qquad + \displaystyle\int_{t_0}^t \phi_0\big((t-\zeta)^2M\big)\hat{f}(\zeta)\mathrm{d}\zeta,
\end{cases}
\tag{7.4}
$$

where $\hat{f}(\zeta) = f(q(\zeta))$.

Approximating the integrals in (7.4) using quadrature formulae leads to ERKN integrators for the system (7.1) (see [29]). Here we only consider explicit ERKN schemes.

Definition 7.1 (*Wu et al.* [29]) An s-stage explicit multi-frequency and multidimensional ERKN integrator for integrating the multi-frequency oscillatory system (7.1) is defined by

$$\begin{cases} Q_{ni} = \phi_0(c_i^2 V)q_n + hc_i\phi_1(c_i^2 V)q_n' + h^2 \sum_{j=1}^{i-1} \bar{a}_{ij}(V)f(Q_{nj}), \quad i = 1, 2, \ldots, s, \\[2em] q_{n+1} = \phi_0(V)q_n + h\phi_1(V)q_n' + h^2 \sum_{i=1}^{s} \bar{b}_i(V)f(Q_{ni}), \\[2em] q_{n+1}' = -hM\phi_1(V)q_n + \phi_0(V)q_n' + h \sum_{i=1}^{s} b_i(V)f(Q_{ni}), \end{cases}$$

$$(7.5)$$

where h is the stepsize, and $b_i(V)$, $\bar{b}_i(V)$ and $\bar{a}_{ij}(V)$, $i, j = 1, 2, \ldots, s$ are matrix-valued functions of $V = h^2 M$.

The scheme (7.5) can also be denoted by the Butcher tableau as

$$\begin{array}{c|c} c & \bar{A}(V) \\ \hline & \bar{b}^\mathsf{T}(V) \\ \hline & b^\mathsf{T}(V) \end{array} = \begin{array}{c|ccccc} c_1 & \mathbf{0}_{d\times d} & \mathbf{0}_{d\times d} & \cdots & \mathbf{0}_{d\times d} \\ c_2 & \bar{a}_{21}(V) & \mathbf{0}_{d\times d} & \cdots & \mathbf{0}_{d\times d} \\ \vdots & \vdots & \vdots & \ddots & \vdots \\ c_s & \bar{a}_{s1}(V) & \bar{a}_{s2}(V) & \cdots & \mathbf{0}_{d\times d} \\ \hline & \bar{b}_1(V) & \bar{b}_2(V) & \cdots & \bar{b}_s(V) \\ \hline & b_1(V) & b_2(V) & \cdots & b_s(V) \end{array}$$

$$(7.6)$$

Remark 7.1 It can be observed that the formula (7.4) does not depend on the decomposition of M, which is different from that proposed by Hairer et al. (see [7, 10, 13, 15]). Accordingly, the ERKN integrators avoid this kind of matrix decompositions. In this sense, (7.4) and (7.5) have broader applications in practice since the class of physical problems usually fall within the scope. Moreover, when $V \to \mathbf{0}_{d\times d}$, an ERKN integrator (7.5) reduces to a classical RKN method.

We define the order of ERKN integrators as follows.

Definition 7.2 An explicit ERKN integrator (7.5) for the system (7.1) has order r if for sufficiently smooth problems (7.1),

$$e_{n+1} := q_{n+1} - q(t_n + h) = \mathcal{O}(h^{r+1}), \quad e_{n+1}' := q_{n+1}' - q'(t_n + h) = \mathcal{O}(h^{r+1}). \quad (7.7)$$

Here $q(t_n + h)$ and $q'(t_n + h)$ are the exact solution of (7.1) and the corresponding derivative at $t_n + h$, respectively, and q_{n+1} and q_{n+1}' are the numerical results obtained in one integration step under the local assumptions: $q_n = q(t_n)$ and $q_n' = q'(t_n)$.

Order conditions for ERKN integrators are presented in [29] and we briefly restate them in the following theorem.

Theorem 7.2 *An s-stage explicit ERKN integrator (7.5) is of order r if and only if*

$$
\begin{cases}
\bar{b}^{\mathsf{T}}(V)\Phi(\tau) = \dfrac{\rho(\tau)!}{\gamma(\tau)}\phi_{\rho(\tau)+1}(V) + \mathcal{O}(h^{r-\rho(\tau)}), & \rho(\tau) = 1, 2, \ldots, r-1, \\[2mm]
b^{\mathsf{T}}(V)\Phi(\tau) = \dfrac{\rho(\tau)!}{\gamma(\tau)}\phi_{\rho(\tau)}(V) + \mathcal{O}(h^{r+1-\rho(\tau)}), & \rho(\tau) = 1, 2, \ldots, r,
\end{cases}
$$

where τ is an extended Nyström tree associated with an elementary differential $\mathscr{F}(\tau)(q_n, q'_n)$ of the function $f(q)$ at q_n.

The set of extended Nyström trees $\mathbb{T}$, the functions $\rho(\tau)$, the *order* of τ, $\alpha(\tau)$, the *number of possible monotonic labelings* of τ, and $\gamma(\tau)$, the *signed density* of τ, are well defined in [33]. The elementary differential $\mathscr{F}$ and the vector $\Phi(\tau) = \big(\Phi_i(\tau)\big)_{i=1}^{s}$ of elementary weights for the internal stages can also be found in [33].

7.2.2 Stability and Phase Properties

Now we turn to the stability and phase properties of explicit ERKN integrators. Following [22], we use a second-order homogeneous linear test model of the form

$$
q''(t) + \omega^2 q(t) = -\varepsilon q(t), \quad \omega^2 + \varepsilon > 0, \tag{7.8}
$$

where ω represents an estimate of the dominant frequency λ, and $\varepsilon = \lambda^2 - \omega^2$ is the error of the estimation. Applying an explicit ERKN integrator (7.5) to the test equation (7.8) yields

$$
\begin{cases}
Q = \phi_0(c^2 V)q_n + \big(c \cdot \phi_1(c^2 V)h\big)q'_n - z\bar{A}(V)Q, & z = \varepsilon h^2, \quad V = h^2\omega^2, \\[1mm]
q_{n+1} = \phi_0(V)q_n + \phi_1(V)hq'_n - z\bar{b}^{\mathsf{T}}(V)Q, \\[1mm]
hq'_{n+1} = -V\phi_1(V)q_n + \phi_0(V)hq'_n - zb^{\mathsf{T}}(V)Q.
\end{cases}
$$

Thus,

$$
\begin{pmatrix} q_{n+1} \\ hq'_{n+1} \end{pmatrix} = S(V, z) \begin{pmatrix} q_n \\ hq'_n \end{pmatrix},
$$

where the stability matrix $S(V, z)$ is determined by

$$
S(V, z) = \begin{pmatrix} \phi_0(V) - z\bar{b}^{\mathsf{T}}(V)N^{-1}\phi_0(c^2 V) & \phi_1(V) - z\bar{b}^{\mathsf{T}}(V)N^{-1}\big(c \cdot \phi_1(c^2 V)\big) \\ -V\phi_1(V) - zb^{\mathsf{T}}(V)N^{-1}\phi_0(c^2 V) & \phi_0(V) - zb^{\mathsf{T}}(V)N^{-1}\big(c \cdot \phi_1(c^2 V)\big) \end{pmatrix}
$$

with $N = I + z\bar{A}(V)$.

The stability of an explicit ERKN integrator is characterized by the spectral radius $\rho(S(V, z))$. We use the two-dimensional region (V, z)-plane to express the stability of an explicit ERKN integrator:

- $R_s = \{(V, z)|\ V > 0 \text{ and } \rho(S) < 1\}$ is called the *stability region* of an explicit ERKN integrator;
- $R_p = \{(V, z)|\ V > 0,\ \rho(S) = 1 \text{ and } \mathrm{tr}(S)^2 < 4\det(S)\}$ is called the *periodicity region* of an explicit ERKN integrator;
- If $R_s = (0, +\infty) \times (-\infty, +\infty)$, the explicit ERKN integrator is said to be *A-stable*;
- If $R_p = (0, +\infty) \times (-\infty, +\infty)$, the explicit ERKN integrator is said to be *P-stable*.

The dispersion error and the dissipation error of an ERKN integrator can be defined in a way similar to those described in [22].

Definition 7.3 Let $H := h\lambda = \sqrt{V + z}$ and

$$\phi := H - \arccos\left(\frac{\mathrm{tr}(S(V, z))}{2\sqrt{\det(S(V, z))}}\right), \quad d := 1 - \sqrt{\det(S(V, z))},$$

where $z = \frac{\varepsilon}{\omega^2 + \varepsilon} H^2$, $V = \frac{\omega^2}{\omega^2 + \varepsilon} H^2$. The Taylor expansions of ϕ and d in H (denoted by $\phi(H)$ and $d(H)$) are called the dispersion error and the dissipation error of an explicit ERKN integrator, respectively. Thus, a method is said to be dispersive of order r if $\phi(H) = \mathscr{O}(H^{r+1})$, and is said to be dissipative of order s if $d(H) = \mathscr{O}(H^{s+1})$. If $\phi(H) = 0$ and $d(H) = 0$, then the method is said to be zero dispersive and zero dissipative, respectively.

7.3 Preliminary Error Analysis

In the following analysis, we restrict ourselves to the case of s-stage explicit ERKN integrators of order p with $s \le 3$ and $p \le 3$.

By Proposition 7.1(*ii*), it is true that $\|\phi_l(V)\|$ ($l = 0, 1, \ldots$, and $V = h^2 M$) is uniformly bounded with respect to a symmetric and positive semi-definite matrix M, and so are $b_i(V)$, $\bar{b}_i(V)$, and $\bar{a}_{ij}(V)$ ($i, j = 1, 2, \ldots, s$), since they are combinations of $\phi_l(V)$ and $\phi_l(c_i^2 V)$. This property is critical in obtaining the required error bounds for ERKN integrators (given in Sect. 7.4) for the important particular case where M is symmetric and positive semi-definite.

7.3.1 Three Elementary Assumptions and a Gronwall's Lemma

Throughout the chapter we use the following three elementary assumptions for the error analysis of explicit ERKN integrators.

Assumption 7.1 We suppose that (7.1) possesses a uniformly bounded and suffi-
ciently smooth solution $q : [t_0, T] \rightarrow X$ with derivatives in X, and that $f(q) : X \rightarrow S$
in (7.1) is Fréchet differentiable sufficiently often in a strip (see [16, 17]) along the
exact solution of (7.1). All occurring derivatives of $f(q)$ are assumed to be uniformly
bounded.

Assumption 7.2 The coefficients of an s-stage explicit ERKN integrator (7.5) satisfy the following assumptions:

$$\sum_{j=1}^{i-1} \bar{a}_{ij}(V) = c_i^2 \phi_2(c_i^2 V), \qquad i = 1, 2, \ldots, s. \tag{7.9}$$

Note that (7.9) implies $c_1 = 0$.

Assumption 7.3 The coefficients $b_i(V)$, $\bar{b}_i(V)$ and $\bar{a}_{ij}(V)$, $i, j = 1, 2, \ldots, s$ of an
s-stage explicit ERKN integrator (7.5) are bounded for any matrix V, and uniformly
bounded for symmetric and positive semi-definite ones.

In the following analysis, we will use a discrete Gronwall lemma (Lemma 2.4 in
[14]). Here, we represent the lemma as follows.

Lemma 7.1 *Let* α, ϕ, ψ, *and* χ *be nonnegative functions defined for* $t_n = n\Delta t$, $n =$
$0, 1, \ldots, N$, *and assume* χ *is nondecreasing. If*

$$\phi_k + \psi_k \leq \chi_k + \Delta t \sum_{n=1}^{k-1} \alpha_n \phi_n, \qquad k = 1, 2, \ldots, N,$$

and if there is a positive constant $\hat{c}$ *such that* $\Delta t \sum_{n=1}^{k-1} \alpha_n \leq \hat{c}$, *then*

$$\phi_k + \psi_k \leq \chi_k e^{\hat{c}k\Delta t}, \qquad k = 1, 2, \ldots, N,$$

where the subscripts k *and* n *mean the functions are evaluated at* $t_k = k\Delta t$ *and*
$t_n = n\Delta t$, *respectively.*

7.3.2   Residuals of ERKN Integrators

Here, we first find an estimate for explicit ERKN integrators.
    Inserting the exact solution of (7.1) into the numerical scheme (7.5) gives

Table 7.1 Stiff order conditions for explicit ERKN integrators

Stiff order conditions	Order
$\sum_{i=1}^{s} b_i(V) = \phi_1(V)$	1
$\sum_{i=1}^{s} b_i(V)c_i = \phi_2(V)$	2
$\sum_{i=1}^{s} \bar{b}_i(V) = \phi_2(V)$	2
$\sum_{j=1}^{i-1} \bar{a}_{ij}(V) = c_i^2\phi_2(c_i^2 V), \ i = 1,2,\ldots,s$	3
$\sum_{i=1}^{s} b_i(V)c_i^2 = 2\phi_3(V)$	3
$\sum_{i=1}^{s} \bar{b}_i(V)c_i = \phi_3(V)$	3

$$\begin{cases} q(t_n + c_ih) = \phi_0(c_i^2 V)q(t_n) + hc_i\phi_1(c_i^2 V)q'(t_n) + h^2 \sum_{j=1}^{i-1} \bar{a}_{ij}(V)\hat{f}(t_n + c_jh) + \triangle_{ni}, \\ \qquad i = 1,2,\ldots,s, \\ q(t_n + h) = \phi_0(V)q(t_n) + h\phi_1(V)q'(t_n) + h^2 \sum_{i=1}^{s} \bar{b}_i(V)\hat{f}(t_n + c_ih) + \delta_{n+1}, \\ q'(t_n + h) = -hM\phi_1(V)q(t_n) + \phi_0(V)q'(t_n) + h \sum_{i=1}^{s} b_i(V)\hat{f}(t_n + c_ih) + \delta'_{n+1}, \end{cases}$$
$$(7.10)$$

where $\triangle_{ni}$, δ_{n+1} and δ'_{n+1} are the discrepancies and $\hat{f}(t) = f(q(t))$. The following lemma gives an estimate of the discrepancies.

Lemma 7.2 *Under Assumptions 7.1 and 7.2, if the stiff order conditions (Table 7.1) of explicit ERKN integrators are satisfied up to order p (p ≤ 3), then*

$$\begin{aligned} \|\triangle_{n1}\| &= 0, \\ \|\triangle_{ni}\| &\le C_1 h^3, \quad i = 2,3, \\ \|\delta_{n+1}\| &\le C_2 h^{p+1}, \\ \|\delta'_{n+1}\| &\le C_3 h^{p+1}. \end{aligned}$$
$$(7.11)$$

Here the constants C_1, C_2 and C_3 depend on $\|M\|$ but are independent of h and n. However, in the important particular case where M is symmetric and positive semi-definite, they are all independent of $\|M\|$.

Proof Expressing $q(t_n + c_i h)$ in (7.10) by the formula (7.4) yields

$$q(t_n + c_i h) = \phi_0(c_i^2 V)q(t_n) + hc_i\phi_1(c_i^2 V)q'(t_n)$$
$$+ h^2 \int_0^{c_i} (c_i - z)\phi_1((c_i - z)^2 V)\hat{f}(t_n + hz)\mathrm{d}z. \tag{7.12}$$

Comparing (7.12) with the first formula in (7.10), we obtain

$$\Delta_{ni} = h^2 \int_0^{c_i} (c_i - z)\phi_1((c_i - z)^2 V)\hat{f}(t_n + hz)\mathrm{d}z - h^2 \sum_{j=1}^{i-1} \bar{a}_{ij}(V)\hat{f}(t_n + c_j h).$$

Expressing $\hat{f}(t_n + hz)$ in the above formula by Taylor series expansions yields

$$\Delta_{ni} = h^2 \int_0^{c_i} (c_i - z)\phi_1((c_i-z)^2 V) \sum_{j=0}^{\infty} \frac{h^j z^j}{j!}\hat{f}^{(j)}(t_n)\mathrm{d}z$$
$$- h^2 \sum_{k=1}^{i-1} \bar{a}_{ik}(V) \sum_{j=0}^{\infty} \frac{c_k^j h^j}{j!}\hat{f}^{(j)}(t_n)$$
$$= \sum_{j=0}^{\infty} h^{j+2} c_i^{j+2} \int_0^1 \frac{(1-\xi)\phi_1(c_i^2(1-\xi)^2 V)\xi^j}{j!}\mathrm{d}\xi \hat{f}^{(j)}(t_n)$$
$$- h^2 \sum_{k=1}^{i-1} \bar{a}_{ik}(V) \sum_{j=0}^{\infty} \frac{c_k^j h^j}{j!}\hat{f}^{(j)}(t_n).$$

By the third property of Proposition 7.1, we have

$$\Delta_{ni} = \sum_{j=0}^{\infty} h^{j+2} c_i^{j+2}\phi_{j+2}(c_i^2 V)\hat{f}^{(j)}(t_n) - h^2 \sum_{k=1}^{i-1} \bar{a}_{ik}(V) \sum_{j=0}^{\infty} \frac{c_k^j h^j}{j!}\hat{f}^{(j)}(t_n)$$
$$= \sum_{j=0}^{\infty} h^{j+2} \left(c_i^{j+2}\phi_{j+2}(c_i^2 V) - \sum_{k=1}^{i-1} \bar{a}_{ik}(V)\frac{c_k^j}{j!} \right)\hat{f}^{(j)}(t_n).$$

It follows from Assumptions 7.1 and 7.2 that

$$\|\Delta_{n1}\| = 0, \quad \|\Delta_{ni}\| \le C_1 h^3, \quad i = 2, 3.$$

Similarly, for δ_{n+1}, we have

$$\delta_{n+1} = h^2 \int_0^1 (1-z)\phi_1\big((1-z)^2 V\big)\hat{f}(t_n + hz)\mathrm{d}z - h^2 \sum_{k=1}^s \bar{b}_k(V)\hat{f}(t_n + c_k h)$$

$$= h^2 \int_0^1 (1-z)\phi_1\big((1-z)^2 V\big) \sum_{j=0}^\infty \frac{h^j z^j}{j!}\hat{f}^{(j)}(t_n)\mathrm{d}z - h^2 \sum_{k=1}^s \bar{b}_k(V) \sum_{j=0}^\infty \frac{c_k^j h^j}{j!}\hat{f}^{(j)}(t_n)$$

$$= \sum_{j=0}^\infty h^{j+2} \int_0^1 \frac{(1-z)\phi_1\big((1-z)^2 V\big)z^j}{j!}\mathrm{d}z \hat{f}^{(j)}(t_n) - h^2 \sum_{k=1}^s \bar{b}_k(V) \sum_{j=0}^\infty \frac{c_k^j h^j}{j!}\hat{f}^{(j)}(t_n)$$

$$= \sum_{j=0}^\infty h^{j+2}\phi_{j+2}(V)\hat{f}^{(j)}(t_n) - h^2 \sum_{k=1}^s \bar{b}_k(V) \sum_{j=0}^\infty \frac{c_k^j h^j}{j!}\hat{f}^{(j)}(t_n)$$

$$= \sum_{j=0}^\infty h^{j+2}\Big(\phi_{j+2}(V) - \sum_{k=1}^s \bar{b}_k(V)\frac{c_k^j}{j!}\Big)\hat{f}^{(j)}(t_n).$$

The third inequality in (7.11) follows immediately from the stiff order conditions for ERKN integrators of order p (Table 7.1).

Likewise, we get

$$\delta'_{n+1} = \sum_{j=0}^\infty h^{j+1}\Big(\phi_{j+1}(V) - \sum_{k=1}^s b_k(V)\frac{c_k^j}{j!}\Big)\hat{f}^{(j)}(t_n),$$

and the fourth inequality in (7.11) holds. $\qquad\square$

Remark 7.2 All the conditions encountered so far in the above proof are collected in Table 7.1. It is easy to check that these conditions can yield the corresponding order conditions (Theorem 7.2). However, these conditions are different from those given in Theorem 7.2, since they use the assumption (7.9) plus some order conditions (Theorem 7.2) with the terms $\mathcal{O}(h^{r-\rho(\tau)})$ and $\mathcal{O}(h^{r+1-\rho(\tau)})$ ignored in these order conditions. The conditions shown in Table 7.1 are regarded as stiff order conditions (following [16, 17]). We will restrict ourselves to the explicit ERKN integrators that fulfill the stiff order conditions up to order p ($p \le 3$) in the remaining parts of this chapter. For the important particular case where M is symmetric and positive semi-definite, it is important to observe that the constants C_1, C_2, and C_3 are all independent of $\|M\|$ since then $\phi_l(V)$, $b_i(V)$, $\bar{b}_i(V)$, and $\bar{a}_{ij}(V)$ ($l = 0, 1, \ldots$ and $i, j = 1, 2, \ldots, s$) are all uniformly bounded.

7.4 Error Bounds

Let e_n denote the difference between numerical and exact solutions at t_n, E_{ni} the difference at $t_n + c_i h$, and e'_n the difference between numerical and exact derivatives at t_n, namely,

$$e_n = q_n - q(t_n), \quad E_{ni} = Q_{ni} - q(t_n + c_i h), \quad e'_n = q'_n - q'(t_n). \tag{7.13}$$

Subtracting (7.10) from (7.5) gives the error recursions

$$
\left\{
\begin{aligned}
& E_{ni} = \phi_0(c_i^2 V)e_n + hc_i\phi_1(c_i^2 V)e_n' + h^2 \sum_{j=1}^{i-1} \bar{a}_{ij}(V)\left(f(Q_{nj}) - \hat{f}(t_n + c_jh)\right) - \Delta_{ni}, \\
& i = 1, 2, \ldots, s, \\
& e_{n+1} = \phi_0(V)e_n + h\phi_1(V)e_n' + h^2 \sum_{i=1}^{s} \bar{b}_i(V)\left(f(Q_{ni}) - \hat{f}(t_n + c_ih)\right) - \delta_{n+1}, \\
& e_{n+1}' = -hM\phi_1(V)e_n + \phi_0(V)e_n' + h \sum_{i=1}^{s} b_i(V)\left(f(Q_{ni}) - \hat{f}(t_n + c_ih)\right) - \delta_{n+1}'.
\end{aligned}
\right.
$$

$$(7.14)$$

We now propose two lemmas from which we present the main result of the analysis in this chapter.

Lemma 7.3 *Under Assumptions 7.1 and 7.2, if the errors e_n and e_n' remain in a neighborhood of* 0, *then*

$$
\begin{aligned}
\left\| f(Q_{ni}) - \hat{f}(t_n + c_ih) \right\| &\le C_4(\|E_{ni}\| + \|E_{ni}\|^2), && i = 1, 2, 3, \\
\|E_{ni}\| &\le C_5(\|e_n\| + h\left\|e_n'\right\| + h^3), && i = 1, 2, 3,
\end{aligned}
$$

$$(7.15)$$

where the constants C_4 and C_5 depend on $\|M\|$ but are independent of h and n. In the important particular case where M is a symmetric and positive semi-definite matrix, C_4 and C_5 are both independent of $\|M\|$.

Proof The Taylor series expansion of $\hat{f}(t_n + c_ih)$ yields

$$
\begin{aligned}
f(Q_{ni}) - \hat{f}(t_n + c_ih) = f(Q_{ni}) - f\left(q(t_n + c_ih)\right) &= \frac{\partial f}{\partial q}\left(q(t_n + c_ih)\right)E_{ni} \\
&+ \int_0^1 (1 - \tau)\frac{\partial^2 f}{\partial q^2}\left(q(t_n + c_ih) + \tau E_{ni}\right)(E_{ni}, E_{ni})d\tau.
\end{aligned}
$$

This formula proves immediately the first inequality of (7.15) by Assumption 7.1.

By setting $i = 1$ in the first formula of (7.14), it is trivial to verify that

$$\|E_{n1}\| = \|e_n\|.$$

$$(7.16)$$

It follows from the first formula of both (7.14) and (7.15) that

$$\|E_{ni}\| \leq C_6 \|e_n\| + C_7 h \left\|e_n'\right\| + C_8 h^2 \sum_{j=1}^{i-1}(\|E_{nj}\| + \|E_{nj}\|^2) + \|\triangle_{ni}\|,$$
$$i = 2, 3. \tag{7.17}$$

By Lemma 7.2, we obtain

$$\|E_{ni}\| \leq C_6 \|e_n\| + C_7 h \left\|e_n'\right\| + C_8 h^2 \sum_{j=1}^{i-1}(\|E_{nj}\| + \|E_{nj}\|^2) + C_1 h^3. \tag{7.18}$$

Thus, the following result holds as long as the error e_n remains in a neighborhood of 0:

$$\begin{aligned}
\|E_{n2}\| &\leq C_6 \|e_n\| + C_7 h \left\|e_n'\right\| + C_8 h^2(\|E_{n1}\| + \|E_{n1}\|^2) + C_1 h^3 \\
&= C_6 \|e_n\| + C_7 h \left\|e_n'\right\| + C_8 h^2(\|e_n\| + \|e_n\|^2) + C_1 h^3 \\
&= (C_6 + C_8 h^2 + C_8 h^2 \|e_n\|) \|e_n\| + C_7 h \left\|e_n'\right\| + C_1 h^3 \\
&\leq \widetilde{C}_2(\|e_n\| + h \left\|e_n'\right\| + h^3).
\end{aligned}$$

In a similar way, we obtain

$$\|E_{n3}\| \leq \widetilde{C}_3(\|e_n\| + h \left\|e_n'\right\| + h^3).$$

Letting $C_5 = \max(1, \widetilde{C}_2, \widetilde{C}_3)$ immediately yields

$$\|E_{ni}\| \leq C_5(\|e_n\| + h \left\|e_n'\right\| + h^3), \quad i = 1, 2, 3.$$

$\square$

Lemma 7.4 *Under Assumptions 7.1 and 7.2, consider an s-stage (s $\leq$ 3) explicit ERKN integrator (7.5) that fulfills the stiff order conditions up to order p (p $\leq$ 3) for the system (7.1). Then, the numerical solution and its derivative satisfy the following inequalities, uniformly on $0 \leq nh \leq T - t_0$:*

$$\begin{cases}
\|e_n\| \leq \overline{C} h^p + \overline{C} \sum_{k=0}^{n-1}(h^2 + h^3)\{\left\|e_k'\right\| + h \left\|e_k'\right\|^2\}, \\
\left\|e_n'\right\| \leq \widetilde{C} h^p + \widetilde{C} \sum_{k=0}^{n-1}(h + h^2)\{\|e_k\| + h^3 \|e_k\| + \|e_k\|^2\},
\end{cases} \tag{7.19}$$

where $\overline{C}$ and $\widetilde{C}$ are constants depending on T and $\|M\|$ but independent of h and n. However, in the important particular case where M is symmetric and positive semi-definite, $\overline{C}$ is independent of $\|M\|$.

Proof From (7.14), we obtain the error recursion

$$\begin{pmatrix} e_{n+1} \\ e'_{n+1} \end{pmatrix} = \mathbf{R} \begin{pmatrix} e_n \\ e'_n \end{pmatrix} + \begin{pmatrix} h^2 \sum_{i=1}^{s} \bar{b}_i(V) \left(f(Q_{ni}) - \hat{f}(t_n + c_i h) \right) - \delta_{n+1} \\ h \sum_{i=1}^{s} b_i(V) \left(f(Q_{ni}) - \hat{f}(t_n + c_i h) \right) - \delta'_{n+1} \end{pmatrix}.$$

Solving this recursion yields

$$\begin{pmatrix} e_n \\ e'_n \end{pmatrix} = \sum_{k=0}^{n-1} \mathbf{R}^{n-k-1} \begin{pmatrix} h^2 \sum_{i=1}^{s} \bar{b}_i(V) \left(f(Q_{ki}) - \hat{f}(t_k + c_i h) \right) - \delta_{k+1} \\ h \sum_{i=1}^{s} b_i(V) \left(f(Q_{ki}) - \hat{f}(t_k + c_i h) \right) - \delta'_{k+1} \end{pmatrix},$$

where $e_0 = \mathbf{0}$, $e'_0 = \mathbf{0}$ are used, and

$$\mathbf{R} = \begin{pmatrix} \phi_0(V) & h\phi_1(V) \\ -hM\phi_1(V) & \phi_0(V) \end{pmatrix}, \quad \mathbf{R}^m = \begin{pmatrix} \phi_0(m^2 V) & mh\phi_1(m^2 V) \\ -mhM\phi_1(m^2 V) & \phi_0(m^2 V) \end{pmatrix}.$$

By $\|m^2 V\| \leq T^2 \|M\|$ and Proposition 7.1,

$$\|\phi_0(m^2 V)\| \leq C \quad \text{and} \quad \|mh\phi_1(m^2 V)\| \leq T \|\phi_1(m^2 V)\| \leq C. \qquad (7.20)$$

Thus,

$$\begin{aligned} \|e_n\| &\leq \sum_{k=0}^{n-1} \{ C(h^2 \sum_{i=1}^{s} \|f(Q_{ki}) - \hat{f}(t_k + c_i h)\| + \|\delta_{k+1}\|) \\ &\quad + C(h \sum_{i=1}^{s} \|f(Q_{ki}) - \hat{f}(t_k + c_i h)\| + \|\delta'_{k+1}\|) \} \\ &\leq C \sum_{k=0}^{n-1} \{ (h^2 + h) \sum_{i=1}^{s} \|f(Q_{ki}) - \hat{f}(t_k + c_i h)\| + \|\delta_{k+1}\| + \|\delta'_{k+1}\| \}. \end{aligned}$$

A direct application of the inequalities in Lemma 7.3 gives

$$\|e_n\| \le C \sum_{k=0}^{n-1} \{(h^2 + h) \sum_{i=1}^{s} C_4 \Big(C_5 (\|e_k\| + h \|e_k'\| + h^3)$$
$$+ (C_5(\|e_k\| + h \|e_k'\| + h^3))^2 \Big) + \|\delta_{k+1}\| + \|\delta_{k+1}'\| \}.$$

It follows from Lemma 7.2 that

$$\|e_n\| \le C \sum_{k=0}^{n-1} \{(h^2 + h)\big(C_6(\|e_k\| + h \|e_k'\| + h^3)$$
$$+ (C_6(\|e_k\| + h \|e_k'\| + h^3))^2\big) + C_2 h^{p+1} + C_3 h^{p+1}\}.$$
$$\le C \sum_{k=0}^{n-1} \{(h^2 + h)\big(C_6 h \|e_k'\| + C_6^2 h^2 \|e_k'\|^2\big) + \check{C} h^{p+1}\} \tag{7.21}$$
$$+ h \sum_{k=0}^{n-1} C(h+1)\big(C_6 + C_6^2 \|e_k\| + 2C_6 h \|e_k'\| + 2C_6 h^3\big) \|e_k\|,$$

where $\check{C}$ is a constant which can easily be chosen such that the above formula is true. Considering Lemma 7.1, we set

$$\phi_n = \|e_n\|, \quad \psi_n \equiv 0,$$
$$\chi_n = C \sum_{k=0}^{n-1} \{(h^2 + h)\big(C_6 h \|e_k'\| + C_6^2 h^2 \|e_k'\|^2\big) + \check{C} h^{p+1}\},$$
$$\alpha_k = C(h+1)\big(C_6 + C_6^2 \|e_k\| + 2C_6 h \|e_k'\| + 2C_6 h^3\big).$$

Therefore, as long as the errors e_n and e_n' remain in a neighborhood of 0, the following result holds:

$$h \sum_{k=0}^{n-1} \alpha_k = h \sum_{k=0}^{n-1} C(h+1)\big(C_6 + C_6^2 \|e_k\| + 2C_6 h \|e_k'\| + 2C_6 h^3\big)$$
$$\le Cnh(h+1)\big(C_6 + C_6^2 C_7 + 2C_6 h C_7 + 2C_6 h^3\big)$$
$$\le CT(h+1)\big(C_6 + C_6^2 C_7 + 2C_6 h C_7 + 2C_6 h^3\big) \le C_8.$$

An application of the discrete Gronwall lemma, Lemma 7.1, to (7.21) gives

$$\|e_n\| \le \widehat{C} \sum_{k=0}^{n-1} h^{p+1} + \widehat{C} \sum_{k=0}^{n-1} (h^2 + h)\{h \left\|e_k'\right\| + h^2 \left\|e_k'\right\|^2\}$$

$$\le \overline{C} h^p + \overline{C} \sum_{k=0}^{n-1} (h^2 + h^3)\{\left\|e_k'\right\| + h \left\|e_k'\right\|^2\}, \tag{7.22}$$

where $\overline{C}$ is a constant depending on T and $\|M\|$ but independent of h and n. However, for the important particular case where M is symmetric and positive semi-definite, $\overline{C}$ is independent of $\|M\|$.

Likewise, from

$$\left\|\phi_0(m^2 V)\right\| \le C \quad \text{and} \quad \left\|-mhM\phi_1(m^2 V)\right\| \le C \|M\|, \tag{7.23}$$

it follows that

$$\left\|e_n'\right\| \le \widetilde{C} h^p + \widetilde{C} \sum_{k=0}^{n-1} (h + h^2)\{\|e_k\| + h^3 \|e_k\| + \|e_k\|^2\}. \tag{7.24}$$

The constant $\widetilde{C}$ depends on $\|M\|$ and T, but is independent of h and n. $\square$

Remark 7.3 It is noted that, by Proposition 7.1 for the important particular case where M is symmetric and positive semi-definite, the constant C in the inequality (7.20) is independent of $\|M\|$. Moreover, in this case, the constants C_i, $i = 1, 2, \ldots, 5$ in the formulae (7.11) and (7.15) are all independent of $\|M\|$. Therefore, when M is symmetric and positive semi-definite, $\overline{C}$ in the above lemma is independent of $\|M\|$, and the fact stated above also results in the conclusion that the error bound of $\|e_n\|$ given in the following theorem is independent of $\|M\|$.

We are now ready to present the main result of this chapter.

Theorem 7.3 *Under the conditions of Lemma 7.4, the explicit ERKN integrator (7.5) converges for $0 \le nh \le T - t_0$. In particular, the numerical solution and its derivative satisfy the following error bounds*

$$\|e_n\| \le \hat{C}_1 h^p,$$
$$\left\|e_n'\right\| \le \hat{C}_2 h^p, \tag{7.25}$$

where the constant $\hat{C}_1 = \overline{C} + 1$ depends on T and $\|M\|$ but is independent of h and n. The constant $\hat{C}_2 = \widetilde{C} + \widetilde{C} T \hat{C}_1 + 1$ depends on $\|M\|$ and T but is independent of h and n. However, in the important particular case where M is symmetric and positive semi-definite, $\hat{C}_1$ is independent of $\|M\|$.

Proof By $e_0 = \mathbf{0}$, $e_0' = \mathbf{0}$ and (7.19), it is easy to verify that $\|e_1\|$ and $\|e_1'\|$ satisfy (7.25). We now apply an induction argument. Suppose the inequalities (7.25) hold for k ($k \le n$). We will prove the result for $n + 1$. By the formula (7.19),

$$\|e_{n+1}\| \le \overline{C}h^p + \overline{C}\sum_{k=0}^{n}(h^2 + h^3)\left(\|e_k'\| + h\|e_k'\|^2\right).$$

By the assumption that the inequalities (7.25) hold for $k \le n$, we obtain

$$
\begin{aligned}
\|e_{n+1}\| &\le \overline{C}h^p + \overline{C}\sum_{k=0}^{n}(h^2 + h^3)\left(\hat{C}_2 h^p + h(\hat{C}_2 h^p)^2\right) \\
&\le \overline{C}h^p + \overline{C}(n+1)(h^2 + h^3)\left(\hat{C}_2 h^p + h(\hat{C}_2 h^p)^2\right) \\
&\le \overline{C}h^p + \overline{C}T(h + h^2)\left(\hat{C}_2 h^p + h(\hat{C}_2 h^p)^2\right) \\
&\le (\overline{C} + 1)h^p.
\end{aligned}
$$

This shows that the first inequality of (7.25) holds for $n + 1$.

In a similar way, we then obtain

$$
\begin{aligned}
\|e_{n+1}'\| &\le \widetilde{C}h^p + \widetilde{C}\sum_{k=0}^{n}(h + h^2)\{\|e_k\| + h^3\|e_k\| + \|e_k\|^2\} \\
&\le \widetilde{C}h^p + \widetilde{C}\sum_{k=0}^{n}(h + h^2)\{\hat{C}_1 h^p + h^3\hat{C}_1 h^p + (\hat{C}_1 h^p)^2\} \\
&\le \widetilde{C}h^p + \widetilde{C}T(1 + h)\{\hat{C}_1 h^p + h^3\hat{C}_1 h^p + (\hat{C}_1 h^p)^2\} \\
&\le (\widetilde{C} + \widetilde{C}T\hat{C}_1 + 1)h^p.
\end{aligned}
$$

This yields the second inequality of (7.25) for $n + 1$. Hence, we have proved the theorem. $\qquad\square$

Remark 7.4 We assume that $h\Upsilon \le 1$, where Υ can be a large positive but finite number since we always consider convergence in the sense of $h \to 0$.

Remark 7.5 The key point to be noted here is that the error analysis of the explicit ERKN integrators derived in this chapter is not dependent on the matrix decomposition of M, unlike the error analysis of Gautschi-type exponential integrators (see [7–9,

13]). Our error analysis for ERKN integrators is applicable not only to symmetric M but also to nonsymmetric M. Our error analysis therefore has wider applicability in general.

Remark 7.6 It has been pointed out in Remark 7.3 that, for the important particular case where M is symmetric and positive semi-definite, $\overline{C}$ in Lemma 7.4 is independent of $\|M\|$. By the formula $\hat{C}_1 = \overline{C} + 1$ given in Theorem 7.3, it is clear that for a symmetric matrix, the error bound of $\|e_n\|$ does not depend on $\|M\|$. Moreover, Table 7.1 presents the stiff order conditions up to order three from which a more effective explicit ERKN integrator for (7.1) is designed in the next section. Therefore, the error analysis for ERKN integrators is not only of theoretical interest but is also insightful.

7.5 An Explicit Third Order Integrator with Minimal Dispersion Error and Dissipation Error

In this section, we propose an explicit three-stage multi-frequency and multidimensional ERKN integrator of order three with minimal dispersion error and dissipation error. First, the scheme (7.5) of a three-stage explicit ERKN integrator can be denoted by the Butcher tableau

$$
\begin{array}{c|ccc}
c_1 & \mathbf{0}_{d \times d} & \mathbf{0}_{d \times d} & \mathbf{0}_{d \times d} \\
c_2 & \bar{a}_{21}(V) & \mathbf{0}_{d \times d} & \mathbf{0}_{d \times d} \\
c_3 & \bar{a}_{31}(V) & \bar{a}_{32}(V) & \mathbf{0}_{d \times d} \\
\hline
 & \bar{b}_1(V) & \bar{b}_2(V) & \bar{b}_3(V) \\
\hline
 & b_1(V) & b_2(V) & b_3(V)
\end{array}
$$

From the stiff order conditions of Table 7.1, three-stage explicit ERKN integrators of order three are given by:

$$
\begin{cases}
b_1(V) + b_2(V) + b_3(V) = \phi_1(V), \\
b_1(V)c_1 + b_2(V)c_2 + b_3(V)c_3 = \phi_2(V), \\
b_1(V)c_1^2 + b_2(V)c_2^2 + b_3(V)c_3^2 = 2\phi_3(V), \\
c_1 = 0, \quad \bar{a}_{21}(V) = c_2^2 \phi_2(c_2^2 V), \\
\bar{a}_{31}(V) + \bar{a}_{32}(V) = c_3^2 \phi_2(c_3^2 V), \\
\bar{b}_1(V) + \bar{b}_2(V) + \bar{b}_3(V) = \phi_2(V), \\
\bar{b}_1(V)c_1 + \bar{b}_2(V)c_2 + \bar{b}_3(V)c_3 = \phi_3(V).
\end{cases} \tag{7.26}
$$

We also consider two more conditions (obtained by modifying the two equations in Theorem 7.2):

$$\begin{cases} \bar{b}_1(V)c_1^2 + \bar{b}_2(V)c_2^2 + \bar{b}_3(V)c_3^2 = 2\phi_4(V), \\ b_3(V)(\bar{a}_{31}(V)c_1 + \bar{a}_{32}(V)c_2) = \phi_4(V). \end{cases} \tag{7.27}$$

Choosing c_2, c_3 as parameters and solving all the equations in (7.26) and (7.27) we obtain

$$\begin{cases} b_1(V) = \dfrac{c_2c_3\phi_1(V) - (c_2 + c_3)\phi_2(V) + 2\phi_3(V)}{c_2c_3}, \\[2mm] b_2(V) = \dfrac{c_3\phi_2(V) - 2\phi_3(V)}{c_2c_3 - c_2^2}, \quad b_3(V) = \dfrac{c_2\phi_2(V) - 2\phi_3(V)}{c_2c_3 - c_3^2}, \\[2mm] \bar{b}_1(V) = \dfrac{c_2c_3\phi_2(V) - (c_2 + c_3)\phi_3(V) + 2\phi_4(V)}{c_2c_3}, \\[2mm] \bar{b}_2(V) = \dfrac{c_3\phi_3(V) - 2\phi_4(V)}{c_2c_3 - c_2^2}, \quad \bar{b}_3(V) = \dfrac{c_2\phi_3(V) - 2\phi_4(V)}{c_2c_3 - c_3^2}, \\[2mm] \bar{a}_{21}(V) = c_2^2\phi_2(c_2^2V), \quad \bar{a}_{31}(V) = c_3^2\phi_2(c_3^2V) - \bar{a}_{32}(V), \\[2mm] \bar{a}_{32}(V) = (c_2 - c_3)c_3\phi_4(V)\Big(c_2(c_2\phi_2(V) - 2\phi_3(V))\Big)^{-1}. \end{cases} \tag{7.28}$$

In what follows we select values c_2 and c_3. Consider (see Definition 7.3)

$$\phi = H - \arccos\left(\frac{\mathrm{tr}(S(V, z))}{2\sqrt{\det(S(V, z))}}\right), \quad d = 1 - \sqrt{\det(S(V, z))}, \tag{7.29}$$

where $z = \frac{\varepsilon}{\omega^2 + \varepsilon}H^2$, $V = \frac{\omega^2}{\omega^2 + \varepsilon}H^2$. Since ω represents an estimate of the dominant frequency λ and $\varepsilon = \lambda^2 - \omega^2$ is the error of the estimation, we now assume that $\varepsilon = 0$. Then, (7.29) only depends on H. The parameters c_2 and c_3 are chosen such that the coefficients of the first terms in the Taylor expansions of $\phi(H)$ and $d(H)$ of the explicit ERKN integrator are minimized simultaneously. This yields

$$c_2 = \frac{6 - \sqrt{6}}{10}, \quad c_3 = \frac{6 + \sqrt{6}}{10}. \tag{7.30}$$

In this case, the formulae (7.28) and (7.30) determine a three-stage explicit multi-frequency and multidimensional ERKN integrator of order three with minimal dispersion error and dissipation error. We denote the above method as MERKN3s3. The Taylor series expansions of these coefficients are

$$b_1(V) = \frac{1}{9}I - \frac{1}{18}V + \frac{31}{7560}V^2 - \frac{1}{8505}V^3 + \cdots,$$

$$b_2(V) = \frac{16 + \sqrt{6}}{36}I + \frac{-8 - 3\sqrt{6}}{144}V + \frac{64 + 29\sqrt{6}}{30240}V^2 + \frac{-88 - 43\sqrt{6}}{2177280}V^3 + \cdots,$$

$$b_3(V) = (\frac{4}{9} - \frac{1}{6\sqrt{6}})I + \frac{-8 + 3\sqrt{6}}{144}V + \frac{64 - 29\sqrt{6}}{30240}V^2 + \frac{-88 + 43\sqrt{6}}{2177280}V^3 + \cdots,$$

$$\bar{b}_1(V) = \frac{1}{9}I - \frac{19}{1080}V + \frac{23}{30240}V^2 - \frac{17}{1088640}V^3 + \cdots,$$

$$\bar{b}_2(V) = \frac{7 + 2\sqrt{6}}{36}I + \frac{26 - 11\sqrt{6}}{2160}V + \frac{19 + 9\sqrt{6}}{60480}V^2 + \frac{-2 - \sqrt{6}}{435456}V^3 + \cdots,$$

$$\bar{b}_3(V) = \frac{7 - 2\sqrt{6}}{36}I + \frac{26 + 11\sqrt{6}}{2160}V + \frac{19 - 9\sqrt{6}}{60480}V^2 + \frac{-2 + \sqrt{6}}{435456}V^3 + \cdots,$$

$$\bar{a}_{21}(V) = -\frac{3(-7 + 2\sqrt{6})}{100}I + \frac{3(-73 + 28\sqrt{6})}{20000}V - \frac{3(-847 + 342\sqrt{6})}{10000000}V^2$$
$$+ \frac{9(-10033 + 4088\sqrt{6})}{28000000000}V^3 + \cdots,$$

$$\bar{a}_{31}(V) = \frac{3 + 8\sqrt{6}}{500}I + \frac{-4809 + 176\sqrt{6}}{300000}V + \frac{-52587 + 25018\sqrt{6}}{42000000}V^2$$
$$+ \frac{-54347217 + 22468088\sqrt{6}}{252000000000}V^3 + \cdots,$$

$$\bar{a}_{32}(V) = \frac{51 + 11\sqrt{6}}{250}I + \frac{381 - 359\sqrt{6}}{75000}V + \frac{39537 - 12943\sqrt{6}}{26250000}V^2$$
$$+ \frac{3345909 - 1424951\sqrt{6}}{15750000000}V^3 + \cdots.$$

With regard to the coefficient $\bar{a}_{32}(V)$ in the method MERKN3s3, we have the following result.

Proposition 7.2 *The coefficient $\bar{a}_{32}(V)$ in the method MERKN3s3 is uniformly bounded over all the symmetric and positive semi-definite matrices V.*

Proof Following the analysis in the proof of Proposition 7.1, we obtain

$$\bar{a}_{32}(V) = P^{\mathsf{T}}\big(\tilde{f}(hW)\big)P,$$

where

$$\tilde{f}(x) = -\big((-39 + 4\sqrt{6})x^2 + 3(-7 + 2\sqrt{6})x^2\cos(x) - 10(-6 + \sqrt{6})x\sin(x)\big)^{-1}$$
$$\big(3(1 + \sqrt{6})(-2 + x^2 + 2\cos(x))\big).$$

From the fact that

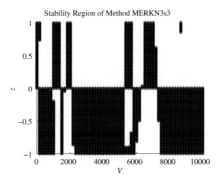

Fig. 7.1 Stability region for the method MERKN3s3

$$\lim_{x \to 0} \tilde{f}(x) = 3(1 + \sqrt{6})(-6 + 16\sqrt{6})^{-1},$$

$$3(1 + \sqrt{6})\left(10(6 - \sqrt{6})\right)^{-1} \leq \lim_{x \to \infty} \tilde{f}(x) \leq 3(1 + \sqrt{6})\left(2(9 + \sqrt{6})\right)^{-1},$$

$$(-39 + 4\sqrt{6})x^2 + 3(-7 + 2\sqrt{6})x^2 \cos(x) - 10(-6 + \sqrt{6})x \sin(x) < 0 \quad \text{for any } x > 0,$$

it follows that $\tilde{f}(hW)$ is uniformly bounded with respect to W. Thus the conclusion is proved. $\qquad\square$

The dispersion error and dissipation error of the method MERKN3s3 are

$$\phi(H) = -\frac{(-7 + 2\sqrt{6})\varepsilon^2}{160(2 + 3\sqrt{6})(\varepsilon + \omega^2)^2} H^5 + \mathcal{O}(H^6);$$

$$d(H) = -\frac{\varepsilon\left(50(-82 + 27\sqrt{6})\varepsilon^2 + 125(-26 + 9\sqrt{6})\varepsilon\omega^2 + 6(2 + 3\sqrt{6})\omega^4\right)}{144000(2 + 3\sqrt{6})(\varepsilon + \omega^2)^3} H^6 + \mathcal{O}(H^7),$$

respectively.

Hence, the method is dispersive of order 4 and dissipative of order 5. The stability region of the method MERKN3s3 is plotted in Fig. 7.1.

7.6 Numerical Experiments

Our aim in this section is to demonstrate the efficiency of the new method MERKN3s3 based on the error analysis. The methods chosen for comparison are:

- A: the symmetric Gautschi's method of order two given in [5];
- B: the symmetric Gautschi's method of order two given in [15];
- C: the symmetric Gautschi's method of order two given in [10];

- ARKN3s4: the three-stage explicit ARKN method of order four given in [27];
- RKN3s4: the three-stage explicit RKN method of order four given in [11];
- W1ERKN3s3: the three-stage explicit ERKN integrator of order three given in [29];
- W2ERKN3s3: the three-stage explicit ERKN integrator of order three given in [29];
- MERKN3s3: the three-stage explicit ERKN integrator of order three derived in this chapter.

Problem 7.1 Consider Problem 6.3 in Chap. 6.

By using second-order symmetric differences, this problem is converted into the ODEs (6.39).

Remark 7.7 The key point to be noted here is that the matrix M determined by (6.50) in this problem is nonsymmetric. We use this problem to show that the error analysis presented in this chapter and the method MERKN3s3 are applicable for (7.1) with nonsymmetric M.

For the parameters in this problem we choose $b = 100$, $g = 9.81$, $d_0 = 10$, $C = 50$. The system is integrated on the interval $[0, 100]$ with $N = 20$ and the stepsizes $h = 0.8/2^j, j = 0, 1, 2, 3$ for the three-stage methods and stepsizes $h = 0.8/(3 \times 2^j), j = 0, 1, 2, 3$ for the two-stage methods. The efficiency curves (accuracy versus the computational cost measured by the number of function evaluations required by each method) are shown in Fig. 7.2 (left). In this experiment we note that for the RKN3s4 method, the errors $\log_{10}(GE)$ are very large when $h = 0.8$ and $h = 0.4$, hence we do not plot the point in Fig. 7.2 (left). Similar situations are encountered in the next two problems and we deal with them in a similar way. In other words, the RKN3s4 method requires a smaller stepsize. Therefore, other methods chosen for this kind of problems are more practical than RKN3s4.

Problem 7.2 Consider the sine-Gordon equation of Chap. 4.

Following the paper [4], we take the initial conditions as

$$U(0) = (\pi)_{i=1}^N, \quad U_t(0) = \sqrt{N} \left(0.01 + \sin(\frac{2\pi i}{N}) \right)_{i=1}^N.$$

Choose $N = 64$, and the problem is integrated on the interval $[0, 10]$ with the stepsizes $h = 0.1/2^j, j = 1, 2, 3, 4$ for the three-stage methods and stepsizes $h = 0.1/(3 \times 2^j), j = 1, 2, 3, 4$ for the two-stage methods. Figure 7.2 (right) shows the error in the positions at $t_{\mathrm{end}} = 10$ versus the computational effort.

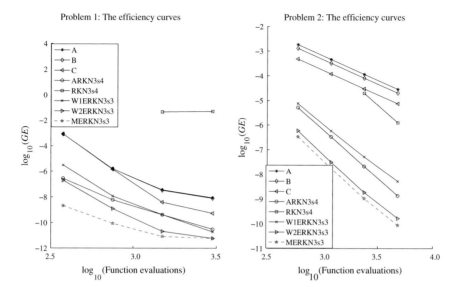

Fig. 7.2 Results for Problem 7.1 (*left*) and Problem 7.2 (*right*): The number of function evaluations (in logarithmic scale) against $\log_{10}(GE)$, the logarithm of the global error over the integration interval

Problem 7.3 Consider the Fermi-Pasta-Ulam problem.
   This problem has been considered in Sect. 2.4.4 of Chap. 2.
   Following [13], we choose

$$m = 3, \; q_1(0) = 1, \; p_1(0) = 1, \; q_4(0) = \frac{1}{\omega}, \; p_4(0) = 1,$$

with zero for the remaining initial values. The system is integrated on the interval $[0, 25]$ with stepsizes $h = 0.02/2^j$, $j = 0, 1, 2, 3$ for the three-stage methods and stepsizes $h = 0.02/(3 \times 2^j)$, $j = 0, 1, 2, 3$ for the two-stage methods. The efficiency curves for different $\omega = 50, 100, 150, 200$ are shown in Fig. 7.3. It can be observed from Fig. 7.3 that although the global errors for the methods A, B and C are independent of ω, the new method MERKN3s3 is much more accurate than these three methods.
   From the numerical results of the three problems, we can conclude that the logarithm of the maximum global error of the MERKN3s3 method is smaller than those of the other methods with the same numbers of function evaluations. This indicates that MERKN3s3 is efficient and highly competitive for multi-frequency oscillatory second-order differential equations of the form (7.1).

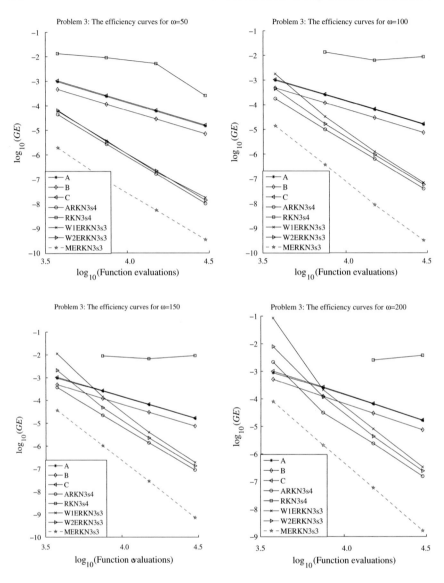

Fig. 7.3 Results for Problem 7.3 with different ω: The number of function evaluations (in logarithmic scale) against $\log_{10}(GE)$, the logarithm of the global error over the integration interval

7.7 Conclusions

In this chapter, we considered the error analysis for explicit multi-frequency and multidimensional ERKN integrators up to stiff order three for systems of multi-frequency oscillatory second-order differential equations (7.1). We analysed and presented the error bounds based on a series of lemmas. *It is important to note that the theoretical analysis in this chapter does not depend on the matrix decomposition of M.* Namely, the error analysis has now been generalized from the work depending on matrix decompositions to the case where the analysis does not depend on matrix decompositions. Here, we have shown that the explicit multi-frequency and multidimensional ERKN integrator fulfilling stiff order p converges with order p, and *for the important particular case where M is a symmetric and positive semi-definite matrix, the error bound of $\|q_n - q(t_n)\|$ is independent of $\|M\|$.* From the main result of our error analysis, we also proposed a novel explicit third order multi-frequency and multidimensional ERKN integrator with minimal dispersion error and dissipation error. The numerical experiments were performed in comparison with some well-known numerical methods in the scientific literature. It follows from the results of the numerical experiments that the explicit third order multi-frequency and multidimensional ERKN integrator derived in this chapter is much more efficient than various other effective methods available in the scientific literature. The error analysis in this chapter clarifies the structure of the error bounds of explicit multi-frequency and multidimensional ERKN integrators. It also provides a powerful means of constructing efficient ERKN integrators for the multi-frequency oscillatory system (7.1).

The material of this chapter is based on the work by Wang et al. [25].

References

1. Cohen D, Hairer E, Lubich C (2005) Numerical energy conservation for multi-frequency oscillatory differential equations. BIT Numer Math 45:287–305
2. Cohen D, Jahnke T, Lorenz K, Lubich C (2006) Numerical integrators for highly oscillatory Hamiltonian systems: a review. In: Mielke A (ed) Analysis, modeling and simulation of multiscale problems. Springer, Berlin, pp 553–576
3. Conte D, Lubich C (2010) An error analysis of the multi-configuration time-dependent Hartree method of quantum dynamics. Math Mod Numer Anal 44:759–780
4. Franco JM (2006) New methods for oscillatory systems based on ARKN methods. Appl Numer Math 56:1040–1053
5. García-Archilla B, Sanz-Serna JM, Skeel RD (1998) Long-time-step methods for oscillatory differential equations. SIAM J Sci Comput 20:930–963
6. Gautschi W (1961) Numerical integration of ordinary differential equations based on trigonometric polynomials. Numer Math 3:381–397
7. Grimm V (2005) On error bounds for the Gautschi-type exponential integrator applied to oscillatory second-order differential equations. Numer Math 100:71–89
8. Grimm V (2005) A note on the Gautschi-type method for oscillatory second-order differential equations. Numer Math 102:61–66

9. Grimm V, Hochbruck M (2006) Error analysis of exponential integrators for oscillatory second-order differential equations. J Phys A Math Gen 39:5495–5507
10. Hairer E, Lubich C (2000) Long-time energy conservation of numerical methods for oscillatory differential equations. SIAM J Numer Anal 38:414–441
11. Hairer E, Nørsett SP, Wanner G (1993) Solving ordinary differential equations I, Nonstiff problems, 2nd edn., Springer series in computational mathematicsSpringer, Berlin
12. Hairer E, Lubich C, Roche M (1988) Error of Runge-Kutta methods for stiff problems studied via differential algebraic equations. BIT Numer Math 28:678–700
13. Hairer E, Lubich C, Wanner G (2006) Geometric numerical integration: structure-preserving algorithms, 2nd edn. Springer, Berlin
14. Hayes LJ (1987) Galerkin alternating-direction methods for nonrectangular regions using patch approximations. SIAM J Numer Anal 18:627–643
15. Hochbruck M, Lubich C (1999) A Gautschi-type method for oscillatory second-order differential equations. Numer Math 83:403–426
16. Hochbruck M, Ostermann A (2005) Explicit exponential Runge-Kutta methods for semilineal parabolic problems. SIAM J Numer Anal 43:1069–1090
17. Hochbruck M, Ostermann A, Schweitzer J (2009) Exponential rosenbrock-type methods. SIAM J Numer Anal 47:786–803
18. Iserles A (2002) On the global error of discretization methods for highly-oscillatory ordinary differential equations. BIT Numer Math 42:561–599
19. Li J, Wang B, You X, Wu X (2011) Two-step extended RKN methods for oscillatory systems. Comput Phys Commun 182:2486–2507
20. Petzold LR, Yen LO, Yen J (1997) Numerical solution of highly oscillatory ordinary differential equations. Acta Numer 7:437–483
21. Shi W, Wu X (2012) On symplectic and symmetric ARKN methods. Comput Phys Commun 183:1250–1258
22. Van der Houwen PJ, Sommeijer BP (1987) Explicit Runge-Kutta(-Nyström) methods with reduced phase errors for computing oscillating solution. SIAM J Numer Anal 24:595–617
23. Vigo-Aguiar J, Simos TE, Ferrándiz JM (2004) Controlling the error growth in long-term numerical integration of perturbed oscillations in one or more frequencies. Proc Roy Soc London Ser A 460:561–567
24. Wang B, Wu X (2012) A new high precision energy-preserving integrator for system of oscillatory second-order differential equations. Phys Lett A 376:1185–1190
25. Wang B, Wu X, Xia J (2013) Error bounds for explicit ERKN integrators for systems of multi-frequency oscillatory second-order differential equations. Appl Numer Math 74:17–34
26. Wang B, Wu X, Zhao H (2013) Novel improved multidimensional Strömer-Verlet formulas with applications to four aspects in scientific computation. Math Comput Model 37:2327–2336
27. Wu X, Wang B (2010) Multidimensional adapted Runge-Kutta-Nyström methods for oscillatory systems. Comput Phys Commun 181:1955–1962
28. Wu X, You X, Xia J (2009) Order conditions for ARKN methods solving oscillatory systems. Comput Phys Commun 180:2250–2257
29. Wu X, You X, Shi W, Wang B (2010) ERKN integrators for systems of oscillatory second-order differential equations. Comput Phys Commun 181:1873–1887
30. Wu X, Wang B, Xia J (2012) Explicit symplectic multidimensional exponential fitting modified Runge-Kutta-Nyström methods. BIT Numer Math 52:773–795
31. Wu X, Wang B, Shi W (2013) Efficient energy-preserving integrators for oscillatory Hamiltonian systems. J Comput Phys 235:587–605
32. Wu X, You X, Wang B (2013) Structure-preserving algorithms for oscillatory differential equations. Springer, Berlin (Jointly published with Science Press Beijing)
33. Yang H, Wu X, You X, Fang Y (2009) Extended RKN-type methods for numerical integration of perturbed oscillators. Comput Phys Commun 180:1777–1794

Chapter 8
Error Analysis of Explicit TSERKN Methods for Highly Oscillatory Systems

The main theme of this chapter is the error analysis for the two-step extended Runge–Kutta–Nyström-type (TSERKN) methods (2011) for the multi-frequency and multi-dimensional oscillatory systems $y'' + My = f(y)$, where high-frequency oscillations in the solutions are generated by the linear part My and $\|M\|$ may be large. TSERKN methods extend the two-step hybrid methods (2003) by modifying both the internal stages and the updates so that they are adapted to the oscillatory properties of the exact solutions. This chapter presents an analysis for a three-stage explicit TSERKN method of order four and derives its global error bound, which is proved to be independent of $\|M\|$ under suitable assumptions. Throughout this chapter, we use the Euclidean norm and its induced matrix norm (spectral norm) denoted by $\|\cdot\|$.

8.1 Motivation

In the last three decades, there has been increasing interest in the numerical integration of the multi-frequency oscillatory second-order initial value problem

$$
\begin{cases}
y'' + My = f(y), & t \in [t_0, T], \\
y(t_0) = y_0, & y'(t_0) = y_0',
\end{cases}
\tag{8.1}
$$

where $M \in \mathbb{R}^{d \times d}$ is a symmetric and positive semi-definite matrix that implicitly contains and preserves the main oscillatory frequencies of the problem. It is particularly attractive when this system is derived from the spatial discretization of wave equations [2, 5, 7, 9] based on the method of lines. In practice, it can be integrated with general-purpose methods or other integrators adapted to the special structure of the problem. However, adapted methods are in general more efficient because they make use of the

© Springer-Verlag Berlin Heidelberg and Science Press, Beijing, China 2015
X. Wu et al., *Structure-Preserving Algorithms for Oscillatory Differential Equations II*, DOI 10.1007/978-3-662-48156-1_8

information contained in the special structure of (8.1) given by the linear term My, and this explains why the ARKN methods (see [4, 16–18]) and ERKN integrators [20] have been explored in recent years. More details on this topic can be found in [19].

After Wu et al. [15] formulated and investigated ERKN integrators based on the matrix-variation-of-constants formula for the system (8.1), Li et al. [12] proposed and studied a new family of TSERKN methods, which incorporate the special structure of the system given by the linear term My into both internal stages and updates, so that they exactly integrate the multi-frequency oscillatory homogeneous second-order differential system $y'' + My = \mathbf{0}$. The corresponding order conditions for TSERKN methods are derived via the B_{BT}-series defined on the set BT of branches and the B_{BWT}-series defined on the subset BWT of BT, where the branches are tree-like graphs consisting of three types of vertex: meagre, fat black and fat white vertices connected by lines. However, the global error analysis for TSERKN methods was not discussed in [12]. In this chapter, we present a global error analysis for TSERKN methods. We will demonstrate our approach to the analysis by constructing a new three-stage explicit TSERKN method of order four and deriving the corresponding global error bound. We will prove that the global error bound for the new method is independent of $\|M\|$ under suitable assumptions. Hence, the new method is suitable for solving highly oscillatory systems (8.1), where $\|M\|$ may be large.

8.2   The Formulation of the New Method

To begin with, we consider the matrix-valued functions defined in Chap. 4:

$$\phi_j(M) = \sum_{k=0}^{\infty} \frac{(-1)^k M^k}{(2k+j)!}, \quad j = 0, 1, \ldots. \tag{8.2}$$

It is known from Li et al. [12] that the exact solution $y(t)$ of (8.1) satisfies

$$
\begin{aligned}
y(t_n + h) = {} & 2\phi_0(V)y(t_n) - y(t_n - h) \\
& + h^2 \int_0^1 (1-z)\phi_1\big((1-z)^2 V\big)\Big(f\big(y(t_n + hz)\big) + f\big(y(t_n - hz)\big)\Big)\,\mathrm{d}z
\end{aligned}
\tag{8.3}
$$

and

$$
\begin{aligned}
y(t_n + c_i h) = {} & (1+c_i)\phi_1^{-1}(V)\phi_1\big((1+c_i)^2 V\big)y(t_n) - c_i\phi_1^{-1}(V)\phi_1(c_i^2 V)y(t_n - h) \\
& + h^2 \int_0^1 c_i(1-z)\Big(\phi_1^{-1}(V)\phi_1(c_i^2 V)\phi_1\big((1-z)^2 V\big)f\big(y(t_n - hz)\big) \\
& \qquad\qquad + c_i\phi_1\big(c_i^2(1-z)^2 V\big)f\big(y(t_n + c_i hz)\big)\Big)\,\mathrm{d}z,
\end{aligned}
\tag{8.4}
$$

where $V = h^2 M$. Here, the integral of a matrix-valued function or vector-valued function is understood componentwise. It is important to observe that when $c_i = 1$, (8.4) reduces to (8.3), as is shown below in Proposition 8.1 of this chapter.

Approximating the integrals in (8.3) and (8.4) with quadrature formulae leads to the following definition for TSERKN method in [12].

Definition 8.1 An s-stage two-step extended Runge–Kutta–Nyström-type method for integrating the oscillatory system (8.1) is defined by

$$
\begin{cases}
Y_{ni} = (1 + c_i)\phi_1^{-1}(V)\phi_1\big((1+c_i)^2 V\big)y_n - c_i\phi_1^{-1}(V)\phi_1(c_i^2 V)y_{n-1} \\
\qquad + h^2 \sum_{j=1}^{s} a_{ij}(V)f(Y_{nj}), \quad i = 1, \ldots, s, \\
y_{n+1} = 2\phi_0(V)y_n - y_{n-1} + h^2 \sum_{i=1}^{s} b_i(V)f(Y_{ni}),
\end{cases}
\tag{8.5}
$$

where $a_{ij}(V)$ and $b_i(V)$, $i, j = 1, \ldots, s$ are matrix-valued functions of $V = h^2 M$.

$a_{ij}(V)$ and $b_i(V)$ can be expressed by

$$
\begin{cases}
a_{ij}(V) = {}^0 a_{ij}I + {}^1 a_{ij}V + {}^2 a_{ij}V^2 + \cdots + {}^m a_{ij}V^m + \cdots, \\
b_i(V) = {}^0 b_i I + {}^1 b_i V + {}^2 b_i V^2 + \cdots + {}^m b_i V^m + \cdots,
\end{cases}
\tag{8.6}
$$

where ${}^m a_{ij}$ and ${}^m b_i$ are real numbers.

The method (8.5) can also be expressed briefly in a Butcher-type tableau as

$$
\begin{array}{c|c|c|ccc}
c_1 & (1+c_1)\phi_1^{-1}(V)\phi_1\big((1+c_1)^2 V\big) & -c_1\phi_1^{-1}(V)\phi_1(c_1^2 V) & a_{11}(V) & \cdots & a_{1s}(V) \\
\vdots & \vdots & \vdots & \vdots & \ddots & \vdots \\
c_s & (1+c_s)\phi_1^{-1}(V)\phi_1\big((1+c_s)^2 V\big) & -c_s\phi_1^{-1}(V)\phi_1(c_s^2 V) & a_{s1}(V) & \cdots & a_{ss}(V) \\
\hline
 & 2\phi_0(V) & -I & b_1(V) & \cdots & b_s(V)
\end{array}
$$

It is easy to see that when $M \to \mathbf{0}_{d \times d}$ ($V \to \mathbf{0}_{d \times d}$), the method (8.5) reduces to the two-step hybrid method proposed in [3].

Order conditions for TSERKN methods are presented in [12], and we briefly restate them here.

Theorem 8.1 *Given exact starting values, the TSERKN method (8.5) is of order p if and only if*

$$
\sum_{i=1}^{s} b_i(V)\Psi_i''(\beta\tau) = \big((-1)^{\rho(\beta\tau)} + 1\big)\rho(\beta\tau)!\phi_{\rho(\beta\tau)}(V) + \mathcal{O}(h^{p+2-\rho(\beta\tau)}),
$$

for all $\beta\tau \in BWT$ with $\rho(\beta\tau) \leq p + 1$.

The set BWT and the function $\Psi''(\beta\tau)$ are defined in [12]. The function $\rho(\beta\tau)$ and the *order* of $\beta\tau$ are defined in [20] as well.

All $\beta\tau \in BWT$ of order up to five and their corresponding Ψ'', ρ and the resulting conditions for order four are listed in Table 8.1. It is noted that different $\beta\tau$ may give the same order conditions.

In [9], Hochbruck and Lubich proposed and analyzed a two-step scheme, named a Gautschi-type method:

$$y_{n+1} = 2y_n - y_{n-1} + h^2\psi(h^2M)\big(-My_n + f(y_n)\big), \tag{8.7}$$

Table 8.1 All $\beta\tau \in BWT$ of order up to five and their corresponding Ψ'', ρ and order conditions

$\beta\tau \in BWT$	ρ	Ψ_i''	Order conditions
	2	2	$\sum_{i=1}^{3} b_i(V) = 2\phi_2(V) + \mathcal{O}(h^4)$
	3	$6c_i$	$\sum_{i=1}^{3} b_i(V)c_i = \mathcal{O}(h^3)$
	4	$12c_i^2$	$\sum_{i=1}^{3} b_i(V)c_i^2 = 4\phi_4(V) + \mathcal{O}(h^2)$
	4	$12c_i^2$	$\sum_{i=1}^{3} b_i(V)c_i^2 = 4\phi_4(V) + \mathcal{O}(h^2)$
	4	$12(-c_i + 2\sum_j {}^0a_{ij})$	$\sum_{i=1}^{3} b_i(V)(-c_i + 2\sum_j {}^0a_{ij}) = 4\phi_4(V) + \mathcal{O}(h^2)$
	5	$20c_i^3$	$\sum_{i=1}^{3} b_i(V)c_i^3 = \mathcal{O}(h)$
	5	$20c_i^3$	$\sum_{i=1}^{3} b_i(V)c_i^3 = \mathcal{O}(h)$
	5	$20c_i^3$	$\sum_{i=1}^{3} b_i(V)c_i^3 = \mathcal{O}(h)$
	5	$20c_i(-c_i + 2\sum_j {}^0a_{ij})$	$\sum_{i=1}^{3} b_i(V)c_i(-c_i + 2\sum_j {}^0a_{ij}) = \mathcal{O}(h)$
	5	$20(c_i + 6\sum_j {}^0a_{ij}c_j)$	$\sum_{i=1}^{3} b_i(V)(c_i + 6\sum_j {}^0a_{ij}c_j) = \mathcal{O}(h)$

where

$$\psi(x^2) = 2\big(1 - \cos(x)\big)/x^2.$$

By the definition of the matrix-valued ϕ-functions (8.2), it is easy to see that $\phi_0(M) = \cos(M^{\frac{1}{2}})$ for any symmetric positive semi-definite matrix M. Thus the scheme (8.7) can also be expressed by

$$y_{n+1} = 2\phi_0(V)y_n - y_{n-1} + h^2\psi(V)f(y_n), \tag{8.8}$$

which is exactly a TSERKN method with $s = 1$, $c_1 = 0$, $a_{11}(V) = \mathbf{0}_{d\times d}$ and $b_1(V) = \psi(V)$.

The authors of [9] have proved that the method (8.7) admits second-order error bounds which are independent of the product of the stepsize and frequencies.

The main purpose of this chapter is to construct TSERKN methods of higher-order with uniform error bounds independent of $\|M\|$.

Before going on to describe the analysis of this chapter, we give the main properties of these matrix-valued functions in the following propositions.

Proposition 8.1 *The matrix-valued functions ϕ_0 and ϕ_1 defined by (8.2) satisfy*

$$\begin{cases} (a+b)\phi_1\big((a+b)^2 M\big) = a\phi_1(a^2 M)\phi_0(b^2 M) + b\phi_1(b^2 M)\phi_0(a^2 M), \\ \phi_0\big((a+b)^2 M\big) = \phi_0(a^2 M)\phi_0(b^2 M) - abM\phi_1(a^2 M)\phi_1(b^2 M), \end{cases} \tag{8.9}$$

where a, $b \in \mathbb{R}$ and $M \in \mathbb{R}^{d\times d}$.

Proof We prove the first formula in (8.9) as follows:

$$\begin{aligned} &a\phi_1(a^2 M)\phi_0(b^2 M) + b\phi_1(b^2 M)\phi_0(a^2 M) \\ &= a\sum_{k=0}^{\infty}\frac{a^{2k}(-1)^k M^k}{(2k+1)!}\sum_{j=0}^{\infty}\frac{b^{2j}(-1)^j M^j}{(2j)!} + b\sum_{k=0}^{\infty}\frac{b^{2k}(-1)^k M^k}{(2k+1)!}\sum_{j=0}^{\infty}\frac{a^{2j}(-1)^j M^j}{(2j)!} \\ &= \sum_{p=0}^{\infty}\left(\sum_{k=0}^{p}\frac{a^{2k+1}(-1)^k M^k}{(2k+1)!}\cdot\frac{b^{2(p-k)}(-1)^{p-k} M^{p-k}}{(2(p-k))!} + \sum_{k=0}^{p}\frac{b^{2k+1}(-1)^k M^k}{(2k+1)!}\cdot\frac{a^{2(p-k)}(-1)^{p-k} M^{p-k}}{(2(p-k))!}\right) \\ &= \sum_{p=0}^{\infty}\left(\sum_{k=0}^{p}\frac{a^{2k+1}b^{2(p-k)}}{(2k+1)!(2p-2k)!} + \sum_{k=0}^{p}\frac{b^{2k+1}a^{2(p-k)}}{(2k+1)!(2p-2k)!}\right)(-1)^p M^p \\ &= \sum_{p=0}^{\infty}\left(\sum_{k=0}^{p}\frac{a^{2k+1}b^{2(p-k)}}{(2k+1)!(2p-2k)!} + \sum_{q=0}^{p}\frac{b^{2(p-q)+1}a^{2q}}{(2p-2q+1)!(2q)!}\right)(-1)^p M^p \\ &= \sum_{p=0}^{\infty}\left(\sum_{k=0}^{2p+1}\frac{a^k b^{2p+1-k}}{k!(2p+1-k)!}\right)(-1)^p M^p = \sum_{p=0}^{\infty}\frac{(a+b)^{2p+1}}{(2p+1)!}(-1)^p M^p \\ &= (a+b)\phi_1\big((a+b)^2 M\big). \end{aligned}$$

The second identity in (8.9) can be obtained in a similar way. $\qquad\square$

Proposition 8.2 *If M is symmetric and positive semi-definite, then*

$$\|\phi_l(M)\| \le \frac{1}{l!}, \quad l = 0, 1, 2, \ldots. \tag{8.10}$$

Proof Since M is symmetric and positive semi-definite, we have $M = P^\mathsf{T}\Omega^2 P$, where $\Omega^2 = \mathrm{diag}(w_1^2, w_2^2, \ldots, w_d^2)$ and P is an orthogonal matrix. Thus

$$
\begin{aligned}
\phi_0(M) &= \sum_{k=0}^{\infty} \frac{(-1)^k M^k}{(2k)!} = \sum_{k=0}^{\infty} \frac{(-1)^k (P^\mathsf{T}\Omega^2 P)^k}{(2k)!} \\
&= P^\mathsf{T} \sum_{k=0}^{\infty} \frac{(-1)^k \Omega^{2k}}{(2k)!} P = P^\mathsf{T}\phi_0(\Omega^2)P,
\end{aligned}
\tag{8.11}
$$

where $\phi_0(\Omega^2) = \mathrm{diag}(\cos w_1, \cos w_2, \ldots, \cos w_d)$. The definition of the spectral norm results in

$$
\|\phi_0(M)\| = \|P^\mathsf{T}\phi_0(\Omega^2)P\| \leq \|\phi_0(\Omega^2)\| \leq 1.
$$

For $m = 0, 1, \ldots$, we have

$$
\|\phi_{m+1}(M)\| = \left\| \int_0^1 \frac{\phi_0((1-\xi)^2 M)\xi^m}{m!} \mathrm{d}\xi \right\| \leq \int_0^1 \left| \frac{\xi^m}{m!} \right| \mathrm{d}\xi = \frac{1}{(m+1)!}.
\tag{8.12}
$$

Thus (8.10) holds.   □

In this chapter, we are only concerned with the explicit TSERKN methods satisfying

$$
\begin{cases}
c_1 = -1, \quad c_2 = 0, \\
a_{ij}(V) = \mathbf{0}_{d \times d}, \quad i = 1, 2, \ j = 1, \ldots, s, \\
a_{ij}(V) = \mathbf{0}_{d \times d}, \quad 3 \leq i \leq s, \ i \leq j \leq s,
\end{cases}
\tag{8.13}
$$

so that $Y_{n1} = y_{n-1}$ and $Y_{n2} = y_n$ (see [13]). In this case, we note that after the starting procedure, the methods only require the evaluations of $f(y_n), f(Y_{n3}), \ldots, f(Y_{ns})$ for each step ($s - 1$ function evaluations).

For the highly oscillatory systems (8.1), where $\|M\|$ may be large, we try to obtain the explicit TSERKN methods for which the global error bounds do not depend on $\|M\|$. To this end, we are looking for c_i ($i = 3, \ldots, s$) satisfying

$$
\left\| (1 + c_i)\phi_1^{-1}(V)\phi_1\big((1 + c_i)^2 V\big) \right\| \leq C, \quad \left\| c_i \phi_1^{-1}(V)\phi_1(c_i^2 V) \right\| \leq C,
$$

where C is independent of $\|M\|$.

Lemma 8.1 *Given a symmetric and positive semi-definite matrix V, if $c \in \mathbb{Z}$, the matrix-valued function*

$$
W(c, V) = c\phi_1^{-1}(V)\phi_1(c^2 V)
$$

satisfies

$$\|W(c, V)\| \le |c|.$$

Proof First of all, we consider $c \in \mathbb{Z}^+$. $\|W(1, V)\| = 1$ is trivial. Assuming that $\|W(n, V)\| \le n$ holds, we can obtain

$$
\begin{aligned}
\|W(n+1, V)\| &= \left\| (n+1)\phi_1^{-1}(V)\phi_1\big((n+1)^2 V\big) \right\| \\
&= \left\| \phi_1^{-1}(V)\big(n\phi_1(n^2 V)\phi_0(V) + \phi_1(V)\phi_0(n^2 V)\big) \right\| \\
&= \left\| W(n, V)\phi_0(V) + \phi_0(n^2 V) \right\| \\
&\le \left\| W(n, V)\phi_0(V) \right\| + \left\| \phi_0(n^2 V) \right\| \\
&\le n+1,
\end{aligned}
\tag{8.14}
$$

from Proposition 8.1. Then it follows that $\|W(-n, V)\| = \|W(n, V)\| \le n$. As for $n = 0$, we have $\|W(0, V)\| = 0$. $\qquad\square$

In this chapter, we only consider the three-stage explicit TSERKN methods with conditions (8.13), since in this way we can avoid the necessity of solving a system of nonlinear equations at each step. In order to obtain a global error bound which is independent of $\|M\|$, we take $c_3 \in \mathbb{Z}$ by Lemma 8.1. The choice of $c_3 = 1$ gives

$$
\begin{cases}
Y_{n3} = 2\phi_0(V)y_n - y_{n-1} + h^2\big(a_{31}(V)f(y_{n-1}) + a_{32}(V)f(y_n)\big), \\
y_{n+1} = 2\phi_0(V)y_n - y_{n-1} + h^2\big(b_1(V)f(y_{n-1}) + b_2(V)f(y_n) + b_3(V)f(Y_{n3})\big),
\end{cases}
\tag{8.15}
$$

where $b_i(V)$, $i = 1, 2, 3$ and $a_{3j}(V)$, $j = 1, 2$ are matrix-valued functions of $V = h^2 M$.

Throughout the chapter, we assume that $f(y(t))$ in (8.1) is sufficiently differentiable with respect to t and all occurring derivatives are uniformly bounded (see [10]).

Inserting the exact solution of (8.1) into the numerical method (8.15) gives

$$
\begin{cases}
y(t_n + h) = 2\phi_0(V)y(t_n) - y(t_n - h) \\
\qquad + h^2\big(a_{31}(V)f\big(y(t_n - h)\big) + a_{32}(V)f\big(y(t_n)\big)\big) + \Delta_{n3}, \\
y(t_n + h) = 2\phi_0(V)y(t_n) - y(t_n - h) \\
\qquad + h^2\big(b_1(V)f\big(y(t_n - h)\big) + b_2(V)f\big(y(t_n)\big) + b_3(V)f\big(y(t_n + h)\big)\big) + \delta_{n+1},
\end{cases}
\tag{8.16}
$$

where Δ_{n3} and δ_{n+1} are the residuals. Comparing (8.4) ($c_3 = 1$) with the first formula of (8.16), we obtain

$$
\begin{aligned}
\Delta_{n3} = {}& h^2 \int_0^1 (1 - z)\phi_1\big((1 - z)^2 V\big)\big(f\big(y(t_n - hz)\big) + f\big(y(t_n + hz)\big)\big)\,dz \\
& - h^2\big(a_{31}(V)f\big(y(t_n - h)\big) + a_{32}(V)f\big(y(t_n)\big)\big).
\end{aligned}
\tag{8.17}
$$

Expressing $\hat{f}(t) = f(y(t))$ in (8.17) using a Taylor series expansion yields

$$
\begin{aligned}
\Delta_{n3} = h^2 \int_0^1 (1-z)\phi_1((1-z)^2 V) \sum_{k=0}^{\infty} \left((-1)^k + 1\right) \tfrac{1}{k!} h^k z^k \hat{f}^{(k)}(t_n) \, dz \\
- h^2 \left(a_{31}(V) \sum_{k=0}^{\infty} \tfrac{1}{k!} h^k (-1)^k \hat{f}^{(k)}(t_n) + a_{32}(V)\hat{f}(t_n)\right).
\end{aligned}
$$

By (7.3) in Sect. 7.2, we have

$$
\begin{aligned}
\Delta_{n3} &= \sum_{k=0}^{\infty} h^{k+2}\left((-1)^k + 1\right)\phi_{k+2}(V)\hat{f}^{(k)}(t_n) \\
&\quad - h^2 \left(a_{31}(V) \sum_{k=0}^{\infty} \frac{1}{k!} h^k (-1)^k \hat{f}^{(k)}(t_n) + a_{32}(V)\hat{f}(t_n)\right) \\
&= \sum_{k=0}^{\infty} h^{k+2}\left((-1)^k \phi_{k+2}(V) + \phi_{k+2}(V) - \frac{1}{k!} a_{31}(V)(-1)^k\right)\hat{f}^{(k)}(t_n) - h^2 a_{32}(V)\hat{f}(t_n).
\end{aligned}
$$
$$(8.18)$$

Similarly, we obtain

$$
\begin{aligned}
\delta_{n+1} = \sum_{k=0}^{\infty} h^{k+2}\left(\phi_{k+2}(V) + (-1)^k \phi_{k+2}(V) - b_1(V)\frac{(-1)^k}{k!} - b_3(V)\frac{1}{k!}\right)\hat{f}^{(k)}(t_n) \\
- h^2 b_2(V)\hat{f}(t_n).
\end{aligned}
$$
$$(8.19)$$

Assuming

$$
\Delta_{n3} = \mathcal{O}(h^4), \quad \delta_{n+1} = \mathcal{O}(h^6), \tag{8.20}
$$

respectively, results in

$$
\begin{cases}
2\phi_2(V) - a_{31}(V) - a_{32}(V) = \mathbf{0}_{d \times d}, & a_{31}(V) = \mathbf{0}_{d \times d}, \\
2\phi_2(V) - b_1(V) - b_2(V) - b_3(V) = \mathbf{0}_{d \times d}, & b_1(V) - b_3(V) = \mathbf{0}_{d \times d}, \\
4\phi_4(V) - b_1(V) - b_3(V) = \mathbf{0}_{d \times d},
\end{cases} \tag{8.21}
$$

which implies that

$$
\begin{cases}
a_{31}(V) = \mathbf{0}_{d \times d}, & a_{32}(V) = 2\phi_2(V), \\
b_1(V) = 2\phi_4(V), & b_2(V) = 2\phi_2(V) - 4\phi_4(V), \; b_3(V) = 2\phi_4(V).
\end{cases} \tag{8.22}
$$

From (8.22) and Proposition 8.2, it can be seen that all $a_{ij}(V)$ and $b_i(V)$ are linear combinations of the ϕ-functions defined by (8.2) and can be bounded by constants which are independent of $\|M\|$.

The method (8.15) determined by (8.22) is denoted by TSERKN3s. It can be verified by a careful calculation that it is of order four by the order conditions for the

TSERKN methods given in Theorem 8.1. In the next section, we will show that the global error bound for this method is independent of $\|M\|$. Hence, there exists C_Δ, independent of $\|M\|$, such that $\|\Delta_{n3}\| \leq C_\Delta h^4$. A similar result is true for δ_{n+1}.

8.3 Error Analysis

In what follows, we suppose that (8.1) possesses a uniformly bounded and sufficiently smooth solution $y : [t_0, T] \to X$ with derivatives in X, and that $f : X \to S$ in (8.1) is Fréchet differentiable sufficiently often in a strip (see, e.g. [10]) along the exact solution of (8.1). All occurring derivatives of $f(y)$ are assumed to be uniformly bounded.

Before continuing, we first mention the discrete Gronwall lemma described by Lemma 7.1 in Chap. 7, which will be useful to our subsequent analysis.

Second, having in mind that Δ_{n3} and δ_{n+1} satisfy (8.20), we next introduce the notation

$$e_n = y_n - y(t_n), \quad E_{n3} = Y_{n3} - y(t_n + h).$$

Subtraction of (8.16) from the numerical method (8.15) gives the error recursions

$$
\begin{cases}
E_{n3} = & 2\phi_0(V)e_n - e_{n-1} + h^2 a_{31}(V)\Big(f(y_{n-1}) - f\big(y(t_n - h)\big)\Big) \\
& + h^2 a_{32}(V)\Big(f(y_n) - f\big(y(t_n)\big)\Big) - \Delta_{n3}, \\
e_{n+1} = & 2\phi_0(V)e_n - e_{n-1} + h^2 b_1(V)\Big(f(y_{n-1}) - f\big(y(t_n - h)\big)\Big) \\
& + h^2 b_2(V)\Big(f(y_n) - f\big(y(t_n)\big)\Big) + h^2 b_3(V)\Big(f(Y_{n3}) - f\big(y(t_n + h)\big)\Big) - \delta_{n+1}.
\end{cases}
\tag{8.23}
$$

Inserting (8.22) into (8.23) leads to

$$
\begin{cases}
E_{n3} = & 2\phi_0(V)e_n - e_{n-1} + 2h^2\phi_2(V)\Big(f(y_n) - f\big(y(t_n)\big)\Big) - \Delta_{n3}, \\
e_{n+1} = & 2\phi_0(V)e_n - e_{n-1} + 2h^2\phi_4(V)\Big(f(y_{n-1}) - f\big(y(t_n - h)\big)\Big) \\
& + 2h^2(\phi_2(V) - 2\phi_4(V))\Big(f(y_n) - f\big(y(t_n)\big)\Big) \\
& + 2h^2\phi_4(V)\Big(f(Y_{n3}) - f\big(y(t_n + h)\big)\Big) - \delta_{n+1}.
\end{cases}
\tag{8.24}
$$

Next we give some lemmas in order to present the main result of this chapter's analysis.

Lemma 8.2 *For the method TSERKN3s with stepsize h, there exists h_0 such that, if $0 < h \leq h_0$ (h is sufficiently small), we have*

$$\|E_{n3}\| \leq C_E(\|e_n\| + \|e_{n-1}\| + h^4), \tag{8.25}$$

where C_E depends on $\|f_y\|$ and is independent of $\|M\|$.

Proof Applying the mean value theorem to f leads to

$$\|f(y_n) - f(y(t_n))\| \leq C_f \|e_n\|,$$

where $C_f := \max \|f_y\|$.

Based on $a_{31}(V) = 0$ and $a_{32}(V) = 2\phi_2(V)$, it follows from (8.24) that

$$\|E_{n3}\| \leq \|2\phi_0(V)e_n\| + \|e_{n-1}\| + h^2 C_f \|e_n\| + \|\Delta_{n3}\|.$$

We then obtain

$$\|E_{n3}\| \leq (2 + C_f h^2)\|e_n\| + \|e_{n-1}\| + C_\Delta h^4.$$

Taking $C_E = \max(2 + C_f h^2, 1, C_\Delta)$ completes the proof. $\square$

Lemma 8.3 *If V is symmetric and positive semi-definite, then*

$$\begin{pmatrix} 2\phi_0(V) & -I \\ I & 0 \end{pmatrix}^n = \begin{pmatrix} W_{n+1} & -W_n \\ -W_n & W_{n-1} \end{pmatrix}, \quad n \geq 0, \tag{8.26}$$

where $W_n = n\phi_1^{-1}(V)\phi_1(n^2 V)$.

Proof For $n = 0$, (8.26) is trivial. Assuming (8.26) holds for $n = k$, we can obtain

$$\begin{pmatrix} 2\phi_0(V) & -I \\ I & 0 \end{pmatrix}^{k+1} = \begin{pmatrix} W_{k+1} & -W_k \\ -W_k & W_{k-1} \end{pmatrix} \begin{pmatrix} 2\phi_0(V) & -I \\ I & 0 \end{pmatrix} = \begin{pmatrix} W_{k+2} & -W_{k+1} \\ -W_{k+1} & W_k \end{pmatrix}.$$

The last equality is obtained by using Proposition 8.1. $\square$

Lemma 8.4 *The global error $e_n = y_n - y(t_n)$ satisfies*

$$e_{n+1} = -W_n e_0 + W_{n+1} e_1 + \sum_{j=1}^{n} W_{n+1-j}(h^2 D_j - \delta_{j+1}), \tag{8.27}$$

where $W_n = n\phi_1^{-1}(V)\phi_1(n^2 V)$, and

$$\begin{aligned} D_j &= b_1(V)\Big(f(y_{j-1}) - f(y(t_j - h))\Big) + b_2(V)\Big(f(y_j) - f(y(t_j))\Big) \\ &\quad + b_3(V)\Big(f(Y_{j3}) - f(y(t_j + h))\Big), \quad j = 1, \ldots, n. \end{aligned} \tag{8.28}$$

Proof The error equation is

$$e_{n+1} = 2\phi_0(V)e_n - e_{n-1} + h^2 D_n - \delta_{n+1},$$

which can be transformed into the one-step form

$$\begin{pmatrix} e_{n+1} \\ e_n \end{pmatrix} = R \begin{pmatrix} e_n \\ e_{n-1} \end{pmatrix} + \begin{pmatrix} h^2 D_n - \delta_{n+1} \\ \mathbf{0} \end{pmatrix}, \tag{8.29}$$

where

$$R = \begin{pmatrix} 2\phi_0(V) & -I \\ I & \mathbf{0} \end{pmatrix}.$$

Solving the Eq. (8.29) yields

$$\begin{pmatrix} e_{n+1} \\ e_n \end{pmatrix} = R^n \begin{pmatrix} e_1 \\ e_0 \end{pmatrix} + \sum_{j=1}^{n} R^{n-j} \begin{pmatrix} h^2 D_j - \delta_{j+1} \\ \mathbf{0} \end{pmatrix}.$$

The result now follows from $(R^n)_{11} = W_{n+1}$ and $(R^n)_{12} = -W_n$ as given by Lemma 8.3. $\qquad\square$

Lemma 8.5 *For the method TSERKN3s with stepsize h satisfying $0 < h < H$, D_n determined by (8.28) satisfies*

$$\|D_n\| \le C_D(\|e_n\| + \|e_{n-1}\| + h^4), \quad n \ge 1, \tag{8.30}$$

where C_D depends on $\|f_y\|$ and is independent of $\|M\|$.

Proof Applying the mean value theorem to f leads to

$$\|D_n\| \le C_f(\|e_{n-1}\| + \|e_n\| + \|E_{n3}\|). \tag{8.31}$$

Inserting (8.25) into (8.31) gives

$$\begin{aligned} \|D_n\| &\le C_f\big(\|e_{n-1}\| + \|e_n\| + C_E(\|e_{n-1}\| + \|e_n\| + h^4)\big) \\ &= (C_f + C_E)\|e_{n-1}\| + (C_f + C_E)\|e_n\| + C_E h^4. \end{aligned}$$

Taking $C_D = C_f + C_E$ completes the proof. $\qquad\square$

Now we are ready to present the main result of this chapter's analysis.

Theorem 8.2 *Suppose that M in the initial value problem (8.1) is symmetric and positive semi-definite and consider the method TSERKN3s with the stepsize h satisfying $0 < h \le h_0$ and the mesh points $t_n = t_0 + nh$, $n = 0, 1, 2, \ldots, N$. Then, if*

$$\|y_1 - y(t_1)\| \le C_0 h^5,$$

where C_0 is independent of $\|M\|$, the error bounds

$$\|e_n\| \le C h^4 \tag{8.32}$$

hold uniformly for $n = 2, \ldots, N$. *The constant* C *depends on* T *and* $\|f_y\|$, *but is independent of* $\|M\|$, n *and* h.

Proof By Lemma 8.1, we know $\|W_n\| = \|W(n, V)\| \leq n$. The expressions (8.27), (8.30) and (8.20) result in

$$
\begin{aligned}
\|e_{n+1}\| &\leq \|W_{n+1}e_1\| + \|\sum_{j=1}^{n} W_{n+1-j}h^2 D_j\| + \|\sum_{j=1}^{n} W_{n+1-j}\delta_{j+1}\| \\
&\leq (n+1)\|e_1\| + \sum_{j=1}^{n} h^2(n+1-j)\|D_j\| + \sum_{j=1}^{n}(n+1-j)\|\delta_{j+1}\| \\
&\leq (n+1)\|e_1\| + \sum_{j=1}^{n} C'nh^2(\|e_j\| + \|e_{j-1}\| + h^4) + \sum_{j=1}^{n} C'nh^6 \\
&\leq \sum_{j=1}^{n} h\bar{C}\|e_j\| + \bar{C}h^4,
\end{aligned}
$$

where $\bar{C}$ depends on T and $\|f_y\|$, but is independent of $\|M\|$, h and n. Thus, we have

$$
\|e_n\| \leq \sum_{j=1}^{n-1} h\bar{C}\|e_j\| + \bar{C}h^4. \tag{8.33}
$$

Considering Lemma 7.1, we set $\phi_n = \|e_n\|$, $\psi_n = 0$, $\chi_n = \bar{C}h^4$, $\alpha_n = \bar{C}$. Therefore, the following result holds:

$$
h\sum_{k=0}^{n-1} \alpha_k = h\sum_{k=0}^{n-1} \bar{C} \leq \hat{C}. \tag{8.34}
$$

An application of the discrete Gronwall lemma to (8.33) gives

$$
\|e_n\| \leq \bar{C}h^4 e^{\hat{C}nh} \leq Ch^4, \tag{8.35}
$$

where C depends on T, $\|f_y\|$ and is independent of $\|M\|$, h and n. $\qquad\square$

8.4 Stability and Phase Properties

In this section, we are concerned with the stability and phase properties of the new method. Consider a second-order modified linear test model [13, 14]

$$
y''(t) + \omega^2 y(t) = -\varepsilon y(t), \qquad \omega^2 + \varepsilon > 0, \tag{8.36}
$$

where ω represents an estimate of the dominant frequency λ and $\varepsilon = \lambda^2 - \omega^2$ is the error of estimation. Applying the TSERKN method (8.15) to (8.36) yields

$$\begin{cases} Y_{n3} = 2\phi_0(V)y_n - y_{n-1} - z\big(a_{31}(V)y_{n-1} + a_{32}(V)y_n\big), \\ y_{n+1} = 2\phi_0(V)y_n - y_{n-1} - z\big(b_1(V)y_{n-1} + b_2(V)y_n + b_3(V)Y_{n3}\big), \end{cases} \tag{8.37}$$

where

$$z = \varepsilon h^2, \quad V = \omega^2 h^2. \tag{8.38}$$

Substituting the first equation of (8.37) into the second one of (8.37), we obtain the recurrence relation

$$y_{n+1} - S(V, z)y_n + P(V, z)y_{n-1} = 0,$$

where $S(V, z)$ and $P(V, z)$ are functions of V and z. The quantities

$$\phi(V, z) = H - \arccos\left(\frac{S(V, z)}{2\sqrt{P(V, z)}}\right), \quad d(V, z) = 1 - \sqrt{P(V, z)}$$

are called the dispersion error and dissipation error, respectively, where $H = h\lambda = \sqrt{V + z}$. The analysis of these errors is similar to that discussed in [13]. Setting

$$V = \frac{\omega^2}{\omega^2 + \varepsilon}H^2, \quad z = \frac{\varepsilon}{\omega^2 + \varepsilon}H^2,$$

we arrive at the following definition.

Definition 8.2   A TSERKN method is said to be dispersive of order q and dissipative of order r, if $\phi(V, z) = \mathcal{O}(H^{q+1})$ and $d(V, z) = \mathcal{O}(H^{r+1})$. A TSERKN method is zero dissipative if $d(V, z) = 0$.

Because $S(V, z)$ and $P(V, z)$ depend on the variables V and z, geometrically, the characterization of stability reduces to the study of a two-dimensional region in the (V, z)-plane for a TSERKN method. Accordingly, we have the following definitions for the stability of a TSERKN method.

- $R_s = \{(V, z)|\ V > 0,\ |P(V, z)| < 1 \text{ and } |S(V, z)| < P(V, z) + 1\}$ is called the *region of stability of a TSERKN method*.
- $R_p = \{(V, z)|\ V > 0\ P(V, z) \equiv 1 \text{ and } |S(V, z)| < 2\}$ is called the *region of periodicity of a TSERKN method*.

In what follows, we give the series expansions of the dispersion error and the dissipation error for the new method TSERKN3s derived in the previous section.

Proposition 8.3   *The method TSERKN3s derived in the previous section is dispersive of order four and zero dissipative. The series expansions of the dispersion error and the dissipation error are*

Fig. 8.1 Region of periodicity for the method TSERKN3s

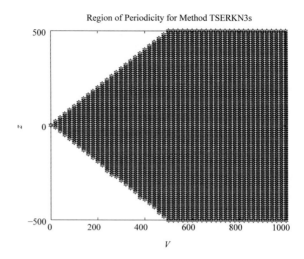

$$\phi(V, z) = \frac{\varepsilon(2\varepsilon - 3\omega^2)H^5}{1440(\varepsilon + \omega^2)^2} + \mathcal{O}(H^7), \quad d(V, z) = 0,$$

respectively.

The region of periodicity for the integrator TSERKN3s is depicted in Fig. 8.1.

8.5 Numerical Experiments

In this section, in order to show the competitiveness and efficiency of the new method compared with some well-known methods in the scientific literature, we use three model problems whose solutions are known to be oscillatory. The criterion used in the numerical comparisons is the decimal logarithm of the global error (GE) versus the logarithm of the number of function evaluations required by each method. Here the global error is computed by the maximum norm of the difference between the numerical solution and the analytical solution. The integrators we select for comparison are

- RKN4s5: the four-stage RKN method of order five given in [8];
- Gautschi2000a: the symmetric Gautschi's method of order two given in [6];
- TS3s4: the three-stage two-step method of order four proposed in [1];
- TSERKN3s: the three-stage TSERKN method of order four proposed in this chapter.

Problem 8.1 We consider the Duffing equation

$$\begin{cases} y'' + 25y = 2k^2 y^3 - k^2 y, & t \in [0, 5000], \\ y(0) = 1, & y'(0) = 0. \end{cases}$$

where $k = 0.03$. The exact solution of this initial value problem is

$$y(t) = sn(wt; k/w),$$

where sn denotes the Jacobian elliptic function. The system is integrated with the stepsizes $h = 1/(3 \times 2^{j-2}), j = 2, 3, 4, 5$ for RKN4s5, $h = 1/(3 \times 2^j), j = 2, 3, 4, 5$ for Gautschi2000a, $h = 1/(3 \times 2^{j-1}), j = 2, 3, 4, 5$ for TS3s4 and TSERKN3s. The numerical results are presented in Fig. 8.2.

Problem 8.2 Consider a nonlinear wave equation

$$\begin{cases} \dfrac{\partial^2 u}{\partial t^2} - \dfrac{\partial^2 u}{\partial x^2} = u^5 - u^3 - 10u, & 0 < x < 1, \ t > 0, \\ u(0, t) = u(1, t) = 0, & u(x, 0) = \dfrac{x(1 - x)}{100}, \quad u_t(x, 0) = 0. \end{cases}$$

By using second-order symmetric differences, this problem is converted to a system of ODEs in time

$$\begin{cases} \dfrac{d^2 u_i}{dt^2} - \dfrac{u_{i+1} - 2u_i + u_{i-1}}{\Delta x^2} = u_i^5 - u_i^3 - 10u_i, & 0 < t \le t_{\text{end}}, \\ u_i(0) = \dfrac{x_i(1 - x_i)}{100}, \quad u_i'(0) = 0, & i = 1, \dots, N - 1, \end{cases}$$

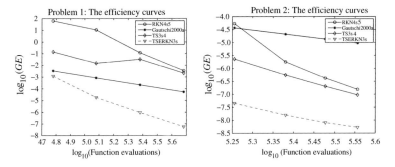

Fig. 8.2 Efficiency curves for Problems 8.1 (*left*) and 8.2 (*right*)

where $\Delta x = 1/N$ is the spatial mesh step and $x_i = i\Delta x$. This semi-discrete oscillatory system has the form

$$
\begin{cases}
\dfrac{d^2 U}{dt^2} + MU = F(U), \quad 0 < t \le t_{end}. \\[2mm]
U(0) = \left(\dfrac{x_1(1 - x_1)}{100}, \ldots, \dfrac{x_{N-1}(1 - x_{N-1})}{100} \right)^{\mathsf{T}}, \quad U'(0) = 0,
\end{cases}
$$

where $U(t) = (u_1(t), \ldots, u_{N-1}(t))^{\mathsf{T}}$ with $u_i(t) \approx u(x_i, t), i = 1, \ldots, N - 1$, and

$$
M = \frac{1}{\Delta x^2}
\begin{pmatrix}
2 & -1 & & & \\
-1 & 2 & -1 & & \\
& \ddots & \ddots & \ddots & \\
& & -1 & 2 & -1 \\
& & & -1 & 2
\end{pmatrix},
$$

$$
F(U) = \left(u_1^5 - u_1^3 - 10u_1, \ldots, u_{N-1}^5 - u_{N-1}^3 - 10u_{N-1} \right)^{\mathsf{T}}.
$$

The system is integrated on the interval $[0, 1000]$ with $N = 20$ and the stepsizes $h = 1/(15 \times j), j = 3, 4, 5, 6$ for RKN4s5, $h = 1/(60 \times j), j = 3, 4, 5, 6$ for Gautschi2000a, $h = 1/(30 \times j), j = 3, 4, 5, 6$ for TS3s4 and TSERKN3s. The numerical results are presented in Fig. 8.2.

Problem 8.3 Consider the Fermi–Pasta–Ulam Problem of Chap. 2.
   Following Hairer et al. [8], we choose

$$
m = 3, \ q_1(0) = 1, \ p_1(0) = 1, \ q_4(0) = \frac{1}{\omega}, \ p_4(0) = 1,
$$

and choose zero for the remaining initial values. The system is integrated on the interval $[0, 25]$ with the stepsizes $h = 0.01/(3 \times 2^{j-3}), j = 2, 3, 4, 5$ for RKN4s5, $h = 0.01 \times 1/(3 \times 2^{j-1}), j = 2, 3, 4, 5$ for Gautschi2000a, $h = 0.01/(3 \times 2^{j-2}), j = 2, 3, 4, 5$ for TS3s4 and TSERKN3s. The efficiency curves for different $\omega = 50, 100, 150, 200$ are shown in Fig. 8.3. In this experiment we note that the error $\log_{10}(GE)$ is very large for RKN4s5 with $h = 0.015$ when $\omega = 150$ or $\omega = 200$, hence we do not plot the corresponding points in Fig. 8.3.

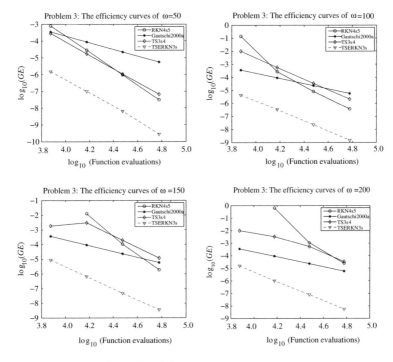

Fig. 8.3 Efficiency curves for Problem 8.3

8.6 Conclusions

In this chapter, we analyzed the error for two-step extended Runge–Kutta–Nyström-type (TSERKN) methods. We derived a three-stage explicit TSERKN method of order four for the multi-frequency oscillatory system (8.1) and presented a rigorous error analysis. The new method is an extension of Gautschi-type methods and shares the favourable property that it exactly integrates the multi-frequency oscillatory homogeneous second-order differential equation $y'' + My = \mathbf{0}$. We showed that the new method has the global error bounds $\|y_n - y(t_n)\| \le Ch^4$, where C is independent of $\|M\|$, h and n. We also gave an analysis of the stability and phase properties. Furthermore, corresponding numerical experiments were presented which confirm that the method derived in this chapter is more efficient when applied to the multi-frequency oscillatory system (8.1) in comparison with the well-known high-quality integrators proposed in the scientific literature.

The material of this chapter is based on the work by Li and Wu [11].

References

1. Chawla MM (1984) Numerov made explicit has better stability. BIT Numer Math 24:117–118
2. Cohen D, Hairer E, Lubich C (2008) Long-time analysis of nonlinearly perturbed wave equations via modulated fourier expansions. Arch Ration Mech Anal 187:341–368
3. Coleman JP (2003) Order conditions for a class of two-step methods for $y'' = f(x, y)$. IMA J Numer Anal 23:197–220
4. Franco JM (2006) New methods for oscillatory systems based on ARKN methods. Appl Numer Math 56:1040–1053
5. Grimm V (2006) On the use of the gautschi-type exponential integrator for wave equation. In: Bermudez de Castro A, Gomez D, Quintela P, Salgado P (eds) Numerical mathematics and advanced applications (ENUMATH2005). Springer, Berlin, pp 557–563
6. Hairer E, Lubich C (2000) Long-time energy conservation of numerical methods for oscillatory differential equations. SIAM J Numer Anal 38:414–441
7. Hairer E, Lubich C (2008) Spectral semi-discretisations of weakly non-linear wave equations over long times. Found Comput Math 8:319–334
8. Hairer E, Nørsett SP, Wanner G (2002) Solving ordinary differential equations i: nonstiff problems, 2nd edn. Springer, Berlin
9. Hochbruck M, Lubich C (1999) A gautschi-type method for oscillatory second-order differential equations. Numer Math 83:403–426
10. Hochbruck M, Ostermann A (2005) Explicit exponential runge-kutta methods for semilineal parabolic problems. SIAM J Numer Anal 43:1069–1090
11. Li J, Wu X (2014) Error analysis of explicit TSERKN methods for highly oscillatory systems. Numer Algo 65:465–483
12. Li J, Wang B, You X, Wu X (2011) Two-step extended RKN methods for oscillatory systems. Comput Phys Commun 182:2486–2507
13. Van de Vyver H (2009) Scheifele two-step methods for perturbed oscillators. J Comput Appl Math 224:415–432
14. Wu X (2012) A note on stability of multidimensional adapted Runge-Kutta-Nyström methods for oscillatory systems. Appl Math Modell 36:6331–6337
15. Wu X, You X, Shi W, Wang B (2010) ERKN integrators for systems of oscillatory second-order differential equations. Comput Phys Commun 181:1873–1887
16. Wu X, You X, Li J (2009) Note on derivation of order conditions for ARKN methods for perturbed oscillators. Comput Phys Commun 180:1545–1549
17. Wu X, You X, Xia J (2009) Order conditions for ARKN methods solving oscillatory systems. Comput Phys Commun 180:2250–2257
18. Wu X, Wang B (2010) Multidimensional adapted Runge-Kutta-Nyström methods for oscillatory systems. Comput Phys Commun 181:1955–1962
19. Wu X, You X, Wang B (2013) Structure-preserving algorithms for oscillatory differential equations. Springer, Berlin
20. Yang H, Wu X, You X, Fang Y (2009) Extended RKN-type methods for numerical integration of perturbed oscillators. Comput Phys Commun 180:1777–1794

Chapter 9
Highly Accurate Explicit Symplectic ERKN Methods for Multi-frequency Oscillatory Hamiltonian Systems

Numerical integration of differential equations having multi-frequency oscillatory solutions is far from being satisfactory and many challenges remain. This is true in particular when one wishes to approximate the underlying multi-frequency oscillatory solution efficiently. This chapter is devoted to exploring explicit symplectic ERKN methods of order five, based on the symplecticity conditions and order conditions of ERKN methods for multi-frequency and multidimensional oscillatory Hamiltonian systems. Two five-stage explicit symplectic ERKN methods of order five with small residuals are derived, and their stability and phase properties are analysed. The numerical results demonstrate and confirm that high-order explicit symplectic ERKN integrators are much more efficient than the existing methods in the scientific literature.

9.1 Motivation

This chapter focuses on highly accurate explicit symplectic ERKN methods for the multi-frequency and multidimensional oscillatory second-order differential equations of the form

$$\begin{cases} q''(t) + Mq(t) = f\big(q(t)\big), & t \in [t_0, T], \\ q(t_0) = q_0, \quad q'(t_0) = q'_0, \end{cases} \tag{9.1}$$

where $M \in \mathbb{R}^{d \times d}$ is a symmetric positive semi-definite matrix implicitly containing and preserving the oscillatory dominant frequencies of the problem, and $f(q)$ is the negative gradient of a real-valued function $U(q)$ whose second derivatives are continuous. Equation (9.1) is in fact identical to the following multi-frequency and multidimensional oscillatory Hamiltonian system:

© Springer-Verlag Berlin Heidelberg and Science Press, Beijing, China 2015
X. Wu et al., *Structure-Preserving Algorithms for Oscillatory
Differential Equations II*, DOI 10.1007/978-3-662-48156-1_9

$$\begin{cases} \dot{p} = -\nabla_q H(p, q), & p(t_0) = p_0, \\ \dot{q} = \nabla_p H(p, q), & q(t_0) = q_0, \end{cases} \tag{9.2}$$

with the Hamiltonian

$$H(p, q) = \frac{1}{2} p^{\mathsf{T}} p + \frac{1}{2} q^{\mathsf{T}} M q + U(q), \tag{9.3}$$

where $p = \dot{q}$. It is of great importance to explore efficient numerical methods for the oscillatory Hamiltonian system (9.2). In fact, the development of numerical integrators for problem (9.2) has received much attention in recent years. For related work, readers are referred to [3–5, 19, 21–23]. In [21], Wu et al. formulated a standard form of ERKN methods for (9.2) using the matrix-variation-of-constants formula, and presented the order conditions using B-series theory based on the set SENT (see e.g. [22]). On the other hand, the importance of geometric numerical integration for the purpose of preserving some structures of differential equations has recently become apparent. For a survey of this field, readers are referred to the book [7]. Ruth [12] was the first to publish the result about canonical numerical integrators, and proposed a three-stage canonical RKN method of order three. Further research about canonical integrators following [12] can be found in [1, 2, 11, 13, 16, 20, 24]. In order to preserve the symplectic structure of (9.2), symplecticity conditions for multi-frequency and multidimensional ERKN methods were derived and some novel explicit symplectic ERKN methods of order up to four were proposed by Wu et al. [20, 22]. However, fifth-order explicit symplectic ERKN methods were not analysed for (9.2). Highly accurate numerical solutions are often needed in applied sciences, such as in the propagation of satellite orbits, where one aims to obtain highly accurate solutions at a low computational cost.

9.2 Preliminaries

In this section, we reformulate the explicit multi-frequency ERKN methods for (9.2) and represent the corresponding order and symplecticity conditions. The stability and phase properties of the explicit ERKN methods are summarized as well.

The discussion commences from defining matrix-valued functions of M given by (4.7) in Chap. 4. Then the following result is a consequence of Theorem 1.1 in Chap. 1:

Theorem 9.1 *If $\nabla U : \mathbb{R}^d \to \mathbb{R}^d$ is continuous in (9.2), then the solution of (9.2) and its derivative satisfy the following equations:*

$$\begin{cases} q(t) = \phi_0\big((t-t_0)^2 M\big)q_0 + (t-t_0)\phi_1\big((t-t_0)^2 M\big)q_0' \\[2mm] \qquad + \int_{t_0}^{t}(t-\zeta)\phi_1\big((t-\zeta)^2 M\big)\hat{f}(\zeta)\mathrm{d}\zeta, \\[4mm] q'(t) = -(t-t_0)M\phi_1\big((t-t_0)^2 M\big)q_0 + \phi_0\big((t-t_0)^2 M\big)q_0' \\[2mm] \qquad + \int_{t_0}^{t}\phi_0\big((t-\zeta)^2 M\big)\hat{f}(\zeta)\mathrm{d}\zeta \end{cases} \tag{9.4}$$

for any $t_0, t \in (-\infty, +\infty)$, where $\hat{f}(\zeta) = -\nabla U(q(\zeta))$.

In order to obtain a highly accurate numerical method for (9.2), we approximate the integrals in (9.4) with quadrature formulae. This leads to the multi-frequency and multidimensional ERKN integrators for (9.2) (see [21]). This chapter is only concerned with explicit ERKN methods.

Definition 9.1 An s-stage explicit ERKN method with stepsize h for solving multi-frequency and multidimensional oscillatory Hamiltonian systems (9.2) is defined by

$$\begin{cases} Q_i = \phi_0(c_i^2 V)q_n + hc_i\phi_1(c_i^2 V)q_n' + h^2 \sum_{j=1}^{i-1}\bar{a}_{ij}(V)\big(-\nabla U(Q_j)\big), \quad i = 1, \ldots, s, \\[4mm] q_{n+1} = \phi_0(V)q_n + h\phi_1(V)q_n' + h^2 \sum_{i=1}^{s}\bar{b}_i(V)\big(-\nabla U(Q_i)\big), \\[4mm] q_{n+1}' = -hM\phi_1(V)q_n + \phi_0(V)q_n' + h \sum_{i=1}^{s}b_i(V)\big(-\nabla U(Q_i)\big), \end{cases} \tag{9.5}$$

where c_i, $i = 1, \ldots, s$, are real constants, $b_i(V)$, $\bar{b}_i(V)$ and $\bar{a}_{ij}(V)$, $i, j = 1, \ldots, s$, are all matrix-valued functions of $V = h^2 M$. The scheme (9.5) can also be characterized by the Butcher tableau

$$\frac{\begin{array}{c|c} c & \bar{A}(V) \\ \hline & \bar{b}^{\mathsf{T}}(V) \\ \hline & b^{\mathsf{T}}(V) \end{array}} = \frac{\begin{array}{c|cccc} c_1 & \mathbf{0}_{d\times d} & \mathbf{0}_{d\times d} & \cdots & \mathbf{0}_{d\times d} \\ c_2 & \bar{a}_{21}(V) & \mathbf{0}_{d\times d} & \cdots & \mathbf{0}_{d\times d} \\ \vdots & \vdots & \vdots & \ddots & \vdots \\ c_s & \bar{a}_{s1}(V) & \bar{a}_{s2}(V) & \cdots & \mathbf{0}_{d\times d} \\ \hline & \bar{b}_1(V) & \bar{b}_2(V) & \cdots & \bar{b}_s(V) \\ \hline & b_1(V) & b_2(V) & \cdots & b_s(V) \end{array}}$$

The following theorem states the order conditions of ERKN methods for multi-frequency and multidimensional oscillatory Hamiltonian systems (9.2).

Theorem 9.2 *A necessary and sufficient condition of an s-stage ERKN method (9.5) for a multi-frequency and multidimensional oscillatory Hamiltonian system (9.2) to be of order r is given by*

$$
\begin{cases}
\bar{b}^{\mathsf{T}}\Phi(\tau) = \dfrac{\rho(\tau)!}{\gamma(\tau)}\phi_{\rho(\tau)+1}(V) + \mathscr{O}(h^{r-\rho(\tau)}), & \rho(\tau) = 1, \ldots, r-1, \\[2ex]
b^{\mathsf{T}}\Phi(\tau) = \dfrac{\rho(\tau)!}{\gamma(\tau)}\phi_{\rho(\tau)}(V) + \mathscr{O}(h^{r+1-\rho(\tau)}), & \rho(\tau) = 1, \ldots, r,
\end{cases}
$$

where τ is the extended Nyström tree associated with the elementary differential $\mathscr{F}(\tau)(q_n, q_n')$ of the function $-\nabla U(q)$ at q_n.

The symplecticity conditions of ERKN methods are derived in the paper [20]. The next theorem presents the symplecticity conditions for explicit multi-frequency ERKN methods when applied to multi-frequency and multidimensional oscillatory Hamiltonian systems (9.2).

Theorem 9.3 *An s-stage explicit multi-frequency ERKN method (9.5) for multi-frequency and multidimensional oscillatory Hamiltonian systems (9.2) is symplectic if its coefficients satisfy*

$$
\begin{cases}
\phi_0(V)b_i + V\phi_1(V)\bar{b}_i = d_i\phi_0(c_i^2 V), \ d_i \in \mathbb{R}, \ i = 1, 2, \ldots, s, \\
\phi_0(V)\bar{b}_i + c_i d_i \phi_1(c_i^2 V) = b_i\phi_1(V), \ i = 1, 2, \ldots, s, \\
\bar{b}_j b_i = \bar{b}_i b_j + d_i a_{ij}, \ i = 1, 2, \ldots, s, \ j = 1, 2, \ldots, i-1,
\end{cases}
\tag{9.6}
$$

where $V = h^2 M$.

The analysis of stability and phase properties is fundamental to demonstrate numerical behaviour of integrators for (9.2). The numerical stability and phase properties of explicit ERKN methods can be analysed using the second-order homogeneous linear test model of the form [15, 17]

$$
q''(t) + \omega^2 q(t) = -\varepsilon q(t), \quad \omega^2 + \varepsilon > 0,
\tag{9.7}
$$

where ω represents an estimation of the dominant frequency λ and $\varepsilon = \lambda^2 - \omega^2$ is the error of that estimation. The procedure of analysis is similar to that described in Chap. 7.

9.3   Explicit Symplectic ERKN Methods of Order Five with Some Small Residuals

We note an important fact that, apart from symplecticity, adaption to oscillation, the algebraic order cannot be ignored when designing efficient numerical methods for solving *multi-frequency oscillatory Hamiltonian systems.*

We consider the fifth-order explicit symplectic ERKN methods. This section will construct two five-stage explicit symplectic ERKN methods of order five with some small residuals for (9.2) and analyse the stability and phase properties for the new highly accurate methods.

For deriving fifth-order explicit symplectic ERKN methods, it seems reasonable not to consider the four-stage explicit ERKN methods, since many order conditions and symplecticity conditions will be involved. However, it also noted that fewer stage number is desirable because it involves fewer function evaluations per step and hence lower computational cost. Based on the facts stated above, five-stage explicit ERKN methods for (9.2) seem appropriate. The scheme (9.5) of a five-stage explicit multi-frequency and multidimensional ERKN method can be denoted by the following Butcher tableau:

$$
\frac{\begin{array}{c|c} c & \bar{A}(V) \end{array}}{\begin{array}{c|c} & \bar{b}^{\mathsf{T}}(V) \\ \hline & b^{\mathsf{T}}(V) \end{array}} =
\frac{\begin{array}{c|ccccc}
c_1 & \mathbf{0}_{d\times d} & \mathbf{0}_{d\times d} & \mathbf{0}_{d\times d} & \mathbf{0}_{d\times d} & \mathbf{0}_{d\times d} \\
c_2 & \bar{a}_{21}(V) & \mathbf{0}_{d\times d} & \mathbf{0}_{d\times d} & \mathbf{0}_{d\times d} & \mathbf{0}_{d\times d} \\
c_3 & \bar{a}_{31}(V) & \bar{a}_{32}(V) & \mathbf{0}_{d\times d} & \mathbf{0}_{d\times d} & \mathbf{0}_{d\times d} \\
c_4 & \bar{a}_{41}(V) & \bar{a}_{42}(V) & \bar{a}_{43}(V) & \mathbf{0}_{d\times d} & \mathbf{0}_{d\times d} \\
c_5 & \bar{a}_{51}(V) & \bar{a}_{52}(V) & \bar{a}_{53}(V) & \bar{a}_{54}(V) & \mathbf{0}_{d\times d}
\end{array}}{\begin{array}{ccccc}
\bar{b}_1(V) & \bar{b}_2(V) & \bar{b}_3(V) & \bar{b}_4(V) & \bar{b}_5(V) \\
\hline
b_1(V) & b_2(V) & b_3(V) & b_4(V) & b_5(V)
\end{array}}
$$

From Theorem 9.3, the symplecticity conditions for the five-stage explicit ERKN method are given by

$$
\begin{cases}
\phi_0(V)b_1(V) + V\phi_1(V)\bar{b}_1(V) = d_1\phi_0(c_1^2 V), \\
\phi_0(V)b_2(V) + V\phi_1(V)\bar{b}_2(V) = d_2\phi_0(c_2^2 V), \\
\phi_0(V)b_3(V) + V\phi_1(V)\bar{b}_3(V) = d_3\phi_0(c_3^2 V), \\
\phi_0(V)b_4(V) + V\phi_1(V)\bar{b}_4(V) = d_4\phi_0(c_4^2 V), \\
\phi_0(V)b_5(V) + V\phi_1(V)\bar{b}_5(V) = d_5\phi_0(c_5^2 V), \\
\phi_0(V)\bar{b}_1(V) + c_1 d_1\phi_1(c_1^2 V)) = b_1(V)\phi_1(V), \\
\phi_0(V)\bar{b}_2(V) + c_2 d_2\phi_1(c_2^2 V)) = b_2(V)\phi_1(V), \\
\phi_0(V)\bar{b}_3(V) + c_3 d_3\phi_1(c_3^2 V)) = b_3(V)\phi_1(V), \\
\phi_0(V)\bar{b}_4(V) + c_4 d_4\phi_1(c_4^2 V)) = b_4(V)\phi_1(V), \\
\phi_0(V)\bar{b}_5(V) + c_5 d_5\phi_1(c_5^2 V)) = b_5(V)\phi_1(V), \\
\bar{b}_1(V)b_2(V) = \bar{b}_2(V)b_1(V) + d_2\bar{a}_{21}(V), \\
\bar{b}_1(V)b_3(V) = \bar{b}_3(V)b_1(V) + d_3\bar{a}_{31}(V), \\
\bar{b}_2(V)b_3(V) = \bar{b}_3(V)b_2(V) + d_3\bar{a}_{32}(V), \\
\bar{b}_1(V)b_4(V) = \bar{b}_4(V)b_1(V) + d_4\bar{a}_{41}(V), \\
\bar{b}_2(V)b_4(V) = \bar{b}_4(V)b_2(V) + d_4\bar{a}_{42}(V), \\
\bar{b}_3(V)b_4(V) = \bar{b}_4(V)b_3(V) + d_4\bar{a}_{43}(V), \\
\bar{b}_1(V)b_5(V) = \bar{b}_5(V)b_1(V) + d_5\bar{a}_{51}(V), \\
\bar{b}_2(V)b_5(V) = \bar{b}_5(V)b_2(V) + d_5\bar{a}_{52}(V), \\
\bar{b}_3(V)b_5(V) = \bar{b}_5(V)b_3(V) + d_5\bar{a}_{53}(V), \\
\bar{b}_4(V)b_5(V) = \bar{b}_5(V)b_4(V) + d_5\bar{a}_{54}(V).
\end{cases}
\tag{9.8}
$$

It is noted that the products of the matrix-valued functions $\phi_0(V)$, $\phi_1(V)$, $b_i(V)$, $\bar{b}_i(V)$, $\bar{a}_{ij}(V)$, $i = 1, \ldots, 5$, $j = 1, \ldots, i - 1$, are commutative since they are all matrix functions of V. Choosing c_i and d_i, $i = 1, \ldots, 5$ as parameters, we then can observe that there are ten coefficients $b_i(V)$, $\bar{b}_i(V)$, $i = 1, 2, \ldots, 5$, in the first ten equations in (9.8). Therefore, choosing c_i and d_i, $i = 1, \ldots, 5$ as parameters and solving the first ten equations in (9.8), we obtain

$$
\begin{cases}
b_1(V) = d_1(\phi_0(V)\phi_0(c_1^2 V) + c_1 V \phi_1(V)\phi_1(c_1^2 V)), \\
b_2(V) = d_2(\phi_0(V)\phi_0(c_2^2 V) + c_2 V \phi_1(V)\phi_1(c_2^2 V)), \\
b_3(V) = d_3(\phi_0(V)\phi_0(c_3^2 V) + c_3 V \phi_1(V)\phi_1(c_3^2 V)), \\
b_4(V) = d_4(\phi_0(V)\phi_0(c_4^2 V) + c_4 V \phi_1(V)\phi_1(c_4^2 V)), \\
b_5(V) = d_5(\phi_0(V)\phi_0(c_5^2 V) + c_5 V \phi_1(V)\phi_1(c_5^2 V)), \\
\bar{b}_1(V) = (b_1(V)\phi_1(V) - c_1 d_1 \phi_1(c_1^2 V))(\phi_0(V))^{-1}, \\
\bar{b}_2(V) = (b_2(V)\phi_1(V) - c_2 d_2 \phi_1(c_2^2 V))(\phi_0(V))^{-1}, \\
\bar{b}_3(V) = (b_3(V)\phi_1(V) - c_3 d_3 \phi_1(c_3^2 V))(\phi_0(V))^{-1}, \\
\bar{b}_4(V) = (b_4(V)\phi_1(V) - c_4 d_4 \phi_1(c_4^2 V))(\phi_0(V))^{-1}, \\
\bar{b}_5(V) = (b_5(V)\phi_1(V) - c_5 d_5 \phi_1(c_5^2 V))(\phi_0(V))^{-1}.
\end{cases}
\tag{9.9}
$$

Then choosing d_i, $b_i(V)$, $\bar{b}_i(V)$, $i = 1, \ldots, 5$ as parameters, it follows from the other ten equations in (9.8) that

$$
\begin{cases}
\bar{a}_{21}(V) = (\bar{b}_1(V)b_2(V) - \bar{b}_2(V)b_1(V))(d_2)^{-1}, \\
\bar{a}_{31}(V) = (\bar{b}_1(V)b_3(V) - \bar{b}_3(V)b_1(V))(d_3)^{-1}, \\
\bar{a}_{32}(V) = (\bar{b}_2(V)b_3(V) - \bar{b}_3(V)b_2(V))(d_3)^{-1}, \\
\bar{a}_{41}(V) = (\bar{b}_1(V)b_4(V) - \bar{b}_4(V)b_1(V))(d_4)^{-1}, \\
\bar{a}_{42}(V) = (\bar{b}_2(V)b_4(V) - \bar{b}_4(V)b_2(V))(d_4)^{-1}, \\
\bar{a}_{43}(V) = (\bar{b}_3(V)b_4(V) - \bar{b}_4(V)b_3(V))(d_4)^{-1}, \\
\bar{a}_{51}(V) = (\bar{b}_1(V)b_5(V) - \bar{b}_5(V)b_1(V))(d_5)^{-1}, \\
\bar{a}_{52}(V) = (\bar{b}_2(V)b_5(V) - \bar{b}_5(V)b_2(V))(d_5)^{-1}, \\
\bar{a}_{53}(V) = (\bar{b}_3(V)b_5(V) - \bar{b}_5(V)b_3(V))(d_5)^{-1}, \\
\bar{a}_{54}(V) = (\bar{b}_4(V)b_5(V) - \bar{b}_5(V)b_4(V))(d_5)^{-1}.
\end{cases}
\tag{9.10}
$$

According to the definition of $\phi_0(V)$ and $\phi_1(V)$ given by (1.7) in Chap. 1 and using Mathematica, we can simplify (9.9) and (9.10) to

$$\begin{cases} b_1(V) = d_1\phi_0((-1+c_1)^2V), \\ b_2(V) = d_2\phi_0((-1+c_2)^2V), \\ b_3(V) = d_3\phi_0((-1+c_3)^2V), \\ b_4(V) = d_4\phi_0((-1+c_4)^2V), \\ b_5(V) = d_5\phi_0((-1+c_5)^2V), \\ \bar{b}_1(V) = d_1(1-c_1)\phi_1((1-c_1)^2V), \\ \bar{b}_2(V) = d_2(1-c_2)\phi_1((1-c_2)^2V), \\ \bar{b}_3(V) = d_3(1-c_3)\phi_1((1-c_3)^2V), \\ \bar{b}_4(V) = d_4(1-c_4)\phi_1((1-c_4)^2V), \\ \bar{b}_5(V) = d_5(1-c_5)\phi_1((1-c_5)^2V), \\ \bar{a}_{21}(V) = -d_1(c_1-c_2)\phi_1((c_1-c_2)^2V), \\ \bar{a}_{31}(V) = -d_1(c_1-c_3)\phi_1((c_1-c_3)^2V), \\ \bar{a}_{32}(V) = -d_2(c_2-c_3)\phi_1((c_2-c_3)^2V), \\ \bar{a}_{41}(V) = -d_1(c_1-c_4)\phi_1((c_1-c_4)^2V), \\ \bar{a}_{42}(V) = -d_2(c_2-c_4)\phi_1((c_2-c_4)^2V), \\ \bar{a}_{43}(V) = -d_3(c_3-c_4)\phi_1((c_3-c_4)^2V), \\ \bar{a}_{51}(V) = -d_1(c_1-c_5)\phi_1((c_1-c_5)^2V), \\ \bar{a}_{52}(V) = -d_2(c_2-c_5)\phi_1((c_2-c_5)^2V), \\ \bar{a}_{53}(V) = -d_3(c_3-c_5)\phi_1((c_3-c_5)^2V), \\ \bar{a}_{54}(V) = -d_4(c_4-c_5)\phi_1((c_4-c_5)^2V). \end{cases} \tag{9.11}$$

On the other hand, from Theorem 9.2, sufficient conditions for a five-stage explicit ERKN method to be of order five are given by

$$\begin{cases} b^{\mathsf{T}}(V)(e \otimes I) = \phi_1(V) + \mathcal{O}(h^5), \\ b^{\mathsf{T}}(V)(c \otimes I) = \phi_2(V) + \mathcal{O}(h^4), \\ b^{\mathsf{T}}(V)(c^2 \otimes I) = 2\phi_3(V) + \mathcal{O}(h^3), \\ b^{\mathsf{T}}(V)(\bar{A}(\mathbf{0})(e \otimes I)) = \phi_3(V) + \mathcal{O}(h^3), \\ b^{\mathsf{T}}(V)(c^3 \otimes I) = 6\phi_4(V) + \mathcal{O}(h^2), \\ b^{\mathsf{T}}(V)(\bar{A}(\mathbf{0})(c \otimes I)) = \phi_4(V) + \mathcal{O}(h^2), \\ b^{\mathsf{T}}(V)((c \otimes I) \cdot (\bar{A}(\mathbf{0})(e \otimes I))) = 3\phi_4(V) + \mathcal{O}(h^2), \\ b^{\mathsf{T}}(V)(c^4 \otimes I) = 24\phi_5(V) + \mathcal{O}(h), \\ b^{\mathsf{T}}(V)(\bar{A}(\mathbf{0})(c^2 \otimes I)) = 2\phi_5(V) + \mathcal{O}(h), \\ b^{\mathsf{T}}(V)(\bar{A}(\mathbf{0})^2(e \otimes I)) = \phi_5(V) + \mathcal{O}(h), \\ b^{\mathsf{T}}(V)((c \otimes I) \cdot (\bar{A}(\mathbf{0})(c \otimes I))) = 4\phi_5(V) + \mathcal{O}(h), \\ b^{\mathsf{T}}(V)(\bar{A}^{(2)}(\mathbf{0})(e \otimes I)) = -2\phi_5(V) + \mathcal{O}(h), \\ b^{\mathsf{T}}(V)((c^2 \otimes I) \cdot (\bar{A}(\mathbf{0})(e \otimes I))) = 12\phi_5(V) + \mathcal{O}(h), \\ b^{\mathsf{T}}(V)(\bar{A}(\mathbf{0})(e \otimes I) \cdot (\bar{A}(\mathbf{0})(e \otimes I))) = 6\phi_5(V) + \mathcal{O}(h), \end{cases} \tag{9.12}$$

and

$$
\begin{cases}
\bar{b}^{\mathsf{T}}(V)(e \otimes I) = \phi_2(V) + \mathcal{O}(h^4), \\
\bar{b}^{\mathsf{T}}(V)(c \otimes I) = \phi_3(V) + \mathcal{O}(h^3), \\
\bar{b}^{\mathsf{T}}(V)(c^2 \otimes I) = 2\phi_4(V) + \mathcal{O}(h^2), \\
\bar{b}^{\mathsf{T}}(V)(\bar{A}(\mathbf{0})(e \otimes I)) = \phi_4(V) + \mathcal{O}(h^2), \\
\bar{b}^{\mathsf{T}}(V)(c^3 \otimes I) = 6\phi_5(V) + \mathcal{O}(h), \\
\bar{b}^{\mathsf{T}}(V)(\bar{A}(\mathbf{0})(c \otimes I)) = \phi_5(V) + \mathcal{O}(h), \\
\bar{b}^{\mathsf{T}}(V)((c \otimes I) \cdot (\bar{A}(\mathbf{0})(e \otimes I))) = 3\phi_5(V) + \mathcal{O}(h),
\end{cases}
\tag{9.13}
$$

where I is the 5×5 identity matrix, $e = (1, 1, 1, 1, 1)^{\mathsf{T}}$ and $A(\mathbf{0})$ denotes the coefficient matrix A when $V \to \mathbf{0}_{d \times d}$.

Substituting (9.11) into some formulae of (9.12)

$$
\begin{cases}
b^{\mathsf{T}}(V)(e \otimes I) = \phi_1(V) + \mathcal{O}(h^5), \\
b^{\mathsf{T}}(V)(c \otimes I) = \phi_2(V) + \mathcal{O}(h^4), \\
b^{\mathsf{T}}(V)(c^2 \otimes I) = 2\phi_3(V) + \mathcal{O}(h^3), \\
b^{\mathsf{T}}(V)(c^3 \otimes I) = 6\phi_4(V) + \mathcal{O}(h^2), \\
b^{\mathsf{T}}(V)(c^4 \otimes I) = 24\phi_5(V) + \mathcal{O}(h).
\end{cases}
\tag{9.14}
$$

Together with (9.11), this yields

$$
\begin{cases}
d_1 = \Big(12 - 15c_4 - 15c_5 + 20c_4c_5 - 5c_3(3 - 4c_4 - 4c_5 + 6c_4c_5) + 5c_2\big(-3 + 4c_4 + 4c_5 \\
\qquad - 6c_4c_5 + 2c_3(2 - 3c_4 - 3c_5 + 6c_4c_5)\big)\Big) \big/ \Big(60(c_1 - c_2)(c_1 - c_3)(c_1 - c_4)(c_1 - c_5)\Big), \\
d_2 = \Big(-12 + 15c_4 + 15c_5 - 20c_4c_5 + 5c_3(3 - 4c_4 - 4c_5 + 6c_4c_5) - 5c_1\big(-3 + 4c_4 + 4c_5 \\
\qquad - 6c_4c_5 + 2c_3(2 - 3c_4 - 3c_5 + 6c_4c_5)\big)\Big) \big/ \Big(60(c_1 - c_2)(c_2 - c_3)(c_2 - c_4)(c_2 - c_5)\Big), \\
d_3 = \Big(12 - 15c_4 - 15c_5 + 20c_4c_5 - 5c_2(3 - 4c_4 - 4c_5 + 6c_4c_5) + 5c_1\big(-3 + 4c_4 + 4c_5 \\
\qquad - 6c_4c_5 + 2c_2(2 - 3c_4 - 3c_5 + 6c_4c_5)\big)\Big) \big/ \Big(60(c_1 - c_3)(c_2 - c_3)(c_3 - c_4)(c_3 - c_5)\Big), \\
d_4 = \Big(-12 + 15c_3 + 15c_5 - 20c_3c_5 + 5c_2(3 - 4c_3 - 4c_5 + 6c_3c_5) - 5c_1\big(-3 + 4c_3 + 4c_5 \\
\qquad - 6c_3c_5 + 2c_2(2 - 3c_3 - 3c_5 + 6c_3c_5)\big)\Big) \big/ \Big(60(c_1 - c_4)(-c_2 + c_4)(-c_3 + c_4)(c_4 - c_5)\Big), \\
d_5 = \Big(12 - 15c_3 - 15c_4 + 20c_3c_4 - 5c_2(3 - 4c_3 - 4c_4 + 6c_3c_4) + 5c_1\big(-3 + 4c_3 + 4c_4 \\
\qquad - 6c_3c_4 + 2c_2(2 - 3c_3 - 3c_4 + 6c_3c_4)\big)\Big) \big/ \Big(60(c_1 - c_5)(c_2 - c_5)(c_3 - c_5)(c_4 - c_5)\Big).
\end{cases}
\tag{9.15}
$$

Thus (9.11) together with (9.15) give the coefficients of the five-stage explicit symplectic multi-frequency and multidimensional ERKN method with parameters c_i, $i = 1, \ldots, 5$.

In order to determine the values of c_i, $i = 1, \ldots, 5$, it is useful to follow a known approach to getting fifth-order explicit symplectic RKN methods. Hairer et al. [8] mentioned the construction of higher-order symplectic RKN methods and referred to

Okunbor et al.'s work. Okunbor et al. [11] constructed five-stage explicit symplectic RKN methods of order five with some small residuals. Meanwhile, it can be observed that as $V \to \mathbf{0}_{d \times d}$, the ERKN method (9.5) reduces to a classical explicit RKN method and the symplecticity conditions (9.6) for explicit ERKN integrators reduce to those for the classical explicit RKN methods. This means that when $V \to \mathbf{0}_{d \times d}$, an explicit symplectic multi-frequency and multidimensional ERKN method reduces to a classical explicit symplectic RKN method. Therefore, we choose the values of c_i, $i = 1, \ldots, 5$ following the paper [11].

Case 1

Following [11], we choose

$$
\begin{cases}
c_1 = 0.96172990014637649292, \quad c_2 = 0.86647581982605526019, \\
c_3 = 0.12704898443392728669, \quad c_4 = 0.75435833521637640775, \qquad (9.16) \\
c_5 = 0.22929655056040595951.
\end{cases}
$$

The formulae (9.16), (9.15) and (9.11) determine a five-stage explicit symplectic multi-frequency and multidimensional ERKN method. It can be verified that these coefficients satisfy

$$
\begin{cases}
b^{\mathsf{T}}(V)(e \otimes I) = \phi_1(V) + \mathcal{O}(h^5), \\
b^{\mathsf{T}}(V)(c \otimes I) = \phi_2(V) + \mathcal{O}(h^4), \\
b^{\mathsf{T}}(V)(c^2 \otimes I) = 2\phi_3(V) + \mathcal{O}(h^3), \\
b^{\mathsf{T}}(V)(\bar{A}(\mathbf{0})(e \otimes I)) = \phi_3(V) + \mathcal{O}(h^3) - 8.7 \times 10^{-15}I, \\
b^{\mathsf{T}}(V)(c^3 \otimes I) = 6\phi_4(V) + \mathcal{O}(h^2), \\
b^{\mathsf{T}}(V)(\bar{A}(\mathbf{0})(c \otimes I)) = \phi_4(V) + \mathcal{O}(h^2) - 7.1 \times 10^{-15}I, \\
b^{\mathsf{T}}(V)((c \otimes I) \cdot (\bar{A}(\mathbf{0})(e \otimes I))) = 3\phi_4(V) + \mathcal{O}(h^2) - 7.1 \times 10^{-15}I, \\
b^{\mathsf{T}}(V)(c^4 \otimes I) = 24\phi_5(V) + \mathcal{O}(h), \\
b^{\mathsf{T}}(V)(\bar{A}(\mathbf{0})(c^2 \otimes I)) = 2\phi_5(V) + \mathcal{O}(h), \\
b^{\mathsf{T}}(V)(\bar{A}(\mathbf{0})^2(e \otimes I)) = \phi_5(V) + \mathcal{O}(h), \\
b^{\mathsf{T}}(V)((c \otimes I) \cdot (\bar{A}(\mathbf{0})(c \otimes I))) = 4\phi_5(V) + \mathcal{O}(h), \\
b^{\mathsf{T}}(V)(\bar{A}(\mathbf{0})(e \otimes I) \cdot (\bar{A}(\mathbf{0})(e \otimes I))) = 6\phi_5(V) + \mathcal{O}(h) - 3.7 \times 10^{-15}I, \\
b^{\mathsf{T}}(V)((c^2 \otimes I) \cdot (\bar{A}(\mathbf{0})(e \otimes I))) = 12\phi_5(V) + \mathcal{O}(h) - 2.9 \times 10^{-15}I, \\
b^{\mathsf{T}}(V)(\bar{A}^{(2)}(\mathbf{0})(e \otimes I)) = -2\phi_5(V) + \mathcal{O}(h) + 1.004 \times 10^{-15}I,
\end{cases}
$$
$$(9.17)$$

and

$$
\begin{cases}
\bar{b}^{\mathsf{T}}(V)(e \otimes I) = \phi_2(V) + \mathcal{O}(h^4), \\
\bar{b}^{\mathsf{T}}(V)(c \otimes I) = \phi_3(V) + \mathcal{O}(h^3), \\
\bar{b}^{\mathsf{T}}(V)(c^2 \otimes I) = 2\phi_4(V) + \mathcal{O}(h^2), \\
\bar{b}^{\mathsf{T}}(V)(\bar{A}(\mathbf{0})(e \otimes I)) = \phi_4(V) + \mathcal{O}(h^2), \qquad (9.18) \\
\bar{b}^{\mathsf{T}}(V)(c^3 \otimes I) = 6\phi_5(V) + \mathcal{O}(h), \\
\bar{b}^{\mathsf{T}}(V)(\bar{A}(\mathbf{0})(c \otimes I)) = \phi_5(V) + \mathcal{O}(h) - 5.8 \times 10^{-15}I, \\
\bar{b}^{\mathsf{T}}(V)((c \otimes I) \cdot (\bar{A}(\mathbf{0})(e \otimes I))) = 3\phi_5(V) + \mathcal{O}(h) - 4.2 \times 10^{-15}I.
\end{cases}
$$

It can be observed that (9.17) and (9.18) are the same as the fifth-order conditions (9.12) and (9.13) except for the presence of some small residuals. The method with coefficients determined by (9.16), (9.15) and (9.11) is denoted by 1SMMERKN5s5. Meanwhile, when $V \rightarrow \mathbf{0}_{d \times d}$, this method reduces to a classical explicit symplectic RKN method (method 2 given in [11]) of order five with some residuals.

Case 2

Following [11], the choice of

$$
\begin{cases}
c_1 = 0.77070344943939539384, \ c_2 = 0.24564166478370674795, \\
c_3 = 0.87295101556657583863, \ c_4 = 0.13352418017438366649, \\
c_5 = 0.03827009985427366062,
\end{cases} \tag{9.19}
$$

also delivers a five-stage explicit symplectic multi-frequency and multidimensional ERKN method with the coefficients determined by (9.19), (9.15) and (9.11). It can be verified that these coefficients satisfy

$$
\begin{cases}
b^\mathsf{T}(V)(e \otimes I) = \phi_1(V) + \mathcal{O}(h^5), \\
b^\mathsf{T}(V)(c \otimes I) = \phi_2(V) + \mathcal{O}(h^4), \\
b^\mathsf{T}(V)(c^2 \otimes I) = 2\phi_3(V) + \mathcal{O}(h^3), \\
b^\mathsf{T}(V)(A(\mathbf{0})(e \otimes I)) = \phi_3(V) + \mathcal{O}(h^3) - 2.122 \times 10^{-13}I, \\
b^\mathsf{T}(V)(c^3 \otimes I) = 6\phi_4(V) + \mathcal{O}(h^2), \\
b^\mathsf{T}(V)(A(\mathbf{0})(c \otimes I)) = \phi_4(V) + \mathcal{O}(h^2) - 1.193 \times 10^{-13}I, \\
b^\mathsf{T}(V)((c \otimes I) \cdot (A(\mathbf{0})(e \otimes I))) = 3\phi_4(V) + \mathcal{O}(h^2) - 1.1927 \times 10^{-13}I, \\
b^\mathsf{T}(V)(c^4 \otimes I) = 24\phi_5(V) + \mathcal{O}(h), \\
b^\mathsf{T}(V)(A(\mathbf{0})(c^2 \otimes I)) = 2\phi_5(V) + \mathcal{O}(h) - 4.85 \times 10^{-14}I, \\
b^\mathsf{T}(V)(A(\mathbf{0})^2(e \otimes I)) = \phi_5(V) + \mathcal{O}(h), \\
b^\mathsf{T}(V)((c \otimes I) \cdot (A(\mathbf{0})(c \otimes I))) = 4\phi_5(V) + \mathcal{O}(h) - 1.258 \times 10^{-14}I, \\
b^\mathsf{T}(V)(A(\mathbf{0})(e \otimes I) \cdot (A(\mathbf{0})(e \otimes I))) = 6\phi_5(V) + \mathcal{O}(h) - 1.36 \times 10^{-14}I, \\
b^\mathsf{T}(V)((c^2 \otimes I) \cdot (A(\mathbf{0})(e \otimes I))) = 12\phi_5(V) + \mathcal{O}(h) - 4.577 \times 10^{-14}I, \\
b^\mathsf{T}(V)(A^{(2)}(\mathbf{0})(e \otimes I)) = -2\phi_5(V) + \mathcal{O}(h) + 2.2126 \times 10^{-14}I,
\end{cases} \tag{9.20}
$$

and

$$
\begin{cases}
\bar{b}^\mathsf{T}(V)(e \otimes I) = \phi_2(V) + \mathcal{O}(h^4), \\
\bar{b}^\mathsf{T}(V)(c \otimes I) = \phi_3(V) + \mathcal{O}(h^3), \\
\bar{b}^\mathsf{T}(V)(c^2 \otimes I) = 2\phi_4(V) + \mathcal{O}(h^2), \\
\bar{b}^\mathsf{T}(V)\big(A(\mathbf{0})(e \otimes I)\big) = \phi_4(V) + \mathcal{O}(h^2) - 9.29 \times 10^{-14}I, \\
\bar{b}^\mathsf{T}(V)(c^3 \otimes I) = 6\phi_5(V) + \mathcal{O}(h), \\
\bar{b}^\mathsf{T}(V)\big(A(\mathbf{0})(c \otimes I)\big) = \phi_5(V) + \mathcal{O}(h) - 1.067 \times 10^{-13}I, \\
\bar{b}^\mathsf{T}(V)\big((c \otimes I) \cdot \big(A(\mathbf{0})(e \otimes I)\big)\big) = 3\phi_5(V) + \mathcal{O}(h) - 7.35 \times 10^{-14}I.
\end{cases} \tag{9.21}
$$

It can be observed that (9.20) and (9.21) are the same as the fifth-order conditions (9.12) and (9.13) except for some small residuals. Therefore, the second explicit symplectic multi-frequency and multidimensional ERKN method of order five with some small residuals is constructed. The method with coefficients determined by (9.19), (9.11) and (9.15) is denoted by 2SMMERKN5s5. Meanwhile, when $V \rightarrow \mathbf{0}_{d \times d}$, this method reduces to a classical explicit symplectic RKN method (method 2 given in [11]) of order five with some residuals.

Remark 9.1 In order to achieve accurate five-stage explicit symplectic ERKN methods of order five, many different values of c_i were tested in an attempt to achieve high accuracy but it proved difficult to obtain an accurate fifth-order explicit symplectic ERKN method. In [11], the authors tried about 10000 different initial guesses and only obtained the five-stage explicit symplectic RKN methods of order five with some residuals in the order conditions. In addition, the authors in [11] speculated that those methods given by them would be the only methods with real coefficients considering the large number of initial guesses they tried. This would mean that the fifth-order methods given in [11] are not accurate fifth-order methods and there are no other five-stage explicit symplectic RKN methods of order five with real coefficients. However, because the norm of the residuals is very small, those RKN methods could be thought of as fifth-order methods for the purposes of practical computation. The situation encountered in the paper [11] is just the same as the situation in this chapter. Five-stage explicit symplectic multi-frequency and multidimensional ERKN methods of order five with some small residuals can be obtained. Moreover, from the results of the numerical experiments in the next section, it can be observed that the two explicit symplectic multi-frequency and multidimensional ERKN methods of order five with some small residuals have better accuracy and preserve the Hamiltonian much better than the fifth-order explicit symplectic RKN methods given in [11] and the third-order explicit symplectic ERKN methods given in [20].

It is known that when RKN-type methods are applied to solve oscillatory differential equations, they usually generate some dispersion and/or dissipation, even though these methods may be of high algebraic order. Therefore, the stability and phase properties of the two new ERKN methods are required to be analysed. The stability regions for the methods 1SMMERKN5s5 and 2SMMERKN5s5 derived in this chapter are depicted in Fig. 9.1. Meanwhile, it can be checked that $\det(S) = 1$ for the two new methods; therefore, both of them are zero dissipative. Their dispersion errors are as follows:

- 1SMMERKN5s5:

$$
\begin{aligned}
\phi(H) = \Bigg(&-0.004712004178867 + \frac{0. \times 10^{-60}\omega^{14}}{(\varepsilon + \omega^2)^7} - \frac{4. \times 10^{-44}\omega^{12}}{(\varepsilon + \omega^2)^6} \\
&+ \frac{0. \times 10^{-20}\omega^{10}}{(\varepsilon + \omega^2)^5} - \frac{0.003542437766148\omega^8}{(\varepsilon + \omega^2)^4} + \frac{0.014802401930786\omega^6}{(\varepsilon + \omega^2)^3} \\
&- \frac{0.023689494741995\omega^4}{(\varepsilon + \omega^2)^2} + \frac{0.017141534756223\omega^2}{\varepsilon + \omega^2} \Bigg) H^7 + \mathcal{O}(H^8);
\end{aligned}
$$

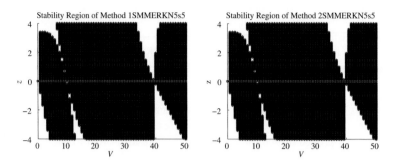

Fig. 9.1 Stability regions for the methods 1SMMERKN5s5 (*left*) and 2SMMERKN5s5 (*right*)

- 2SMMERKN5s5:

$$
\phi(H) = \left(-1.6 \times 10^{-13} + \frac{0. \times 10^{-30}\omega^6}{(\varepsilon + \omega^2)^3} - \frac{1.06 \times 10^{-13}\omega^4}{(\varepsilon + \omega^2)^2} + \frac{2.12 \times 10^{-13}\omega^2}{\varepsilon + \omega^2} \right) H^3
$$
$$
+ \left(5 \times 10^{-16} + \frac{0. \times 10^{-43}\omega^{10}}{(\varepsilon + \omega^2)^5} + \frac{5.6 \times 10^{-27}\omega^8}{(\varepsilon + \omega^2)^4} + \frac{2.16 \times 10^{-14}\omega^6}{(\varepsilon + \omega^2)^3} \right.
$$
$$
\left. - \frac{4.3 \times 10^{-14}\omega^4}{(\varepsilon + \omega^2)^2} + \frac{2.1 \times 10^{-14}\omega^2}{\varepsilon + \omega^2} \right) H^5
$$
$$
+ \left(-0.0047120041788978 + \frac{0. \times 10^{-56}\omega^{14}}{(\varepsilon + \omega^2)^7} - \frac{6. \times 10^{-40}\omega^{12}}{(\varepsilon + \omega^2)^6} + \frac{9.8 \times 10^{-28}\omega^{10}}{(\varepsilon + \omega^2)^5} \right.
$$
$$
- \frac{0.0035424377661716\omega^8}{(\varepsilon + \omega^2)^4} + \frac{0.014802401930890\omega^6}{(\varepsilon + \omega^2)^3} - \frac{0.023689494742163\omega^4}{(\varepsilon + \omega^2)^2}
$$
$$
\left. + \frac{0.017141534756343\omega^2}{\varepsilon + \omega^2} \right) H^7 + \mathcal{O}(H^8).
$$

These two expressions show that the method 1SMMERKN5s5 is dispersive of order six, and 2SMMERKN5s5 is dispersive of order six with small residuals since the coefficients of H^3 and H^5 in $\phi(H)$ of the method W2ESRKN5s5 contain terms with very small constants (smaller than 10^{-12}).

9.4   Numerical Experiments

In this section, three problems are used to show the quantitative behaviour of the new methods compared with some existing methods in the literature. The methods selected are

- C: The symmetric and symplectic method of order two given in [17] of Chap. 6;
- E: The symmetric method of order two given in [6];
- SRKN3s4: The three-stage symplectic RKN method of order four given in [8, 14];
- ARKN4s5: The four-stage explicit ARKN method of order five given in [19];

- M2SRKN5s5: The method 2 (five-stage explicit symplectic RKN method of order five with some small residuals) given in [11];
- M4SRKN5s5: The method 4 (five-stage explicit symplectic RKN method of order five with some small residuals) given in [11];
- SMMERKN3s3: The three-stage explicit symplectic multi-frequency and multi-dimensional ERKN method of order three given in [20];
- 1SMMERKN5s5: The five-stage explicit symplectic multi-frequency and multi-dimensional ERKN method of order five with some small residuals given in this chapter;
- 2SMMERKN5s5: The five-stage explicit symplectic multi-frequency and multi-dimensional ERKN method of order five with some small residuals given in this chapter.

The numerical experiments were carried out on a PC and the algorithm implemented using MATLAB-R2009a.

Problem 9.1 Consider the model for stellar orbits in a galaxy in Problem 2.4 of Chap. 2.

The system is integrated on the interval $[0, 1000]$ with the stepsizes $h = 1/(5 \times 2^j)$ for C and E, $h = 3/(5 \times 2^j)$ for SERKN3s3 and SRKN3s4, $h = 4/(5 \times 2^j)$ for ARKN4s5, and $h = 1/2^j$ for the other methods, where $j = 1, \ldots, 4$. The efficiency curves (accuracy versus the computational cost measured by the number of function evaluations required by each method) are presented in Fig. 9.2a. Then, we solve this problem on the interval $[0, 1000]$ with the stepsizes $h = 1/2^j$, $j = 3, \ldots, 6$, and plot the global errors versus the CPU time in Fig. 9.2b. Finally, we integrate this problem with the stepsize $h = 1/4$ on the intervals $[0, t_{end}]$ with $t_{end} = 10^i$ for $i = 1, \ldots, 4$. The global error for the energy (the logarithm of the maximum global error of the Hamiltonian GEH versus the logarithm of time) is shown in Fig. 9.2c.

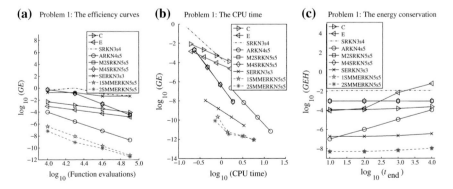

Fig. 9.2 Results for Problem 9.1. **a** The logarithm of the global error (GE) against the number of function evaluations. **b** The logarithm of the global error (GE) against the CPU time. **c** The logarithm of the maximum global error of Hamiltonian $GEH = \max |H_n - H_0|$ against $\log_{10}(t_{end})$

Problem 9.2 Consider the nonlinear Klein-Gordon equation [9]

$$\frac{\partial^2 u}{\partial t^2} - \frac{\partial^2 u}{\partial x^2} = -u^3 - u, \quad 0 < x < L, \quad t > 0,$$

$$u(x, 0) = A\left(1 + \cos\left(\frac{2\pi}{L}x\right)\right), \quad u_t(x, 0) = 0, \quad u(0, t) = u(L, t),$$

where $L = 1.28, A = 0.9$. A semi-discretization of the spatial variable using second-order symmetric differences yields the following system of second-order ODEs in time:

$$\frac{\mathrm{d}^2 U}{\mathrm{d}t^2} + MU = F(U), \quad 0 < t \le t_{\text{end}},$$

where $U(t) = \left(u_1(t), \ldots, u_d(t)\right)^{\mathsf{T}}$ with $u_i(t) \approx u(x_i, t), i = 1, \ldots, d$,

$$M = \frac{1}{\Delta x^2}\begin{pmatrix} 2 & -1 & & & -1 \\ -1 & 2 & -1 & & \\ & \ddots & \ddots & \ddots & \\ & & -1 & 2 & -1 \\ -1 & & & -1 & 2 \end{pmatrix}_{d \times d}$$

with $\Delta x = L/d$, $x_i = i\Delta x$ and $F(U) = \left(-u_1^3 - u_1, \ldots, -u_d^3 - u_d\right)^{\mathsf{T}}$. The corresponding Hamiltonian of this system is

$$H(U', U) = \frac{1}{2}U'^{\mathsf{T}}U' + \frac{1}{2}U^{\mathsf{T}}MU + \frac{1}{4}\sum_{i=1}^{d}\left(2u_i^2 + u_i^4\right).$$

Take the initial conditions

$$U(0) = \left(0.9\left(1 + \cos\left(\frac{2\pi i}{d}\right)\right)\right)_{i=1}^{d}, \quad U'(0) = (0)_{i=1}^{d}$$

with $d = 32$. The problem is integrated on the interval $[0, 100]$ with the stepsizes $h = 1/(5 \times 2^j)$ for C and E, $h = 3/(5 \times 2^j)$ for SERKN3s3 and SRKN3s4, $h = 4/(5 \times 2^j)$ for ARKN4s5, and $h = 1/2^j$ for the other methods, where $j = 5, \ldots, 8$. Figure 9.3a shows the results. Then this problem is solved on the interval $[0, 100]$ with the stepsizes $h = 1/2^j$, $j = 5, \ldots, 8$ and the global errors versus the CPU time are presented in Fig. 9.3b. Finally, this problem is integrated with the stepsize $h = 1/20$ on the intervals $[0, 10^i]$ with $i = 0, \ldots, 3$. The global errors for the energy are shown in Fig. 9.3c. It is noted that some of the errors of SRKN3s4 in Fig. 9.3c are very large and we do not plot the corresponding points.

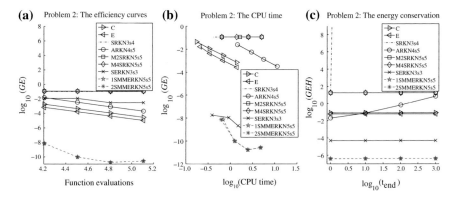

Fig. 9.3 Results for Problem 9.2. **a** The logarithm of the global error (GE) against the number of function evaluations. **b** The logarithm of the global error (GE) against the CPU time. **c** The logarithm of the maximum global error of Hamiltonian $GEH = \max |H_n - H_0|$ against $\log_{10}(t_{\text{end}})$

Problem 9.3 Consider the Fermi–Pasta–Ulam Problem of Chap. 2.

We choose $m = 3$, $\omega = 100$ with the initial conditions

$$q_1(0) = 1, \ p_1(0) = 1, \ q_4(0) = \frac{1}{\omega}, \ p_4(0) = 1,$$

and zero for the remaining initial values.

Figure 9.4a displays the efficiency curves on the interval $[0, 100]$ with the stepsizes $h = 1/(500 \times 2^j)$ for C and E, $h = 3/(500 \times 2^j)$ for SERKN3s3 and SRKN3s4, $h = 4/(500 \times 2^j)$ for ARKN4s5, and $h = 1/(100 \times 2^j)$ for the other methods, where $j = 0, \ldots, 3$. This problem is solved on the interval $[0, 100]$ with the stepsizes

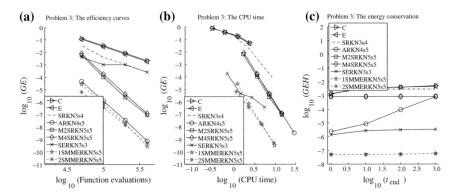

Fig. 9.4 Results for Problem 9.3. **a** The logarithm of the global error (GE) against the number of function evaluations. **b** The logarithm of the global error (GE) against the CPU time. **c** The logarithm of the maximum global error of Hamiltonian $GEH = \max |H_n - H_0|$ against $\log_{10}(t_{\text{end}})$

$h = 1/(100 \times 2^j)$, $j = 0, \ldots, 3$ and the global errors versus the CPU time are shown in Fig. 9.4b. Finally, we integrate this problem with the stepsize $h = 1/100$ on the intervals $[0, 10^i]$ with $i = 0, \ldots, 3$. The global errors of the Hamiltonian are shown in Fig. 9.4c.

It may be concluded from the results of the three numerical experiments carried out in this chapter that the two new methods for the same numbers of function evaluations have better accuracy and that the global errors GEH nearly do not increase as t_{end} increases. It also can be observed that the values of the global errors GEH of the two new methods are much smaller than those of the other methods. Consequently, the new methods preserve the symplectic structure of the Hamiltonian system (9.2) and approximately preserve the energy very well.

9.5   Conclusions and Discussions

In this chapter, the symplecticity conditions and fifth-order conditions of ERKN methods were used to construct two five-stage explicit symplectic ERKN methods of order five having small residuals for the multi-frequency and multidimensional oscillatory Hamiltonian system (9.2). The stability and phase properties of the two new explicit symplectic ERKN methods were also analysed. Numerical experiments were conducted to compare these methods with existing methods in the scientific literature. The numerical results demonstrate that the new explicit symplectic multi-frequency ERKN methods of order five with some small residuals are very efficient and conserve energy well.

It seems that in using the order conditions, we can achieve higher-order explicit symplectic and symmetric multi-frequency ERKN integrators based directly on the coupled conditions derived in Sect. 2.5. However, it should be noted that, generally speaking, it is not so easy to achieve the purpose since so many algebraic conditions (order, symplectic and symmetric conditions) must be satisfied simultaneously. It is noted that high-order composition methods based on the ARKN methods and ERKN methods may be worth consideration and an investigation can be found in a very recent paper by Liu et al. [10].

The material of this chapter is based on the work by Wang and Wu [18].

References

1. Calvo MP, Sanz-Serna JM (1992) Order conditions for canonical Runge-Kutta-Nyström methods. BIT Numer Math 32:131–142
2. Cohen D, Hairer E, Lubich C (2005) Numerical energy conservation for multi-frequency oscillatory differential equations. BIT Numer Math 45:287–305
3. Franco JM (2004) Exponentially fitted explicit Runge-Kutta-Nyström methods. J Comput Appl Math 167:1–19

4. Franco JM (2006) New methods for oscillatory systems based on ARKN methods. Appl Numer Math 56:1040–1053

5. Hochbruck M, Lubich C (1999) A Gautschi-type method for oscillatory second-order differential equations. Numer Math 83:403–426

6. Hairer E, Lubich C (2000) Long-time energy conservation of numerical methods for oscillatory differential equations. SIAM J Numer Anal 38:414–441

7. Hairer E, Lubich C, Wanner G (2006) Geometric numerical integration: structure-preserving algorithms for odinary differential equations, 2nd edn. Springer, Berlin

8. Hairer E, Nørsett SP, Wanner G (1993) Solving ordinary differential equations I: nonstiff problems. Springer, Berlin

9. Jimánez S, Vázquez L (1990) Analysis of four numerical schemes for a nonlinear Klein-Gordon equation. Appl Math Comput 35:61–93

10. Liu K, Wu X, High-order symplectic and symmetric composition methods for multifrequency and multi-dimensional oscillatory Hamiltonian systems. J Comput Math [in press]

11. Okunbor D, Skeel RD (1994) Canonical Runge-Kutta-Nyström methods of order 5 and 6. J Comput Appl Math 51:375–382

12. Ruth RD (1983) A canonical integration technique. IEEE Trans Nuclear Sci NS-30(4): 2669–2671

13. Simos TE, Vigo-Aguiar J (2003) Exponentially fitted symplectic integrator. Phys Rev E 67:016701–016707

14. Sanz-Serna JM, Calvo MP (1994) Numerical Hmiltonian problem. Applied mathematics and matematical computation, vol 7. Chapman and Hall, London

15. Van der Houwen PJ, Sommeijer BP (1987) Explicit Runge-Kutta(-Nyström) methods with reduced phase errors for computing oscillating solution. SIAM J Numer Anal 24:595–617

16. Van de Vyver H (2005) A symplectic exponentially fitted modified Runge-Kutta-Nyström methods for the numerical integration of orbital problems. New Astron 10:261–269

17. Van de Vyver H (2009) Scheifele two-step methods for perturbed oscillators. J Comput Appl Math 224:415–432

18. Wang B, Wu X (2014) A highly accurate explicit symplectic ERKN method for multi-frequency and multidimensional oscillatory Hamiltonian systems. Numer Algo 65:705–721

19. Wu X, Wang B (2010) Multidimensional adapted Runge-Kutta-Nyström methods for oscillatory systems. Comput Phys Commun 181:1955–1962

20. Wu X, Wang B, Xia J (2012) Explicit symplectic multidimensional exponential fitting modified Runge-Kutta-Nyström methods. BIT Numer Math 52:773–795

21. Wu X, You X, Shi W, Wang B (2010) ERKN integrators for systems of oscillatory second-order differential equations. Comput Phys Commun 181:1873–1887

22. Wu X, You X, Wang B (2013) Structure-preserving algorithms for oscillatory differential equations. Springer, Berlin

23. Wu X, You X, Xia J (2009) Order conditions for ARKN methods solving oscillatory systems. Comput Phys Commun 180:2250–2257

24. Yoshida H (1990) Construction of high order symplectic integrators. Phys Lett A 150:262–268

Chapter 10
Multidimensional ARKN Methods for General Multi-frequency Oscillatory Second-Order IVPs

Based on B-series theory, the order conditions of the multidimensional ARKN methods were presented by Wu et al. [18] for general multi-frequency oscillatory second-order initial value problems where the functions on right-hand side *depend on both position and velocity*. The class of physical problems which fall within its scope is broader. These multidimensional ARKN methods integrate unperturbed oscillators exactly. This chapter focuses mainly on the analysis of concrete multidimensional ARKN methods.

10.1 Motivation

The multiscale simulation of differential equations with multi-frequency oscillatory solutions have received a great deal of attention in the last few years. These problems are frequently encountered in many fields of applied sciences and engineering, such as celestial mechanics, theoretical physics, chemistry, electronics and control engineering. Some progress in theoretical and numerical research has been made on the modelling and simulation of these oscillations (see, e.g. [1–3, 9, 11]). A typical topic is the numerical integration of a multi-frequency and multidimensional oscillatory system associated with an initial value problem of the form

$$\begin{cases} y'' + My = f(y, y'), & t \in [t_0, T], \\ y(t_0) = y_0, & y'(t_0) = y_0', \end{cases} \tag{10.1}$$

where $M \in \mathbb{R}^{d \times d}$ is a positive semi-definite matrix (stiffness matrix) that implicitly preserves the dominant frequencies of the oscillatory problem, and $f : \mathbb{R}^d \times \mathbb{R}^d \to \mathbb{R}^d$, $y \in \mathbb{R}^d$, $y' \in \mathbb{R}^d$.

© Springer-Verlag Berlin Heidelberg and Science Press, Beijing, China 2015   211
X. Wu et al., *Structure-Preserving Algorithms for Oscillatory
Differential Equations II*, DOI 10.1007/978-3-662-48156-1_10

It has been widely believed in numerical simulation that, for special classes of problems with a structure of particular interest, numerical algorithms should respect that structure. An outstanding advantage of multidimensional ARKN methods for (10.1) is that their updates incorporate the special structure of (10.1) inherent in the linear term My, so that they naturally integrate the multi-frequency and multidimensional unperturbed problem $y'' + My = 0$. For research on this topic, readers are referred to [4–7, 12, 14, 16–21] as well as the references contained therein.

Multidimensional ARKN methods were constructed in [17] for oscillatory systems where the functions on the right-hand side are independent of y'. This chapter presents the analysis of concrete multidimensional ARKN methods for multi-frequency oscillatory systems with right-hand-side functions depending on both y and y'. This is accomplished using the order conditions for the ARKN methods derived by Wu et al. [19] and based on the Nyström tree theory. The construction of multidimensional ARKN methods for the multi-frequency and multidimensional oscillatory system (10.1) will be discussed in detail. An analysis of the stability and phase properties will also be made as well.

10.2   Multidimensional ARKN Methods and the Corresponding Order Conditions

This chapter begins with the matrix-variation-of-constants formula.

If $f : \mathbb{R}^d \times \mathbb{R}^d \to \mathbb{R}^d$ in (10.1) is continuous, from Theorem 1.1 in Chap. 1, the exact solution and its derivative for the multi-frequency oscillatory system (10.1) satisfy the matrix-variation-of-constants formula:

$$
\begin{cases}
y(t) = \phi_0\big((t-t_0)^2 M\big) y_0 + (t-t_0)\phi_1\big((t-t_0)^2 M\big) y_0' \\
\qquad + \displaystyle\int_{t_0}^t (t-\zeta)\phi_1\big((t-\zeta)^2 M\big) \hat{f}(\zeta) \mathrm{d}\zeta, \\
y'(t) = -(t-t_0) M \phi_1\big((t-t_0)^2 M\big) y_0 + \phi_0\big((t-t_0)^2 M\big) y_0' \\
\qquad + \displaystyle\int_{t_0}^t \phi_0\big((t-\zeta)^2 M\big) \hat{f}(\zeta) \mathrm{d}\zeta,
\end{cases}
\tag{10.2}
$$

for any real number t_0, t, where $\hat{f}(\zeta) = f\big(y(\zeta), y'(\zeta)\big)$.

To obtain a numerical integrator for (10.1), we approximate the integrals by quadrature formulae. This leads to the following multidimensional ARKN methods proposed in [19] for (10.1).

Definition 10.1 An s-stage multidimensional ARKN scheme for the numerical integration of the initial value problem (10.1) is defined by

$$
\begin{cases}
Y_i = y_n + hc_i y_n' + h^2 \sum_{j=1}^{s} \bar{a}_{ij} \big(f(Y_j, Y_j') - MY_j \big), & i = 1, \ldots, s, \\[2mm]
Y_i' = y_n' + h \sum_{j=1}^{s} a_{ij} \big(f(Y_j, Y_j') - MY_j \big), & i = 1, \ldots, s, \\[2mm]
y_{n+1} = \phi_0(V) y_n + h\phi_1(V) y_n' + h^2 \sum_{i=1}^{s} \bar{b}_i(V) f(Y_i, Y_i'), \\[2mm]
y_{n+1}' = \phi_0(V) y_n' - hM\phi_1(V) y_n + h \sum_{i=1}^{s} b_i(V) f(Y_i, Y_i'),
\end{cases}
\tag{10.3}
$$

where $\bar{a}_{ij}$ and a_{ij} are real constants, the weight functions $b_i(V) : \mathbb{R}^{d \times d} \to \mathbb{R}^{d \times d}$ and $\bar{b}_i(V) : \mathbb{R}^{d \times d} \to \mathbb{R}^{d \times d}$, $i = 1, \ldots, s$, in the updates are matrix-valued functions of $V = h^2 M$. The scheme (10.3) can also be denoted by the following Butcher tableau:

$$
\begin{array}{c|c|c}
c & A & \bar{A} \\ \hline
& b^{\mathsf{T}}(V) & \bar{b}^{\mathsf{T}}(V)
\end{array}
=
\begin{array}{c|ccc|ccc}
c_1 & a_{11} & \cdots & a_{1s} & \bar{a}_{11} & \cdots & \bar{a}_{1s} \\
\vdots & \vdots & \ddots & \vdots & \vdots & \ddots & \vdots \\
c_s & a_{s1} & \cdots & a_{ss} & \bar{a}_{s1} & \cdots & \bar{a}_{ss} \\ \hline
& b_1(V) & \cdots & b_s(V) & \bar{b}_1(V) & \cdots & \bar{b}_s(V)
\end{array}
$$

In the special case where the right-hand-side function of (10.1) is independent of y', the scheme (10.3) reduces to the corresponding ARKN method for the special oscillatory system [17],

$$
\begin{cases}
y'' + My = f(y), & t \in [t_0, T], \\
y(t_0) = y_0, & y'(t_0) = y_0',
\end{cases}
\tag{10.4}
$$

namely,

$$
\begin{cases}
Y_i = y_n + hc_i y_n' + h^2 \sum_{j=1}^{s} \bar{a}_{ij} \big(f(Y_j) - MY_j \big), & i = 1, \ldots, s, \\[2mm]
y_{n+1} = \phi_0(V) y_n + h\phi_1(V) y_n' + h^2 \sum_{i=1}^{s} \bar{b}_i(V) f(Y_i), \\[2mm]
y_{n+1}' = \phi_0(V) y_n' - hM\phi_1(V) y_n + h \sum_{i=1}^{s} b_i(V) f(Y_i).
\end{cases}
\tag{10.5}
$$

The scheme (10.5) can also be denoted by the following Butcher tableau:

$$
\begin{array}{c|c}
c & \bar{A} \\
\hline
\bar{b}^{\mathsf{T}}(V) \\
\hline
b^{\mathsf{T}}(V)
\end{array}
\;=\;
\begin{array}{c|ccc}
c_1 & \bar{a}_{11} & \cdots & \bar{a}_{1s} \\
\vdots & \vdots & \ddots & \vdots \\
c_s & \bar{a}_{s1} & \cdots & \bar{a}_{ss} \\
\hline
& \bar{b}_1(V) & \cdots & \bar{b}_s(V) \\
\hline
& b_1(V) & \cdots & b_s(V)
\end{array}
$$

A multidimensional ARKN method (10.3) is said to be of *order* p if, for a sufficiently smooth problem (10.1), the conditions

$$
\begin{aligned}
e_{n+1} &:= y_{n+1} - y(t_n + h) = \mathcal{O}(h^{p+1}) \\
e'_{n+1} &:= y'_{n+1} - y'(t_n + h) = \mathcal{O}(h^{p+1})
\end{aligned}
\tag{10.6}
$$

are satisfied simultaneously, where $y(t_n + h)$ and $y'(t_n + h)$ are the respective exact solution and its derivative of (10.1) at $t_n + h$, whereas y_{n+1} and y'_{n+1} are the one-step numerical approximations obtained by the method from the local assumptions: $y_n = y(t_n)$ and $y'_n = y'(t_n)$. The order conditions, established in [19] for the method (10.3), are given below.

Theorem 10.1 (Wu et al. [19]) *A necessary and sufficient condition for a multidimensional ARKN method (10.3) to be of order p is given by*

$$
\begin{cases}
\left(\Phi(\tau)^{\mathsf{T}} \otimes I_m\right)\bar{b}(V) - \dfrac{\rho(\tau)!}{\gamma(\tau)}\phi_{\rho(\tau)+1}(V) = \mathcal{O}(h^{p-\rho(\tau)}), & \rho(\tau) = 1, \ldots, p-1, \\[2ex]
\left(\Phi(\tau)^{\mathsf{T}} \otimes I_m\right)b(V) - \dfrac{\rho(\tau)!}{\gamma(\tau)}\phi_{\rho(\tau)}(V) = \mathcal{O}(h^{p+1-\rho(\tau)}), & \rho(\tau) = 1, \ldots, p,
\end{cases}
$$

where τ is the Nyström tree associated with the elementary differential $\mathscr{F}(\tau)(y_n, y'_n)$ of the function $\tilde{f}(y, y') = f(y, y') - My$ at (y_n, y'_n), the entries $\Phi_j(\tau)$, $j = 1, \ldots, s$, of $\Phi(\tau)$ are weight functions defined in [8], and $\phi_{\rho(\tau)}(V)$, $\rho(\tau) = 1, \ldots, p$, are given by (4.7).

10.3 ARKN Methods for General Multi-frequency and Multidimensional Oscillatory Systems

This section constructs three explicit multidimensional ARKN schemes for the general multi-frequency and multidimensional oscillatory system (10.1) and analyses the stability and phase properties of the schemes.

10.3.1 Construction of Multidimensional ARKN Methods

First, consider an explicit three-stage ARKN method of order three. Since f depends on both y and y' in (10.1), at least three stages are required to satisfy all the order conditions. The explicit three-stage ARKN methods for (10.1) can be expressed by the following Butcher tableau:

$$
\begin{array}{c|cc}
c & A & \bar{A} \\
\hline
& b^{\mathsf{T}}(V) & \bar{b}^{\mathsf{T}}(V)
\end{array}
=
\begin{array}{c|cccccc}
c_1 & 0 & 0 & 0 & 0 & 0 & 0 \\
c_2 & a_{21} & 0 & 0 & \bar{a}_{21} & 0 & 0 \\
c_3 & a_{31} & a_{32} & 0 & \bar{a}_{31} & \bar{a}_{32} & 0 \\
\hline
& b_1(V) & b_2(V) & b_3(V) & \bar{b}_1(V) & \bar{b}_2(V) & \bar{b}_3(V)
\end{array}
$$

From Theorem 10.1, a three-stage ARKN method is of order three if its coefficients satisfy

$$
\begin{cases}
(e^{\mathsf{T}} \otimes I)b(V) = \phi_1(V) + \mathcal{O}(h^3), & (c^{\mathsf{T}} \otimes I)b(V) = \phi_2(V) + \mathcal{O}(h^2), \\
((c^2)^{\mathsf{T}} \otimes I)b(V) = 2\phi_3(V) + \mathcal{O}(h), & (e^{\mathsf{T}} \otimes I)\bar{b}(V) = \phi_2(V) + \mathcal{O}(h^2), \\
(c^{\mathsf{T}} \otimes I)\bar{b}(V) = \phi_3(V) + \mathcal{O}(h), & ((\bar{A}e)^{\mathsf{T}} \otimes I)b(V) = \phi_3(V) + \mathcal{O}(h), \\
((Ae)^{\mathsf{T}} \otimes I)\bar{b}(V) = \phi_3(V) + \mathcal{O}(h), & ((Ae)^{\mathsf{T}} \otimes I)b(V) = \phi_2(V) + \mathcal{O}(h^2), \\
((Ac)^{\mathsf{T}} \otimes I)b(V) = \phi_3(V) + \mathcal{O}(h), & ((c \cdot Ae)^{\mathsf{T}} \otimes I)b(V) = 2\phi_3(V) + \mathcal{O}(h), \\
((A^2e)^{\mathsf{T}} \otimes I)b(V) = \phi_3(V) + \mathcal{O}(h), & ((Ae \cdot Ae)^{\mathsf{T}} \otimes I)b(V) = 2\phi_3(V) + \mathcal{O}(h),
\end{cases}
\tag{10.7}
$$

where $e = (1, 1, 1)^{\mathsf{T}}$.

Choosing $c_1 = 0$, $c_2 = \dfrac{1}{2}$, $c_3 = 1$, and solving all the equations in (10.7), we obtain

$$
\begin{cases}
a_{21} = \dfrac{1}{2}, \quad a_{31} = -1, \quad a_{32} = 2, \\[2mm]
\bar{a}_{21} = \dfrac{1}{8}, \quad \bar{a}_{31} = \dfrac{1}{2}, \quad \bar{a}_{32} = 0,
\end{cases}
\tag{10.8}
$$

and

$$
\begin{cases}
b_1(V) = \phi_1(V) - 3\phi_2(V) + 4\phi_3(V), & \bar{b}_1(V) = \phi_2(V) - \dfrac{3}{2}\phi_3(V), \\[2mm]
b_2(V) = 4\phi_2(V) - 8\phi_3(V), & \bar{b}_2(V) = \phi_3(V), \\[2mm]
b_3(V) = -\phi_2(V) + 4\phi_3(V), & \bar{b}_3(V) = \dfrac{1}{2}\phi_3(V).
\end{cases}
\tag{10.9}
$$

Equations (10.8)–(10.9) give a three-stage ARKN method of order three. The Taylor expansions of the coefficients $b_i(V)$ and $\bar{b}_i(V)$ of this method are given by

$$
\begin{cases}
b_1(V) = \dfrac{1}{6}I - \dfrac{3}{40}V + \dfrac{5}{1008}V^2 - \dfrac{7}{51840}V^3 + \dfrac{1}{492800}V^4 + \cdots, \\[2mm]
b_2(V) = \dfrac{2}{3}I - \dfrac{1}{10}V + \dfrac{1}{252}V^2 - \dfrac{1}{12960}V^3 + \dfrac{1}{1108800}V^4 + \cdots, \\[2mm]
b_3(V) = \dfrac{1}{6}I + \dfrac{1}{120}V - \dfrac{1}{1680}V^2 + \dfrac{1}{72576}V^3 - \dfrac{1}{5702400}V^4 + \cdots, \\[2mm]
\bar{b}_1(V) = \dfrac{1}{4}I - \dfrac{7}{240}V + \dfrac{11}{10080}V^2 - \dfrac{1}{48348}V^3 + \dfrac{19}{79833600}V^4 + \cdots, \\[2mm]
\bar{b}_2(V) = \dfrac{1}{6}I - \dfrac{1}{120}V + \dfrac{1}{5040}V^2 - \dfrac{1}{362880}V^3 + \dfrac{1}{39916800}V^4 + \cdots, \\[2mm]
\bar{b}_3(V) = \dfrac{1}{12}I - \dfrac{1}{240}V + \dfrac{1}{10080}V^2 - \dfrac{1}{725760}V^3 + \dfrac{1}{79833600}V^4 + \cdots.
\end{cases}
$$
$$(10.10)$$

This method is denoted by ARKN3s3.

We next consider the explicit multidimensional ARKN methods of order four for (10.1). In this case, four stages are needed. An explicit four-stage ARKN method can be denoted by the following Butcher tableau:

$$
\begin{array}{c|c c}
c & A & \bar{A} \\
\hline
 & b^{\mathsf{T}}(V) & \bar{b}^{\mathsf{T}}(V)
\end{array}
=
\begin{array}{c|cccc|cccc}
c_1 & 0 & 0 & 0 & 0 & 0 & 0 & 0 & 0 \\
c_2 & a_{21} & 0 & 0 & 0 & \bar{a}_{21} & 0 & 0 & 0 \\
c_3 & a_{31} & a_{32} & 0 & 0 & \bar{a}_{31} & \bar{a}_{32} & 0 & 0 \\
c_4 & a_{41} & a_{42} & a_{43} & 0 & \bar{a}_{41} & \bar{a}_{42} & \bar{a}_{43} & 0 \\
\hline
 & b_1(V) & b_2(V) & b_3(V) & b_4(V) & \bar{b}_1(V) & \bar{b}_2(V) & \bar{b}_3(V) & \bar{b}_4(V)
\end{array}
$$

The simplifying conditions $Ae = c$, $\bar{A} = A^2$ are used to reduce the order to those listed below,

$$
\begin{cases}
(e^{\mathsf{T}} \otimes I)b(V) = \phi_1(V) + \mathcal{O}(h^4), & (c^{\mathsf{T}} \otimes I)b(V) = \phi_2(V) + \mathcal{O}(h^3), \\
((c^2)^{\mathsf{T}} \otimes I)b(V) = 2\phi_3(V) + \mathcal{O}(h^2), & ((Ac)^{\mathsf{T}} \otimes I)b(V) = \phi_3(V) + \mathcal{O}(h^2), \\
((A^2c)^{\mathsf{T}} \otimes I)b(V) = \phi_4(V) + \mathcal{O}(h), & ((Ac^2)^{\mathsf{T}} \otimes I)b(V) = 3\phi_4(V) + \mathcal{O}(h), \\
((c^3)^{\mathsf{T}} \otimes I)b(V) = 6\phi_4(V) + \mathcal{O}(h), & ((c \cdot Ac)^{\mathsf{T}} \otimes I)b(V) = 2\phi_4(V) + \mathcal{O}(h), \\
(e^{\mathsf{T}} \otimes I)\bar{b}(V) = \phi_2(V) + \mathcal{O}(h^3), & (c^{\mathsf{T}} \otimes I)\bar{b}(V) = \phi_3(V) + \mathcal{O}(h^2), \\
((c^2)^{\mathsf{T}} \otimes I)\bar{b}(V) = 2\phi_4(V) + \mathcal{O}(h), & ((Ac)^{\mathsf{T}} \otimes I)\bar{b}(V) = \phi_4(V) + \mathcal{O}(h),
\end{cases}
$$
$$(10.11)$$

where $e = (1, 1, 1, 1)^{\mathsf{T}}$.

Choosing $c_1 = 0$, $c_2 = \dfrac{1}{2}$, $c_3 = \dfrac{1}{2}$, $c_4 = 1$, and solving the above equations, we obtain

$$
\begin{cases}
a_{21} = \dfrac{1}{2}, \ a_{31} = 0, \ a_{32} = \dfrac{1}{2}, \ a_{41} = 0, \ a_{42} = 0, \ a_{43} = 1, \\[2mm]
\bar{a}_{21} = 0, \ \bar{a}_{31} = \dfrac{1}{4}, \ \bar{a}_{32} = 0, \ \bar{a}_{41} = 0, \ \bar{a}_{42} = \dfrac{1}{2}, \ \bar{a}_{43} = 0,
\end{cases}
\tag{10.12}
$$

and

$$
\begin{cases}
b_1(V) = \phi_1(V) - 3\phi_2(V) + 4\phi_3(V), \ \bar{b}_1(V) = \phi_2(V) - 3\phi_3(V) + 4\phi_4(V), \\
b_2(V) = 2\phi_2(V) - 4\phi_3(V), \quad\quad\quad\ \ \bar{b}_2(V) = 2\phi_3(V) - 4\phi_4(V), \\
b_3(V) = 2\phi_2(V) - 4\phi_3(V), \quad\quad\quad\ \ \bar{b}_3(V) = 2\phi_3(V) - 4\phi_4(V), \\
b_4(V) = -\phi_2(V) + 4\phi_3(V), \quad\quad\quad\ \bar{b}_4(V) = -\phi_3(V) + 4\phi_4(V).
\end{cases}
\tag{10.13}
$$

Equations (10.12)–(10.13) give a four-stage ARKN method of order four. The Taylor expansions of the coefficients $b_i(V)$ and $\bar{b}_i(V)$ of this method are

$$
\begin{cases}
b_1(V) = \dfrac{1}{6}I - \dfrac{3}{40}V + \dfrac{5}{1008}V^2 - \dfrac{7}{51840}V^3 + \cdots, \\[2mm]
b_2(V) = \dfrac{1}{3}I - \dfrac{1}{20}V + \dfrac{1}{504}V^2 - \dfrac{1}{25920}V^3 + \cdots, \\[2mm]
b_3(V) = \dfrac{1}{3}I - \dfrac{1}{20}V + \dfrac{1}{504}V^2 - \dfrac{1}{25920}V^3 + \cdots, \\[2mm]
b_4(V) = \dfrac{1}{6}I + \dfrac{1}{120}V - \dfrac{1}{1680}V^2 + \dfrac{1}{72576}V^3 + \cdots, \\[2mm]
\bar{b}_1(V) = \dfrac{1}{6}I - \dfrac{1}{45}V + \dfrac{1}{1120}V^2 - \dfrac{1}{56700}V^3 + \cdots, \\[2mm]
\bar{b}_2(V) = \dfrac{1}{6}I - \dfrac{1}{90}V + \dfrac{1}{3360}V^2 - \dfrac{1}{226800}V^3 + \cdots, \\[2mm]
\bar{b}_3(V) = \dfrac{1}{6}I - \dfrac{1}{90}V + \dfrac{1}{3360}V^2 - \dfrac{1}{226800}V^3 + \cdots, \\[2mm]
\bar{b}_4(V) = \dfrac{1}{360}V - \dfrac{1}{10080}V^2 + \dfrac{1}{604800}V^3 + \cdots.
\end{cases}
\tag{10.14}
$$

This method is denoted by ARKN4s4.

In what follows, the construction of the ARKN methods of order five for (10.1) is considered. In this case, at least six stages are required. Thus, consider the following Butcher tableau:

c_1	0	0	0	0	0	0	0	0	0	0	0	0
c_2	a_{21}	0	0	0	0	0	$\bar{a}_{21}$	0	0	0	0	0
c_3	a_{31}	a_{32}	0	0	0	0	$\bar{a}_{31}$	$\bar{a}_{32}$	0	0	0	0
c_4	a_{41}	a_{42}	a_{43}	0	0	0	$\bar{a}_{41}$	$\bar{a}_{42}$	$\bar{a}_{43}$	0	0	0
c_5	a_{51}	a_{52}	a_{53}	a_{54}	0	0	$\bar{a}_{51}$	$\bar{a}_{52}$	$\bar{a}_{53}$	$\bar{a}_{54}$	0	0
c_6	a_{61}	a_{62}	a_{63}	a_{64}	a_{65}	0	$\bar{a}_{61}$	$\bar{a}_{62}$	$\bar{a}_{63}$	$\bar{a}_{64}$	$\bar{a}_{65}$	0
	$b_1(V)$	$b_2(V)$	$b_3(V)$	$b_4(V)$	$b_5(V)$	$b_6(V)$	$\bar{b}_1(V)$	$\bar{b}_2(V)$	$\bar{b}_3(V)$	$\bar{b}_4(V)$	$\bar{b}_5(V)$	$\bar{b}_6(V)$

The order conditions up to order five are imposed using the simplifying assumptions $Ae = c$, $\bar{A} = A^2$ and the coefficients must then satisfy

$$
\begin{cases}
(e^\mathsf{T} \otimes I)b(V) = \phi_1(V) + \mathcal{O}(h^5), & (c^\mathsf{T} \otimes I)b(V) = \phi_2(V) + \mathcal{O}(h^4), \\
((c^2)^\mathsf{T} \otimes I)b(V) = 2\phi_3(V) + \mathcal{O}(h^3), & ((Ac)^\mathsf{T} \otimes I)b(V) = \phi_3(V) + \mathcal{O}(h^3), \\
((A^2c)^\mathsf{T} \otimes I)b(V) = \phi_4(V) + \mathcal{O}(h^2), & ((Ac^2)^\mathsf{T} \otimes I)b(V) = 2\phi_4(V) + \mathcal{O}(h^2), \\
((c^3)^\mathsf{T} \otimes I)b(V) = 6\phi_4(V) + \mathcal{O}(h^2), & ((c \cdot Ac)^\mathsf{T} \otimes I)b(V) = 3\phi_4(V) + \mathcal{O}(h^2), \\
((A^3c)^\mathsf{T} \otimes I)b(V) = 6\phi_5(V) + \mathcal{O}(h), & ((c \cdot A^2c)^\mathsf{T} \otimes I)b(V) = 4\phi_5(V) + \mathcal{O}(h), \\
((c^4)^\mathsf{T} \otimes I)b(V) = 24\phi_5(V) + \mathcal{O}(h), & ((c^2 \cdot Ac)^\mathsf{T} \otimes I)b(V) = 12\phi_5(V) + \mathcal{O}(h), \\
((Ac \cdot Ac)^\mathsf{T} \otimes I)b(V) = 6\phi_5(V) + \mathcal{O}(h), & ((c \cdot Ac^2)^\mathsf{T} \otimes I)b(V) = 8\phi_5(V) + \mathcal{O}(h), \\
((A^2c^2)^\mathsf{T} \otimes I)b(V) = 2\phi_5(V) + \mathcal{O}(h), & ((A(c \cdot Ac))^\mathsf{T} \otimes I)b(V) = 3\phi_5(V) + \mathcal{O}(h), \\
((Ac^3)^\mathsf{T} \otimes I)b(V) = 6\phi_5(V) + \mathcal{O}(h),
\end{cases}
$$

$$(10.15)$$

and

$$
\begin{cases}
(e^\mathsf{T} \otimes I)\bar{b}(V) = \phi_2(V) + \mathcal{O}(h^4), & (c^\mathsf{T} \otimes I)\bar{b}(V) = \phi_3(V) + \mathcal{O}(h^3), \\
((c^2)^\mathsf{T} \otimes I)\bar{b}(V) = 2\phi_4(V) + \mathcal{O}(h^2), & ((Ac)^\mathsf{T} \otimes I)\bar{b}(V) = \phi_4(V) + \mathcal{O}(h^2), \\
((A^2c)^\mathsf{T} \otimes I)\bar{b}(V) = \phi_5(V) + \mathcal{O}(h), & ((Ac^2)^\mathsf{T} \otimes I)\bar{b}(V) = 2\phi_5(V) + \mathcal{O}(h), \\
((c^3)^\mathsf{T} \otimes I)\bar{b}(V) = 6\phi_5(V) + \mathcal{O}(h), & ((c \cdot Ac)^\mathsf{T} \otimes I)\bar{b}(V) = 3\phi_5(V) + \mathcal{O}(h),
\end{cases}
$$

$$(10.16)$$

where $e = (1, 1, 1, 1, 1, 1)^\mathsf{T}$.

Choosing $c_1 = 0$, $c_2 = \dfrac{1}{6}$, $c_3 = \dfrac{1}{3}$, $c_4 = \dfrac{1}{2}$, $c_5 = \dfrac{2}{3}$ and $c_6 = 1$, and solving the above equations, we obtain

$$
\begin{cases}
a_{21} = \dfrac{1}{6}, & a_{31} = 0, & a_{32} = \dfrac{1}{3}, & a_{41} = -\dfrac{1}{4}, & a_{42} = \dfrac{3}{4}, \\[2ex]
a_{43} = 0, & a_{51} = -\dfrac{1}{27}, & a_{52} = \dfrac{2}{9}, & a_{53} = \dfrac{1}{3}, & a_{54} = \dfrac{4}{27}, \\[2ex]
a_{61} = -\dfrac{2}{11}, & a_{62} = \dfrac{3}{11}, & a_{63} = \dfrac{27}{11}, & a_{64} = -4, & a_{65} = \dfrac{27}{11}, \\[2ex]
\bar{a}_{21} = 0, & \bar{a}_{31} = \dfrac{1}{18}, & \bar{a}_{32} = 0, & \bar{a}_{41} = \dfrac{1}{8}, & \bar{a}_{42} = 0, \\[2ex]
\bar{a}_{43} = 0, & \bar{a}_{51} = 0, & \bar{a}_{52} = \dfrac{2}{9}, & \bar{a}_{53} = 0, & \bar{a}_{54} = 0, \\[2ex]
\bar{a}_{61} = \dfrac{21}{22}, & \bar{a}_{62} = -\dfrac{18}{11}, & \bar{a}_{63} = \dfrac{9}{11}, & \bar{a}_{64} = \dfrac{4}{11}, & \bar{a}_{65} = 0,
\end{cases}
$$

$$(10.17)$$

and

$$
\begin{cases}
b_1(V) = \phi_1(V) - \dfrac{15}{2}\phi_2(V) + 40\phi_3(V) - 135\phi_4(V) + 216\phi_5(V), \\[2mm]
b_2(V) = 0, \\[2mm]
b_3(V) = 27\big(\phi_2(V) - 9\phi_3(V) + 39\phi_4(V) - 72\phi_5(V)\big), \\[2mm]
b_4(V) = -32\big(\phi_2(V) - 11\phi_3(V) + 54\phi_4(V) - 108\phi_5(V)\big), \\[2mm]
b_5(V) = \dfrac{27}{2}\big(\phi_2(V) - 12\phi_3(V) + 66\phi_4(V) - 144\phi_5(V)\big), \\[2mm]
b_6(V) = -\phi_2(V) + 13\phi_3(V) - 81\phi_4(V) + 216\phi_5(V), \\[2mm]
\bar{b}_1(V) = \phi_2(V) - 5\phi_3(V) + \dfrac{64}{5}\phi_4(V) - 13\phi_5(V), \\[2mm]
\bar{b}_2(V) = 0, \\[2mm]
\bar{b}_3(V) = 9\phi_3(V) - \dfrac{171}{5}\phi_4(V) + 45\phi_5(V), \\[2mm]
\bar{b}_4(V) = -4\phi_3(V) + \dfrac{64}{5}\phi_4(V) - 16\phi_5(V), \\[2mm]
\bar{b}_5(V) = \dfrac{54}{5}\phi_4(V) - 27\phi_5(V), \\[2mm]
\bar{b}_6(V) = -\dfrac{11}{5}\phi_4(V) + 11\phi_5(V).
\end{cases}
\tag{10.18}
$$

Equations (10.17)–(10.18) give a six-stage ARKN method of order five. The Taylor expansions of the coefficients (10.18) are given by

$$
\begin{cases}
b_1(V) = \dfrac{11}{120}I - \dfrac{3}{70}V + \dfrac{25}{8064}V^2 - \dfrac{259}{2851200}V^3 + \cdots, \\[2mm]
b_2(V) = 0, \\[2mm]
b_3(V) = \dfrac{27}{40}I - \dfrac{99}{560}V + \dfrac{9}{896}V^2 - \dfrac{17}{70400}V^3 + \cdots, \\[2mm]
b_4(V) = -\dfrac{8}{15}I + \dfrac{4}{35}V - \dfrac{1}{126}V^2 + \dfrac{19}{89100}V^3 + \cdots, \\[2mm]
b_5(V) = \dfrac{27}{40}I - \dfrac{9}{140}V + \dfrac{3}{896}V^2 - \dfrac{3}{35200}V^3 + \cdots, \\[2mm]
b_6(V) = \dfrac{11}{120}I + \dfrac{1}{336}V - \dfrac{1}{4480}V^2 + \dfrac{47}{7983360}V^3 + \cdots, \\[2mm]
\bar{b}_1(V) = \dfrac{11}{120}I - \dfrac{383}{25200}V + \dfrac{1231}{1814400}V^2 - \dfrac{2839}{199584000}V^3 + \cdots, \\[2mm]
\bar{b}_2(V) = 0, \\[2mm]
\bar{b}_3(V) = \dfrac{9}{20}I - \dfrac{51}{1400}V + \dfrac{107}{100800}V^2 - \dfrac{61}{3696000}V^3 + \cdots, \\[2mm]
\bar{b}_4(V) = -\dfrac{4}{15}I + \dfrac{59}{3150}V - \dfrac{59}{113400}V^2 + \dfrac{197}{24948000}V^3 + \cdots, \\[2mm]
\bar{b}_5(V) = \dfrac{9}{40}I - \dfrac{27}{2280}V + \dfrac{13}{67200}V^2 - \dfrac{17}{7392000}V^3 + \cdots, \\[2mm]
\bar{b}_6(V) = \dfrac{11}{12600}V - \dfrac{11}{453600}V^2 + \dfrac{1}{3024000}V^3 + \cdots.
\end{cases}
\tag{10.19}
$$

This method is denoted by ARKN6s5.

10.3.2   Stability and Phase Properties of Multidimensional ARKN Methods

This section is concerned with the stability and phase properties of multidimensional ARKN methods (10.3). In the case of the classical RKN methods, the stability properties are analysed using the second-order homogeneous linear test equation

$$y''(t) = -\lambda^2 y(t), \quad \text{with } \lambda > 0. \tag{10.20}$$

Since the ARKN methods integrate $y'' + My = 0$ exactly, it is pointless to consider the stability and phase properties of ARKN methods on the basis of the conventional linear test equation (10.20).

For the stability analysis of multidimensional ARKN methods, we use the following revised test equation [13, 16]:

$$y''(t) + \omega^2 y(t) = -\varepsilon y(t), \quad \text{with } \varepsilon + \omega^2 > 0, \tag{10.21}$$

where ω represents an estimate of the dominant frequency λ and $\varepsilon = \lambda^2 - \omega^2$ is the error of that estimation. Applying an ARKN integrator to (10.21), we obtain

$$\begin{cases} Y = ey_n + chy_n' - (V+z)AY, \quad z = \varepsilon h^2, \quad V = h^2\omega^2, \\ y_{n+1} = \phi_0(V)y_n + \phi_1(V)hy_n' - z\bar{b}^\mathsf{T}(V)Y, \\ hy_{n+1}' = -V\phi_1(V)y_n + \phi_0(V)hy_n' - zb^\mathsf{T}(V)Y. \end{cases} \tag{10.22}$$

It follows from (10.22) that

$$\begin{pmatrix} y_{n+1} \\ hy_{n+1}' \end{pmatrix} = R(V, z) \begin{pmatrix} y_n \\ hy_n' \end{pmatrix},$$

where the stability matrix $R(V, z)$ is given by

$$R(V, z) = \begin{pmatrix} \phi_0(V) - z\bar{b}^\mathsf{T}(V)N^{-1}e & \phi_1(V) - z\bar{b}^\mathsf{T}(V)N^{-1}c \\ -V\phi_1(V) - zb^\mathsf{T}(V)N^{-1}e & \phi_0(V) - zb^\mathsf{T}(V)N^{-1}c \end{pmatrix},$$

with $N = I + (V+z)A$ and $e = (1, 1, \dots, 1)^\mathsf{T}$.

The spectral radius $\rho(R(V, z))$ represents the stability of an ARKN method. Since the stability matrix $R(V, z)$ depends on the variables V and z, the characterization of stability is determined by two-dimensional regions in the (V, z)-plane. Accordingly, we have the following definitions of stability for an ARKN method:

(i) $R_s = \{(V, z) \mid V > 0, \ z > 0 \text{ and } \rho(R) < 1\}$ is called the *stability region of an ARKN method*.

(ii) $R_p = \{(V, z) \mid V > 0, \ z > 0, \ \rho(R) = 1 \text{ and } \text{tr}(R)^2 < 4\det(R)\}$ is called the *periodicity region of an ARKN method*.

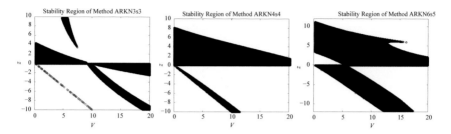

Fig. 10.1 Stability regions of the methods ARKN3s3 (*left*), ARKN4s4 (*middle*), and ARKN6s5 (*right*)

(iii) If $R_s = (0, \infty) \times (0, \infty)$, the ARKN method is called *A-stable*.
(iv) If $R_p = (0, \infty) \times (0, \infty)$, the ARKN method is called *P-stable*.

The stability regions based on the test equation (10.21) for the methods derived in this section are depicted in Fig. 10.1.

For the integration of oscillatory problems, it is common practice to consider the phase properties (dispersion order and dissipation order) of the numerical methods.

Definition 10.2 The quantities

$$\phi(H) = H - \arccos\left(\frac{\mathrm{tr}(R)}{2\sqrt{\det(R)}}\right), \quad d(H) = 1 - \sqrt{\det(R)}$$

are, respectively, called the dispersion error and the dissipation error of ARKN methods, where $H = \sqrt{V + z}$. Then, a method is said to be dispersive of order q and dissipative of order r, if $\phi(H) = \mathscr{O}(H^{q+1})$ and $d(H) = \mathscr{O}(H^{r+1})$. If $\phi(H) = 0$ and $d(H) = 0$, then the method is said to be zero dispersive and zero dissipative.

The dissipation errors and the dispersion errors for the methods derived in Sect. 10.3.1 are

- ARKN3s3:
$$d(H) = \frac{\varepsilon}{96(\varepsilon + \omega^2)} H^4 + \mathscr{O}(H^6),$$

$$\phi(H) = -\frac{\varepsilon}{480(\varepsilon + \omega^2)} H^5 + \mathscr{O}(H^7);$$

- ARKN4s4:
$$d(H) = \frac{\varepsilon(4\varepsilon + 3\omega^2)}{576(\varepsilon + \omega^2)^2} H^6 + \mathscr{O}(H^8),$$

$$\phi(H) = \frac{\varepsilon}{120(\varepsilon + \omega^2)} H^5 + \mathscr{O}(H^7);$$

• ARKN6s5:

$$d(H) = -\frac{\varepsilon(154\varepsilon + 129\omega^2)}{110880(\varepsilon + \omega^2)^2}H^6 + \mathcal{O}(H^8),$$

$$\phi(H) = \frac{\varepsilon(44\varepsilon + 27\omega^2)}{36960(\varepsilon + \omega^2)^2}H^7 + \mathcal{O}(H^9).$$

10.4   Numerical Experiments

In order to show the remarkable efficiency of the ARKN methods derived in Sect. 10.3 in comparison with some existing methods in the scientific literature, four problems are considered. The methods we select for comparison are

• ARKN3s3: the three-stage ARKN method of order three;
• ARKN4s4: the four-stage ARKN method of order four;
• ARKN6s5: the six-stage ARKN method of order five;
• RKN4s4: the classical four-stage RKN method of order four (see, e.g. II.14 of [8]);
• RKN6s5: the six-stage RKN method of order five obtained from ARKN6s5 with $V = \mathbf{0}$.

For each experiment, we display the efficiency curves of accuracy versus computational cost as measured by the number of function evaluations required by each method.

Problem 10.1 The linear problem:

$$\begin{cases} y''(t) + \omega^2 y(t) = -\delta y'(t), \\ y(0) = 1, \quad y'(0) = -\dfrac{\delta}{2}. \end{cases}$$

The analytic solution of this initial value problem is given by

$$y(t) = \exp\left(-\frac{\delta}{2}t\right)\cos\left(\sqrt{\omega^2 - \frac{\delta^2}{4}}\, t\right).$$

In this test, the parameter values $\omega = 1$, $\delta = 10^{-3}$ are chosen. The problem is integrated on the interval $[0, 100]$ with the stepsizes $h = 1/2^j$, $j = 1, 2, 3, 4$. The numerical results are displayed in Fig. 10.2 (left).

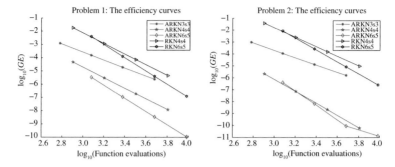

Fig. 10.2 Results for Problems 10.1 and 10.2: The log–log plot of maximum global error against number of function evaluations

Problem 10.2 The van de Pol equation:

$$
\begin{cases}
y''(t) + y(t) = \delta(1 - y(t)^2)y'(t), \\[2mm]
y(0) = 2 + \dfrac{1}{96}\delta^2 + \dfrac{1033}{552960}\delta^4 + \dfrac{1019689}{55738368000}\delta^6, \quad y'(0) = 0,
\end{cases}
$$

with $\delta = 0.8 \times 10^{-4}$. Integrate the problem on the interval $[0, 100]$ with the stepsizes $h = 1/2^j$, $j = 1, 2, 3, 4$. In order to evaluate the error for each method, a reference numerical solution is obtained by the method RKN4s4 in II.14 of [8] with a very small stepsize. The numerical results are displayed in Fig. 10.2 (right).

Problem 10.3 The oscillatory initial value problem

$$
\begin{cases}
y''(t) + \begin{pmatrix} 13 & -12 \\ -12 & 13 \end{pmatrix} y(t) = \dfrac{12\varepsilon}{5} \begin{pmatrix} 3 & 2 \\ -2 & -3 \end{pmatrix} y'(t) + \varepsilon^2 \begin{pmatrix} f_1(t) \\ f_2(t) \end{pmatrix}, \quad 0 < t \le t_{\text{end}}, \\[3mm]
y(0) = \begin{pmatrix} \varepsilon \\ \varepsilon \end{pmatrix}, \quad y'(0) = \begin{pmatrix} -4 \\ 6 \end{pmatrix},
\end{cases}
$$

with

$$
f_1(t) = \dfrac{36}{5}\sin(t) + 24\sin(5t), \quad f_2(t) = -\dfrac{24}{5}\sin(t) - 36\sin(5t).
$$

The analytic solution to this problem is given by

$$
y(t) = \begin{pmatrix} \sin(t) - \sin(5t) + \varepsilon\cos(t) \\ \sin(t) + \sin(5t) + \varepsilon\cos(5t) \end{pmatrix}.
$$

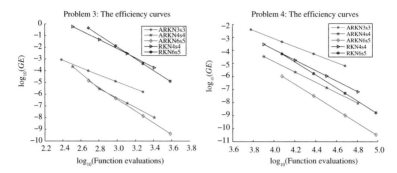

Fig. 10.3 Results for Problems 10.3 and 10.4: The log–log plot of maximum global error against number of function evaluations

In this test, the problem is integrated on the interval $[0, 20]$ with the stepsizes $h = 1/2^j$, $j = 2, 3, 4, 5$ and $\varepsilon = 10^{-3}$. The numerical results are displayed in Fig. 10.3 (left).

Problem 10.4 The damped wave equation with periodic boundary conditions

$$\begin{cases} \dfrac{\partial^2 u}{\partial t^2} + \delta \dfrac{\partial u}{\partial t} - \dfrac{\partial^2 u}{\partial x^2} = f(u), \quad -1 < x < 1, \ t > 0, \\ u(-1, t) = u(1, t). \end{cases}$$

This problem models the wave propagation in a medium [15]. A semi-discretization in the spatial variable by second-order symmetric differences leads to the following system of second-order ordinary differential equations in time:

$$U'' + MU = F(U, U'), \quad 0 < t \le t_{\text{end}},$$

where $U(t) = \big(u_1(t), \ldots, u_N(t)\big)^{\mathsf{T}}$ with $u_i(t) \approx u(x_i, t), i = 1, \ldots, N,$

$$M = \frac{1}{\Delta x^2} \begin{pmatrix} 2 & -1 & & & -1 \\ -1 & 2 & -1 & & \\ & \ddots & \ddots & \ddots & \\ & & -1 & 2 & -1 \\ -1 & & & -1 & 2 \end{pmatrix}$$

with $\Delta x = 2/N$ and $x_i = -1 + i\Delta x$, and $F(U,\ U') = \big(f(u_1) - \delta u_1', \ldots, f(u_N) - \delta u_N'\big)^{\mathsf{T}}$.

In this numerical experiment, $f(u) = -\sin u$, $\delta = 0.8$ and the initial conditions are

$$U(0) = \big(\pi\big)_{i=1}^N, \quad U'(0) = \sqrt{N}\big(0.01 + \sin(\frac{2\pi i}{N})\big)_{i=1}^N,$$

with $N = 40$. Integrate the system on the interval $[0, 100]$ with the stepsizes $h = 0.1/2^j$, $j = 1, 2, 3, 4$. The reference numerical solution is obtained by the method RKN4s4 with a very small stepsize. The numerical results are shown in Fig. 10.3 (right).

10.5 Conclusions and Discussions

This chapter has presented further discussion and analysis of multidimensional ARKN methods proposed in [19] for the general multi-frequency oscillatory system (10.1) whose right-hand-side functions *depend on both y and y'*. Three multidimensional ARKN schemes for the oscillatory system (10.1) were constructed based on the order conditions presented in [19]. The stability and phase properties of the schemes were analysed, based on the revised linear model (10.21). The multidimensional ARKN methods were implemented and numerical comparisons were made with some existing methods. The results show that the new methods perform much more efficiently than the existing methods, and are therefore more suitable for the numerical integration of the general oscillatory system (10.1).

The stability analysis based on (10.21) may be reconsidered since the right-hand-side functions depend on both y and y' in (10.1). Therefore, the stability properties can be analysed using the new linear test equation involving y' of the form

$$y''(t) = -\omega^2 y(t) - \mu y'(t), \tag{10.23}$$

with $\omega > \mu/2 > 0$, where ω is called the (undamped) natural frequency of the system and μ is called the damping. The new stability analysis for the methods ARKN3s3, ARKN4s4 and ARKN6s5 is presented below.

Applying an ARKN integrator to (10.23), we obtain

$$
\begin{cases}
Y_i = y_n + hc_i y'_n + h^2 \sum_{j=1}^{s} \bar{a}_{ij}\left(-\omega^2 Y_j - \mu Y'_j\right), & i = 1, \ldots, s, \\[2mm]
hY'_i = hy'_n + h^2 \sum_{j=1}^{s} a_{ij}\left(-\omega^2 Y_j - \mu Y'_j\right), & i = 1, \ldots, s, \\[2mm]
y_{n+1} = \phi_0(v^2) y_n + h\phi_1(v^2) y'_n + h^2 \sum_{i=1}^{s} \bar{b}_i(v^2)\left(-\mu Y'_i\right), \\[2mm]
hy'_{n+1} = \phi_0(v^2) hy'_n - v^2 \phi_1(v^2) y_n + h^2 \sum_{i=1}^{s} b_i(v^2)\left(-\mu Y'_i\right),
\end{cases}
\tag{10.24}
$$

or, in the compact form

$$
\begin{cases}
Y = e \otimes y_n + c \otimes h y_n' + \bar{A}\big(-v^2 Y - \sigma h Y'\big), \\
h Y' = e \otimes h y_n' + A\big(-v^2 Y - \sigma h Y'\big), \\
y_{n+1} = \phi_0 y_n + \phi_1 h y_n' - \sigma \bar{b}^\mathsf{T} h Y', \\
h y_{n+1}' = \phi_0 h y_n' - v^2 \phi_1 y_n - \sigma b^\mathsf{T} h Y',
\end{cases}
\tag{10.25}
$$

where $v = h\omega$, $\sigma = h\mu$ and the argument (v^2) has been suppressed. It follows from (10.25) that

$$
\begin{pmatrix} y_{n+1} \\ h y_{n+1}' \end{pmatrix} = R(v, \sigma) \begin{pmatrix} y_n \\ h y_n' \end{pmatrix},
$$

where the stability matrix $R(v, \sigma)$ is given by

$$
R(v, \sigma) = \begin{pmatrix} \phi_0 + \sigma v^2 \bar{b}^\mathsf{T} A C^{-1} e & \phi_1 + \sigma v^2 \bar{b}^\mathsf{T} A C^{-1} c + \sigma^2 \bar{b}^\mathsf{T} A C^{-1} e - \sigma \bar{b}^\mathsf{T} e \\ -v^2 \phi_1 + \sigma v^2 b^\mathsf{T} A C^{-1} e & \phi_0 + \sigma v^2 b^\mathsf{T} A C^{-1} c + \sigma^2 b^\mathsf{T} A C^{-1} e - \sigma b^\mathsf{T} e \end{pmatrix}.
$$

The spectral radius $\rho(R(v, \sigma))$ represents the stability of an ARKN method. Since the stability matrix $R(v, \sigma)$ depends on the variables v and σ, the stability is characterized by the regions in the (v, σ) plane. This results in the following new definitions of stability for an ARKN method based on the new linear test equation (10.23).

(i) $R_s = \{(v, \sigma) \mid v > 0, \ \sigma > 0 \text{ and } \rho(R) < 1\}$ is called the *stability region* of the ARKN method.

(ii) If $R_s = (0, \infty) \times (0, \infty)$, the ARKN method is called *A-stable*.

The stability regions based on the new linear test equation (10.23) for the methods derived in this chapter are depicted in Fig. 10.4.

The material of this chapter is based on the work by Liu and Wu [10].

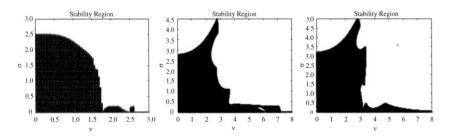

Fig. 10.4 Stability regions for the methods ARKN3s3 (*left*), ARKN4s4 (*middle*), and ARKN6s5 (*right*)

References

1. Ascher UM, Reich S (1999) On some difficulties in integrating highly oscillatory Hamiltonian systems. In: Proceedings of the computational molecular dynamics, Springer Lecture Notes, pp 281-296
2. Cohen D, Hairer E, Lubich C (2003) Modulated Fourier expansions of highly oscillatory differential equations. Found Comput Math 3:327–450
3. Garcá-Archilla B, Sanz-Serna J, Skeel R (1998) Long-time-step methods for oscillatory differential equations. SIAM J Sci Comput 30:930–963
4. Franco J (2006) New methods for oscillatory systems based on ARKN methods. Appl Numer Math 56:1040–1053
5. Fang Y, Wu X (2007) A new pair of explicit ARKN methods for the numerical integration of general perturbed oscillators. Appl Numer Math 57:166–175
6. Fang Y, Wu X (2008) A trigonometrically fitted explicit Numerov-type method for second-order initial value problems with oscillating solutions. Appl Numer Math 58:341–351
7. García-Alonso F, Reyes J, Ferrádiz J, Vigo-Aguiar J (2009) Accurate numerical integration of perturbed oscillatory systems in two frequencies. ACM Trans Math Softw 36:21–34
8. Hairer E, Nørsett SP, Wanner G (1993) Solving ordinary differnetial equations I, Nonstiff problems, 2nd edn., Springer series in computational mathematicsSpringer, Berlin
9. Iserles A (2002) Think globally, act locally: solving highly-oscillatory ordinary differential equations. Appl Numer Math 43:145–160
10. Liu K, Wu X (2014) Multidimensional ARKN methods for general oscillatory second-order initial value problems. Comput Phys Commun 185:1999–2007
11. Petzold L, Jay L, Yen J (1997) Numerical solution of highly oscillatory ordinary differential equations. Acta Numer 6:437–484
12. Shi W, Wu X (2012) On symplectic and symmetric ARKN methods. Comput Phys Commun 183:1250–1258
13. Van der Houwen PJ, Sommeijer BP (1987) Explicit Runge-Kutta(-Nyström) methods with reduced phase errors for computing oscillating solution. SIAM J Numer Anal 24:595–617
14. Van de Vyver H (2005) Stability and phase-lag analysis of explicit Runge-Kutta methods with variable coefficients for oscillatory problems. Comput Phys Commun 173:115–130
15. Weinberger HF (1965) A first course in partial differential equations with complex variables and transform methods. Dover Publications, Inc, New York
16. Wu X (2012) A note on stability of multidimensional adapted Runge-Kutta-Nyström methods for oscillatory systems. Appl Math Model 36:6331–6337
17. Wu X, Wang B (2010) Multidimensional adapted Runge-Kutta-Nyström methods for oscillatory systems. Comput Phys Commun 181:1955–1962
18. Wu X, You X, Li J (2009) Note on derivation of order conditions for ARKN methods for perturbed oscillators. Comput Phys Commun 180:1545–1549
19. Wu X, You X, Xia J (2009) Order conditions for ARKN methods solving oscillatory systems. Comput Phys Commun 180:2250–2257
20. Wu X, You X, Wang B (2013) Structure-preserving integrators for oscillatory ordinary differential equations. Springer, Heidelberg (jointly published with Science Press Beijing)
21. Yang H, Wu X (2008) Trigonometrically-fitted ARKN methods for perturbed oscillators. Appl Numer Math 58:1375–1395

Chapter 11
A Simplified Nyström-Tree Theory for ERKN Integrators Solving Oscillatory Systems

In the study of extended Runge-Kutta-Nyström (ERKN) methods for the integration of multi-frequency oscillatory systems, a quite complicated set of algebraic conditions arises which must be satisfied for a method to achieve some specified order. A theory of tri-coloured tree was proposed by Yang et al. [36] for achieving the order conditions of ERKN methods. However, the tri-coloured tree theory for the order conditions is not completely satisfactory due to the existence of redundant trees. In this chapter, a simplified tri-coloured tree theory for the order conditions for ERKN integrators is developed. This leads to a novel Nyström-tree theory for the order conditions of ERKN methods without any redundant trees.

11.1 Motivation

The purpose of this chapter is to investigate and analyse a simplification of the Nyström-tree theory for ERKN methods for solving a system of multi-frequency oscillatory second-order initial value problems of the form

$$\begin{cases} \mathbf{y}''(t) + M\mathbf{y}(t) = \mathbf{f}\big(\mathbf{y}(t)\big), & t \in [t_0, T], \\ \mathbf{y}(t_0) = \mathbf{y}_0, \quad \mathbf{y}'(t_0) = \mathbf{y}'_0, \end{cases} \tag{11.1}$$

where M is a $d \times d$ positive semi-definite constant matrix containing implicitly the frequencies of the problem, and $\mathbf{f} : \mathbb{R}^d \to \mathbb{R}^d$ is a vector-valued function. Such problems are frequently encountered in celestial mechanics, theoretical physics and chemistry, and electronics. The simplified theory will make the construction of a numerical integrator for the oscillatory system (11.1) much more efficient. Since by appending $t'' = 0$, we can transform the following non-autonomous form

$$\begin{cases} \mathbf{y}''(t) + M\mathbf{y}(t) = \mathbf{f}\big(t, \mathbf{y}(t)\big), & t \in [t_0, T], \\ \mathbf{y}(t_0) = \mathbf{y}_0, \quad \mathbf{y}'(t_0) = \mathbf{y}'_0, \end{cases} \tag{11.2}$$

© Springer-Verlag Berlin Heidelberg and Science Press, Beijing, China 2015
X. Wu et al., *Structure-Preserving Algorithms for Oscillatory
Differential Equations II*, DOI 10.1007/978-3-662-48156-1_11

into an autonomous form (11.1). Thus, we consider only the autonomous form (11.1) for the simplified Nyström-tree theory.

The Nyström method (see [18]) is the well-known classical numerical integrator for second-order differential equations. The Nyström-tree theory is important for the analysis of order conditions [8, 9]. In [36], based on a set of special extended Nyström trees (SENT) and some real-valued mappings on it, a tri-coloured tree theory and order conditions for ERKN methods were established. Regretfully, the tri-coloured tree theory could not be satisfactory due to the existence of redundant special extended Nyström trees (SEN-trees). Therefore, in this chapter, we create a simplified Nyström-tree theory for the order conditions of ERKN methods without any redundant trees. This leads to a great reduction in the number of the order conditions for ERKN methods in [36].

In general, the nth order derivative of the vector-valued function $f(y)$ would help us gain an insight into the set of rooted trees and the theory of B-series for the order conditions of ERKN methods. The nth order derivative of $f(y)$ with respect to y will be denoted by $f^{(n)}$. Just as shown in [36], for the second-order differential equation $y'' = -My + f(y)$, the term $f^{(1)}(-My)$ appears in the second derivative $\frac{d^2 f(y)}{dt^2}$. In fact, we have

$$\frac{d^2 f(y)}{dt^2} = f^{(2)}(y', y') + f^{(1)} f + f^{(1)}(-My).$$

The authors in [36] supplemented the set of classical special Nyström trees with a tri-coloured tree ⎡, and then produced the set SENT. At first sight, the tri-coloured tree theory looks beautiful. However, it can be observed from Tables 2, 3 and 4 in [36] that the results are neither elegant nor perfect yet for two reasons: First, the quantities of the weight $\Phi_i(\cdot)$ and the density $\gamma(\cdot)$ of this tree are exactly the same as those of the Nyström tree ⎤. This means that this tri-coloured tree can be regarded as a redundant tree for the order conditions. Second, the number of redundant trees is huge.

It is really a regrettable story that the introduced tri-coloured tree ⎡ is a redundant tree in the original paper [36]. The tree plays a key part in constructing of the tri-coloured tree theory, whereas it contributes little towards the order conditions for the ERKN methods. It can be noticed that the differences between the two trees ⎤ and ⎡ are the elementary differential $\mathscr{F}(\cdot)(y)$ and the symmetry $\alpha(\cdot)$, which have nothing to do with the order conditions. In this chapter, the redundant tree ⎡ is removed by introducing an "extended" elementary differential, which contains not one function but a composition of several functions and their derivatives. For example, we introduce an "extended" elementary differential $D_h^2 f$ as the sum of $f^{(2)}(y', y')$ and $f^{(1)}(-My)$, namely,

$$D_h^2 f = f^{(2)}\left(y', y'\right) + f^{(1)}\left(-My\right),$$

where $D_h^2 f$ denotes the second derivative of $f\left(\phi_0(h^2 M)y + \phi_1(h^2 M)hy'\right)$ with respect to h, at $h = 0$. In this way, it will be observed that all the redundant trees in [36] are eliminated due to the use of the extended elementary differentials. For order 5, for example, from the new theory there remain only 7 trees out of the original 13 trees in [36]. It turns out that, in light of the simplified Nyström-tree theory based on the extended elementary differentials, almost one half of the tri-coloured trees in [36] can be removed.

11.2   ERKN Methods and Related Issues

In this section, using the matrix-variation-of-constants formula, we represent ERKN methods for the multi-frequency oscillatory system (11.1). The ERKN methods make full use of the special structure of the equation generated by the linear term My.

From the matrix-variation-of-constants formula (1.6) in Chap. 1, the solution of the IVP (11.1) and its derivative, at $t + \mu h$, satisfy the following integral equations:

$$
\begin{cases}
y(t + \mu h) = \phi_0(\mu^2 V)y(t) + \mu h \phi_1(\mu^2 V)y'(t) \\
\qquad + h^2 \displaystyle\int_0^\mu (\mu - z)\phi_1\big((\mu - z)^2 V\big) f\big(y(t + hz)\big)dz, \\
y'(t + \mu h) = \phi_0(\mu^2 V)y'(t) - h\mu M \phi_1(\mu^2 V)y(t) \\
\qquad + h \displaystyle\int_0^\mu \phi_0\big((\mu - z)^2 V\big) f\big(y(t + hz)\big)dz,
\end{cases}
\tag{11.3}
$$

where $V = h^2 M$ and the matrix-valued functions $\phi_i(V)$ are defined by (4.7) in Chap. 4. The matrix-valued functions $\phi_i(V)$ have the following properties:

$$
\int_0^1 \frac{(1 - z)\phi_1\big((1 - z)^2 V\big)z^j}{j!}dz = \phi_{j+2}(V), \qquad \int_0^1 \frac{\phi_0\big((1 - z)^2 V\big)z^j}{j!}dz = \phi_{j+1}(V).
\tag{11.4}
$$

It is noted that the matrix-variation-of-constants formula (11.3) does not require the decomposition of M. This point is significant since M is not necessarily diagonal nor symmetric in (11.1). Approximating the integrals in (11.3) by suitable quadrature formulae results in the following ERKN methods derived in [34, 36].

Definition 11.1 An s-stage *ERKN method* for the numerical integration of (11.1) is defined by the following scheme:

$$
\begin{cases}
Y_i = \phi_0(c_i^2 V)y_n + hc_i\phi_1(c_i^2 V)y_n' + h^2 \sum_{j=1}^{s} \bar{a}_{ij}(V)f(Y_j), \qquad i = 1, \ldots, s, \\[2ex]
y_{n+1} = \phi_0(V)y_n + h\phi_1(V)y_n' + h^2 \sum_{i=1}^{s} \bar{b}_i(V)f(Y_i), \\[2ex]
y_{n+1}' = -hM\phi_1(V)y_n + \phi_0(V)y_n' + h \sum_{i=1}^{s} b_i(V)f(Y_i),
\end{cases}
$$

(11.5)

where c_i, $i = 1, \ldots, s$ *are real constants,* $b_i(V)$, $\bar{b}_i(V)$, $i = 1, \ldots, s$ *and* $\bar{a}_{ij}(V)$, $i, j = 1, \ldots, s$ *are matrix-valued functions of* $V = h^2 M$, *which are assumed* to have the following series expansions

$$
\bar{a}_{ij}(V) = \sum_{k=0}^{\infty} \frac{\bar{a}_{ij}^{(2k)}}{(2k)!} V^k, \quad \bar{b}_i(V) = \sum_{k=0}^{\infty} \frac{\bar{b}_i^{(2k)}}{(2k)!} V^k, \quad b_i(V) = \sum_{k=0}^{\infty} \frac{b_i^{(2k)}}{(2k)!} V^k,
$$

with real constant coefficients $\bar{a}_{ij}^{(2k)}$, $\bar{b}_i^{(2k)}$, $b_i^{(2k)}$, $k = 0, 1, 2, \ldots$.

In essence, ERKN methods incorporate the particular structure of the differential equation (11.1) into both the internal stages and the updates.

ERKN methods have been widely studied. The relation between ERKN methods and exponentially or trigonometric fitted methods [4–7, 13, 14, 20–23, 35] has been investigated. Moreover, multidimensional ERKN methods and multidimensional exponentially fitted methods have been constructed [27, 32, 34]. For oscillatory Hamiltonian systems, the symplecticity and symmetry conditions for ERKN methods are presented in [26, 28, 30–33]. ERKN methods are extended to two-step hybrid methods in [15], to Falkner-type methods in [16], to energy-preserving methods in [24, 29], to asymptotic methods for highly oscillatory problems in [25] and to multisymplectic methods for Hamiltonian partial differential equations in [19].

However, the tri-coloured tree theory in [36], as an important approach in analysing the algebraic theory of order conditions for ERKN methods, has not been simplified yet. It is noted that when the tri-coloured tree theory is used to find the order conditions, one has to consider all rooted trees to check whether or not the tree under consideration is a redundant tree. As a result, it is neither convenient nor efficient to use the original tri-coloured tree theory to find the order conditions for ERKN methods. Thus, in what follows, a simplified and efficient approach to dealing with the problem will be presented.

11.3 Higher Order Derivatives of Vector-Valued Functions

11.3.1 Taylor Series of Vector-Valued Functions

In this section, we introduce multivariate Taylor series expansions which are required in the analysis and presentation of the main results in this chapter.

Suppose that $f : \mathbb{R}^d \to \mathbb{R}^d$ is a vector-valued function of the $d \times 1$ vector y and is infinitely differentiable in an open neighbourhood of y_0. The Taylor series of $f(y)$ can be then expressed by

$$f(y) = f|_{y=y_0} + \sum_{i=1}^{\infty} \frac{1}{i!} f^{(i)}\Big|_{y=y_0} (y - y_0)^{\otimes i}, \tag{11.6}$$

where the higher order derivatives are given by

$$f^{(n)} = \frac{\partial f^{(n-1)}}{\partial y^{\mathsf{T}}} = \Big(\frac{\partial}{\partial y_1}, \frac{\partial}{\partial y_2}, \cdots, \frac{\partial}{\partial y_d} \Big) \otimes f^{(n-1)}, \tag{11.7}$$

and in general, the dimension of $f^{(n)}$ is $d \times d^n$, the nth power of $(y - y_0)$ is denoted by

$$(y - y_0)^{\otimes n} = \underbrace{(y - y_0) \otimes (y - y_0) \otimes \cdots \otimes (y - y_0)}_{n\text{-fold}}. \tag{11.8}$$

We then have the following Taylor series expansion of $f^{(n)} a \in \mathbb{R}^d$

$$f^{(n)} a = \sum_{k \geq 0} \frac{1}{k!} f^{(n+k)}\Big|_{y=y_0} \Big((y - y_0)^{\otimes k} \otimes a \Big), \tag{11.9}$$

where a is a constant vector. Equation (11.9) follows from the property of the derivative of matrix products

$$\frac{\partial (F \cdot b)}{\partial y^{\mathsf{T}}} = \frac{\partial F}{\partial y^{\mathsf{T}}} \Big(I_d \otimes b \Big), \tag{11.10}$$

and the property of the Kronecker product $\otimes$

$$\Big(I_d \otimes b \Big)(y - y_0) = (y - y_0) \otimes b, \tag{11.11}$$

where F is a vector-valued function of y and b is a constant vector in (11.10).

A detailed analysis of multivariate Taylor series expansions can be found in [1].

11.3.2 Kronecker Inner Product

In this section, the definition and the properties of the Kronecker inner product are introduced. An example, which works in the proofs of two B-series theorems, Theorems 11.1 and 11.2, will be given to make this chapter more accessible.

Definition 11.2 The *Kronecker inner product* of vectors $a_i \in \mathbb{R}^{m_i \times 1}$, $i = 1, \ldots, n$, denoted by $(a_1, a_2, \ldots, a_n)$ is defined as

$$(a_1, a_2, \ldots, a_n) = \frac{1}{\beta} \sum a_{i_1} \otimes a_{i_2} \otimes \cdots \otimes a_{i_n},$$

where the Kronecker inner product is a sum of the possible permutations of $\{a_1, a_2, \ldots, a_n\}$ and β is the number of the permutations.

The following proposition gives interesting properties of the Kronecker inner product.

Proposition 11.1 *The Kronecker inner product defined by Definition 11.2 satisfies the properties as follows.*

- *The dimension of Kronecker inner product, in Definition 11.2, is $m_1 m_2 \cdots m_n \times 1$.*
- *Symmetry:*
$$(a_1, a_2, \ldots, a_n) = (a_{i_1}, a_{i_2}, \ldots, a_{i_n}),$$

where $\{a_{i_1}, a_{i_2}, \ldots, a_{i_n}\}$ is one possible permutation of $\{a_1, a_2, \ldots, a_n\}$. In particular, if a, b and c are three different vectors, we have

$$(a, b) = (b, a),$$

$$(a, b, c) = (a, c, b) = (b, a, c) = (b, c, a) = (c, a, b) = (c, b, a).$$

- *Linearity:*
$$(k_1 a_1, k_2 a_2, \ldots, k_n a_n) = k_1 k_2 \ldots k_n (a_1, a_2, \ldots, a_n),$$

for any real numbers $k_1, k_2, \ldots, k_n$, and

$$(a_1, \ldots, a_i + b, \ldots, a_n) = (a_1, \ldots, a_i, \ldots, a_n) + (a_1, \ldots, b, \ldots, a_n),$$

for $i = 1, 2, \ldots, n$.

Example: Let b be a linear combination of different vectors $a_1, a_2, a_3, \ldots \in \mathbb{R}^{d \times 1}$:

$$b = k_1 a_1 + k_2 a_2 + k_3 a_3 + \cdots.$$

The term associated with $k_1 k_2 k_3$ in $b \otimes b \otimes b$ is given by

$$a_1 \otimes a_2 \otimes a_3 + a_1 \otimes a_3 \otimes a_2 + a_2 \otimes a_1 \otimes a_3$$
$$+ a_2 \otimes a_3 \otimes a_1 + a_3 \otimes a_1 \otimes a_2 + a_3 \otimes a_2 \otimes a_1.$$

This is exactly $6(a_1, a_2, a_3)$, where the number 6 is the number of permutations of $\{a_1, a_2, a_3\}$. The terms associated with $k_1^2 k_2$ in $b \otimes b \otimes b$ is

$$a_1 \otimes a_1 \otimes a_2 + a_1 \otimes a_2 \otimes a_1 + a_2 \otimes a_1 \otimes a_1.$$

This is exactly $3(a_1, a_1, a_2)$, where the number 3 is the number of permutations of $\{a_1, a_1, a_2\}$, and the terms associated with k_1^3 in $b \otimes b \otimes b$ is

$$a_1 \otimes a_1 \otimes a_1.$$

This is exactly (a_1, a_1, a_1), where the number 1 is the number of permutations of $\{a_1, a_1, a_1\}$.

11.3.3 The Higher Order Derivatives and Kronecker Inner Product

In this section, we will give the following result:

$$f^{(n)} \cdot a_1 \otimes a_2 \otimes \cdots \otimes a_n = f^{(n)} \cdot (a_1, a_2, \ldots, a_n), \tag{11.12}$$

where $a_1, a_2, \ldots, a_n \in \mathbb{R}^{d \times 1}$. Equation (11.12) comes from the symmetry of derivative. In fact, the case of $n = 2$ gives

$$f^{(2)} = \begin{pmatrix} \dfrac{\partial^2 f_1}{\partial y_1 \partial y_1} & \dfrac{\partial^2 f_1}{\partial y_1 \partial y_2} & \cdots & \dfrac{\partial^2 f_1}{\partial y_1 \partial y_d} & \cdots & \dfrac{\partial^2 f_1}{\partial y_d \partial y_1} & \dfrac{\partial^2 f_1}{\partial y_d \partial y_2} & \cdots & \dfrac{\partial^2 f_1}{\partial y_d \partial y_d} \\[2ex] \dfrac{\partial^2 f_2}{\partial y_1 \partial y_1} & \dfrac{\partial^2 f_2}{\partial y_1 \partial y_2} & \cdots & \dfrac{\partial^2 f_2}{\partial y_1 \partial y_d} & \cdots & \dfrac{\partial^2 f_2}{\partial y_d \partial y_1} & \dfrac{\partial^2 f_2}{\partial y_d \partial y_2} & \cdots & \dfrac{\partial^2 f_2}{\partial y_d \partial y_d} \\[2ex] \vdots & \vdots & \vdots & \vdots & \vdots & \vdots & \vdots & \vdots & \vdots \\[2ex] \dfrac{\partial^2 f_d}{\partial y_1 \partial y_1} & \dfrac{\partial^2 f_d}{\partial y_1 \partial y_2} & \cdots & \dfrac{\partial^2 f_d}{\partial y_1 \partial y_d} & \cdots & \dfrac{\partial^2 f_d}{\partial y_d \partial y_1} & \dfrac{\partial^2 f_d}{\partial y_d \partial y_2} & \cdots & \dfrac{\partial^2 f_d}{\partial y_d \partial y_d} \end{pmatrix}.$$

The dimension of $f^{(2)} \cdot a_1 \otimes a_2$ is $d \times 1$, and the ith entry is as follows:

$$\frac{\partial^2 f_i}{\partial y_1 \partial y_1} a_{11} a_{21} + \frac{\partial^2 f_i}{\partial y_1 \partial y_2} a_{11} a_{22} + \cdots + \frac{\partial^2 f_i}{\partial y_1 \partial y_d} a_{11} a_{2d}$$

$$+ \frac{\partial^2 f_i}{\partial y_2 \partial y_1} a_{12} a_{21} + \frac{\partial^2 f_i}{\partial y_2 \partial y_2} a_{12} a_{22} + \cdots + \frac{\partial^2 f_i}{\partial y_2 \partial y_d} a_{12} a_{2d}$$

$$+ \cdots$$

$$+ \frac{\partial^2 f_i}{\partial y_d \partial y_1} a_{1d} a_{21} + \frac{\partial^2 f_i}{\partial y_d \partial y_2} a_{1d} a_{22} + \cdots + \frac{\partial^2 f_i}{\partial y_d \partial y_d} a_{1d} a_{2d}.$$

Because of the symmetry of partial derivatives, we then have

$$f^{(2)} \cdot a_1 \otimes a_2 = f^{(2)} \cdot a_2 \otimes a_1,$$

and

$$f^{(2)} \cdot a_1 \otimes a_2 = f^{(2)} \cdot (a_1, a_2).$$

In similar way, we have

$$f^{(n)} \cdot a_1 \otimes a_2 \otimes \ldots \otimes a_n = f^{(n)} \cdot (a_1, a_2, \ldots, a_n).$$

11.3.4 A Definition Associated with the Elementary Differentials

In this section, we will give the definition of $D_h^m f^{(n)}$ which is associated with the elementary differentials and plays an important role in the presentation of extended elementary differentials $\mathscr{F}(\tau)(y)$ introducing in Sect. 11.4.

The nth order derivative of $f(y)$ evaluated at

$$\hat{y} = \phi_0(V) y + \phi_1(V) h y',$$

denoted by $f^{(n)} \big|_{\hat{y}}$, is a $d \times d^n$ matrix. For each entry of this matrix, being a function of h, with the following form

$$\frac{\partial^n f_i}{\partial y_{i_1} \cdots \partial y_{i_n}} \bigg|_{\hat{y}},$$

we can calculate the mth order derivative with respect to h and evaluated at $h = 0$:

$$\frac{\mathbf{d}^m}{\mathbf{d}h^m}\left(\left.\frac{\partial^n f_i}{\partial y_{i_1}\cdots\partial y_{i_n}}\right|_{\hat{y}}\right)\Bigg|_{h=0}.$$

We now turn to the following definition.

Definition 11.3 The *mth order derivative* of function $\left.f^{(n)}\right|_{\hat{y}}$ with respect to h at $h = 0$ is denoted by $D_h^m f^{(n)}$.

Remark 11.1 The dimension of $D_h^m f^{(n)}$, in Definition 11.3, is $d \times d^n$.

To make Definition 11.3 clear and easy to understand, we will give a more detailed analysis about $D_h^m f^{(n)}$. It is natural to start with the Taylor series expansion. On one hand, the Taylor series expansion of the matrix-valued function $\left.f^{(n)}\right|_{\hat{y}}$ at $h = 0$ is given by

$$\left.f^{(n)}\right|_{\hat{y}} = \sum_{m\geq 0}\frac{1}{m!}h^m D_h^m f^{(n)}, \tag{11.13}$$

which comes from the fact that the function $\left.f^{(n)}\right|_{\hat{y}}$ is a matrix depending on h, which is denoted by $g(h)$, namely,

$$g(h) := \left.f^{(n)}\right|_{\hat{y}}. \tag{11.14}$$

On the other hand, the Taylor series expansion of vector-valued function $\left.f^{(n)}\right|_{\hat{y}}z^{\otimes n} \in \mathbb{R}^{d\times 1}$ with respect to y, is given by

$$\left.f^{(n)}\right|_{\hat{y}}z^{\otimes n} = \sum_{k\geq 0}\frac{1}{k!}f^{(n+k)}\left((\hat{y}-y)^{\otimes k}\otimes z^{\otimes n}\right), \tag{11.15}$$

where the $d \times 1$ vector z is constant. The term $\hat{y} - y$ in Eq. (11.15) has the following series expansion with respect to h:

$$\hat{y} - y = hy' + \frac{h^2}{2!}(-M)y + \frac{h^3}{3!}(-M)y' + \frac{h^4}{4!}M^2 y + \cdots. \tag{11.16}$$

A careful calculation can obtain the relationship between $D_h^m f^{(n)}$ and $\{f^{(i)}, -M, y, y'\}$, from (11.15) and (11.16), but the detailed presentation is not our interest in this chapter. Accordingly, we just give the first few terms for $D_h^m f^{(n)}z^{\otimes n}$:

$$D_h^1 f = f^{(1)} y',$$

$$D_h^2 f = f^{(2)}(y', y') + f^{(1)}(-M y),$$

$$D_h^3 f = f^{(3)}(y', y', y') + 3 f^{(2)}(-M y, y') + f^{(1)}(-M y'),$$

$$D_h^4 f = f^{(4)}(y', y', y', y') + 6 f^{(3)}(-M y, y', y') + 3 f^{(2)}(-M y, -M y)$$
$$+ 4 f^{(2)}(-M y', y') + f^{(1)}((-M)^2 y),$$

and

$$D_h^1 f^{(1)} z = f^{(2)}(y', z), \tag{11.17}$$

$$D_h^2 f^{(1)} z = f^{(3)}(y', y', z) + f^{(2)}(-M y, z), \tag{11.18}$$

where Eqs. (11.17) and (11.18) come from the symmetry of the derivative (see Sect. 11.3.3).

Remark 11.2 If M is the null matrix, with $\phi_i(0) = I$ and $\hat{y} = y + h y'$, it follows from Definition 11.3 that

$$D_h^m f^{(n)} z = f^{(m+n)}(\underbrace{y', \ldots, y'}_{m\text{-fold}}, z), \tag{11.19}$$

in which the symmetry of derivatives and the properties of Kronecker inner product are used.

11.4 The Set of Simplified Special Extended Nyström Trees

In this section, we will define a set of rooted trees named as the set of simplified special extended Nyström trees (SSENT). The set SSENT is exactly an optimal extension of the set of SN-trees since one new tree is appended to the fifth-order classical SN-trees, two new trees to sixth-order classical SN-trees and seven new trees to seventh-order classical SN-trees, and so on.

11.4.1 Tree Set SSENT and Related Mappings

In this section, we define recursively the set SSENT and six mappings on it. To make it more accessible, we will give an explanatory example and some comments on the six mappings on a tree $\tau \in$ SSENT.

Fig. 11.1 Figure of tree $W_+B_+(b_+B_+)^p(\tau)$

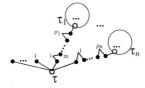

Fig. 11.2 The mode of trees in SSENT

Definition 11.4 The set *SSENT* is recursively defined as follows:

(i) o and $\overset{\bullet}{\circ}$ belong to SSENT,
(ii) $\tau_1 \times \cdots \times \tau_\mu$ belongs to SSENT, if τ_1, ..., τ_μ belong to SSENT,
(iii) $W_+B_+(b_+B_+)^p(\tau)$ belongs to SSENT, if τ belongs to SSENT, $\forall p = 0, 1, 2, \ldots,$

where '$\times$' is the merging product (Butcher [2]; Murua [17]), and $W_+B_+(b_+ B_+)^p(\tau)$ is a mapping by graphing the root of tree τ, to a new black fat node first and then to a new meagre node (p times), and then to a new black fat node, and last to a new white node (see in Fig. 11.1).

Each tree $\tau \in$ SSENT can be denoted by

$$\tau := \underbrace{\tau_* \times \cdots \times \tau_*}_{m\text{-fold}} \times \left(W_+B_+(b_+B_+)^{p_1}(\tau_1)\right) \times \cdots \times \left(W_+B_+(b_+B_+)^{p_n}(\tau_n)\right),$$

(11.20)

where $\tau_* = \overset{\bullet}{\circ}$. Figure 11.2 gives the mode of trees in SSENT.

From Definition 11.4, the following rules for forming a tree $\tau \in$ SSENT can be obtained straightforwardly:

(i) The root of a tree is always a fat white vertex.
(ii) Upwards pointing branches of a white vertex must lead to fat black vertices.
(iii) A fat black vertex has at most one upwards pointing branch, and this branch (if exists) cannot lead to a fat black vertex.
(iv) A meagre vertex must have one upwards pointing branch which leads to a fat black vertex, and the fat black vertex must link to a white vertex.

Definition 11.5 The *order* $\rho(\tau)$, the *extended elementary differential* $\mathscr{F}(\tau)(y)$, the *symmetry* $\alpha(\tau)$, the *weight* $\Phi_i(\tau)$, the *density* $\gamma(\tau)$ and the *sign* $\kappa(\tau)$ on SSENT are recursively defined as follows:

(i) $\rho(\mathbf{o}) = 1$, $\mathscr{F}(\mathbf{o})(y) = f$, $\alpha(\mathbf{o}) = 1$, $\Phi_i(\mathbf{o}) = 1$, $\gamma(\mathbf{o}) = 1$, and $\kappa(\mathbf{o}) = 1$.

(ii) for $\tau \in$ SSENT denoted by (11.20),

- $\rho(\tau) = 1 + m + \sum_{i=1}^{n} \big(1 + 2p_i + \rho(\tau_i)\big),$

- $\mathscr{F}(\tau)(y) = D_h^m f^{(n)}\Big((-M)^{p_1}\mathscr{F}(\tau_1)(y), \ldots, (-M)^{p_n}\mathscr{F}(\tau_n)(y)\Big),$

- $\alpha(\tau) = \dfrac{(\rho(\tau) - 1)!}{m! \prod\limits_{k=1}^{n}\big(1 + 2p_k + \rho(\tau_k)\big)!} \cdot \dfrac{1}{J_1! J_2! \ldots} \cdot \prod\limits_{k=1}^{n} \alpha(\tau_k),$

- $\Phi_i(\tau) = c_i^m \prod\limits_{k=1}^{n}\Big(\sum\limits_{j=1}^{s} \bar{a}_{ij}^{(2p_k)}\Phi_j(\tau_k)\Big),$

- $\gamma(\tau) = \rho(\tau) \prod\limits_{k=1}^{n}\Big(\dfrac{(1 + 2p_k + \rho(\tau_k))!}{(2p_k)! \rho(\tau_k)!}\gamma(\tau_k)\Big),$

- $\kappa(\tau) = \prod\limits_{k=1}^{n}(-1)^{p_k}\kappa(\tau_k),$

where $\sum\limits_{k=1}^{0} = 0$, $\prod\limits_{k=1}^{0} = 1$ and $J_1, J_2, \ldots$, count equal the same branches.

Remark 11.3 The dimension of $\mathscr{F}(\tau)(y)$, in Definition 11.5, is $d \times 1$.

Remark 11.4 In particular, we have

$$\mathscr{F}(\underbrace{\tau_* \times \cdots \times \tau_*}_{m\text{-fold}})(y) = D_h^m f,$$

where $\tau_* = \overset{\bullet}{\mathbf{o}}$, namely, when $n = 0$, the extended elementary differential is exactly the mth order derivative of the function $f\big(\phi_0(h^2 M)y + \phi_1(h^2 M)hy'\big)$ with respect to h at $h = 0$.

Definition 11.6 The set $SSENT_m$ is defined as

$$\mathrm{SSENT}_m = \big\{\tau : \rho(\tau) = m, \ \tau \in \mathrm{SSENT}\big\}.$$

Example: For the tree τ with the following structure,

$$\tau =$$

the order, extended elementary differential, symmetry, weights, density and sign are shown below:

- The order $\rho(\tau)$ is 7.
- The extended elementary differential $\mathscr{F}(\tau)(\mathbf{y})$ is $D_h^0 f^{(2)}\left(-M D_h^0 f^{(0)}, D_h^0 f^{(0)}\right)$, namely,

$$\mathscr{F}(\tau)(\mathbf{y}) = f^{(2)}\left(-M f, f\right).$$

- The symmetry $\alpha(\tau)$ is $\dfrac{6!}{4! \times 2!} = 15$.
- The weights $\Phi_i(\tau)$ are $\sum_{j,k} \bar{a}_{ij}^{(2)} \bar{a}_{ik}^{(0)}$.
- The density $\gamma(\tau)$ is $7 \times (\dfrac{1}{2!} \times 4! \times 1) \times (2! \times 1) = 168$.
- The sign $\kappa(\tau)$ is -1.

The order $\rho(\tau)$ is the number of the tree τ's vertices.

The extended elementary differential $\mathscr{F}(\tau)(\mathbf{y})$ is a product of $(-M)^p$ (p is the number of meagre vertices between a white vertex and the next coming white vertex) and $D_h^m f^{(n)}$ (m is the number of $\tau_* = \overset{\uparrow}{\circ}$, and n is the difference between the ramifications number and the number m).

The symmetry $\alpha(\tau)$ is the number of possible different monotonic labellings of τ. In Table 11.1, we list all the 15 possible different monotonic labellings for the tree.

Table 11.1 The possible labelling of one seventh-order tree

The weights $\Phi_i(\tau)$ are sums over the indices of all white vertices and of all end vertices. The general term of the sum is a product of $\bar{a}_{ij}^{(2p)}$ (p is the number of the meagre vertices between the white vertices i and j), and c_i^m (m is the number of end vertices from the white vertex i).

The density $\gamma(\tau)$ is a product of the density of a tree by overlooking the differences between vertices, and of $(2p)!$ (p is the number of the meagre vertices between two white vertices).

The sign $\kappa(\tau)$ is 1 if the number of the meagre vertices is even, and -1 if the number of the meagre vertices is odd.

11.4.2   The Set SSENT and the Set of Classical SN-Trees

In this section, we can see that one special fifth-order tree and seven seventh-order trees are appended to the classical special Nyström-tree set. We will see that when M is the null matrix, the set SSENT with the six mappings is exactly the set of the classical SN-trees with the corresponding six mappings.

In Table 11.2, we present all simplified special extended Nyström trees (SSEN-trees) up to order 5 and their corresponding orders $\rho(\cdot)$, extended elementary differentials $\mathscr{F}(\cdot)(\mathbf{y})$, symmetries $\alpha(\cdot)$, weights $\Phi_i(\cdot)$, signs $\kappa(\cdot)$ and densities $\gamma(\cdot)$. If the entries $\bar{a}_{ij}(V)$, $b_i(V)$ and $\bar{b}_i(V)$ in the method are all assumed to be constants, the orders $\rho(\cdot)$, the symmetries $\alpha(\cdot)$, the weights $\Phi_i(\cdot)$ and the densities $\gamma(\cdot)$ on SSENT are all the same as those for the corresponding classical SN-trees, except the (extended) elementary differentials $\mathscr{F}(\cdot)(\mathbf{y})$, and in this case, $\kappa(\cdot) = 1$.

It can be also seen from Table 11.2 that the set SSENT is really an extension of the classical SN-tree set. This special and new appended fifth-order tree is one with a meagre vertex, while its weights Φ_i are functions of $\bar{a}_{ij}^{(2)}$, functions of the second derivative of $\bar{a}(V)$ with respect to h, at $h = 0$, where $V = h^2 M$.

When M is the null matrix, the set SSENT is exactly the same as that of classical SN-trees and it can be seen clearly from Table 11.2. In fact, if M is the null matrix, with the disappearing of all meagre vertices, the rules for forming the tree set are straightforwardly reduced to those of the classical SN-tree set.

When M is the null matrix, from Definition 11.5, it is easy to see that the order $\rho(\cdot)$, the symmetry $\alpha(\cdot)$, the weights $\Phi_i(\cdot)$ and the density $\gamma(\cdot)$ are exactly the same as those of the classical SN-tree. Moreover, the extended elementary differential $\mathscr{F}(\cdot)(\mathbf{y})$ shares the same result. In fact, in the case where the meagre vertices do not appear, from Definition 11.5 and Remark 11.2, we can observe that the extended elementary differential on SSENT is exactly the elementary differential on the set of classical SN-trees. This fact can also be seen clearly from Tables 11.2, 11.4 and 11.5.

In Table 11.3, we give all new trees which are appended to the classical SN-tree set to form the set SSENT. There are two sixth-order trees and seven seventh-order trees. All of these new trees are those with at least one meagre vertex. There is no

Table 11.2 Trees in SSENT$_m$, $m \le 5$, and the corresponding classical SN-trees

$\tau \in$ SSENT	ρ	α	$\mathscr{F}$	κ	γ	Φ_i	$\tau \in$ SNT	$\mathscr{F}$
○	1	1	f	1	1	1	○	f
	2	1	$D_h^1 f$	1	2	c_i		$f'y'$
	3	1	$D_h^2 f$	1	3	c_i^2		$f''(y',y')$
	3	1	$f^{(1)} f$	1	6	$\sum_j \bar{a}_{ij}^{(0)}$		$f'f$
	4	1	$D_h^3 f$	1	4	c_i^3		$f^{(3)}(y',y',y')$
	4	3	$D_h^1 f^{(1)} f$	1	8	$c_i \sum_j \bar{a}_{ij}^{(0)}$		$f''(y',f)$
	4	1	$f^{(1)} D_h^1 f$	1	24	$\sum_j \bar{a}_{ij}^{(0)} c_j$		$f'f'y'$
	5	1	$D_h^4 f$	1	5	c_i^4		$f^{(4)}(y',y',y',y')$
	5	6	$D_h^2 f^{(1)} f$	1	10	$c_i^2 \sum_j \bar{a}_{ij}^{(0)}$		$f^{(3)}(y',y',f)$
	5	3	$f^{(2)}(f,f)$	1	20	$\sum_{j,k} \bar{a}_{ij}^{(0)} \bar{a}_{ik}^{(0)}$		$f''(f,f)$
	5	4	$D_h^1 f^{(1)} D_h^1 f$	1	30	$c_i \sum_j \bar{a}_{ij}^{(0)} c_j$		$f''(f'y',y')$
	5	1	$f^{(1)} D_h^2 f$	1	60	$\sum_j \bar{a}_{ij}^{(0)} c_j^2$		$f'f''(y',y')$
	5	1	$f^{(1)} f^{(1)} f$	1	120	$\sum_{j,k} \bar{a}_{ij}^{(0)} \bar{a}_{jk}^{(0)}$		$f'f'f$
	5	1	$f^{(1)}(-Mf)$	-1	60	$\sum_j \bar{a}_{ij}^{(2)}$	– –	– –

corresponding classical SN-tree for each of them. In fact, it is a direct result of the fact that the entries $\bar{a}_{ij}(V)$ in classical SN-tree theory are all constant coefficients, whereas the weights Φ_i in Table 11.3 are all functions of $\bar{a}_{ij}^{(2k)}$, i.e. functions of higher order derivatives of $\bar{a}_{ij}(V)$ with respect to h, at $h = 0$.

From the above analysis, it can be concluded that this extension of the classical SN-tree set is significant and can be thought of as an optimal extension. The main

Table 11.3 Trees with meagre vertex in SSENT_m with $5 < m \le 7$

$\tau \in \text{SSENT}$	ρ	α	$\mathscr{F}$	κ	γ	Φ_i
	6	5	$D_h^1 f^{(1)}(-M)f$	-1	72	$c_i \sum_j \bar{a}_{ij}^{(2)}$
	6	1	$f^{(1)}(-M)D_h^1 f$	-1	360	$\sum_j \bar{a}_{ij}^{(2)} c_j$
	7	1	$f^{(1)}(-M)D_h^2 f$	-1	1260	$\sum_j \bar{a}_{ij}^{(2)} c_j^2$
	7	6	$D_h^1 f^{(1)}(-M)D_h^1 f$	-1	420	$c_i \sum_j \bar{a}_{ij}^{(2)} c_j$
	7	15	$D_h^2 f^{(1)}(-M)f$	-1	84	$c_i^2 \sum_j \bar{a}_{ij}^{(2)}$
	7	1	$f^{(1)}(-M)f^{(1)}f$	-1	2520	$\sum_{j,k} \bar{a}_{ij}^{(2)} \bar{a}_{jk}^{(0)}$
	7	1	$f^{(1)}(-M)^2 f$	1	210	$\sum_j \bar{a}_{ij}^{(4)}$
	7	15	$f^{(2)}\big(f,(-M)f\big)$	-1	168	$\sum_{j,k} \bar{a}_{ij}^{(2)} \bar{a}_{ik}^{(0)}$
	7	1	$f^{(1)}f^{(1)}(-M)f$	-1	2520	$\sum_{j,k} \bar{a}_{ij}^{(0)} \bar{a}_{jk}^{(2)}$

result of this chapter, i.e. there is no longer any redundant tree in the set SSENT, has been shown clearly, from Tables 11.2 and 11.3, since different trees correspond to different weights $\Phi_i(\cdot)$.

Table 11.4 Trees in SSENT$_m$, $m \leq 5$, and the corresponding SENT trees—(a)

$\tau \in$ SSENT$_m$	α	$\mathscr{F}$	$\tau \in$ SENT$_m$	α	$\mathscr{F}$
	1	f		1	f
	1	$f^{(1)}y'$		1	$f'y'$
	1	$f^{(2)}\left(y',y'\right)$		1	$f''(y',y')$
		$+f^{(1)}\left(-My\right)$		1	$f'(-My)$
	1	$f^{(1)}f$		1	$f'f$
	1	$f^{(3)}\left(y',y',y'\right)$		1	$f^{(3)}(y',y',y')$
		$+f^{(1)}\left(-My'\right)$		1	$f'(-M)y'$
		$+3f^{(2)}\left(-My,y'\right)$		3	$f''(-My,y')$
	3	$f^{(2)}(y',f)$		3	$f''(y',f)$
	1	$f^{(1)}\left(f^{(1)}y'\right)$		1	$f'f'y'$
	1	$f^{(4)}\left(y',y',y',y'\right)$		1	$f^{(4)}(y',y',y',y')$
		$+6f^{(3)}\left(-My,y',y'\right)$		6	$f'''(-My,y',y')$
		$+3f^{(2)}\left(-My,-My\right)$		3	$f''(-My,-My)$
		$+4f^{(2)}\left(y',-My'\right)$		4	$f''(y',-My')$
		$+f^{(1)}\left((-M)^2y\right)$		1	$f'(-M)^2y$
$\vdots$	$\vdots$	$\vdots$	$\vdots$	$\vdots$	$\vdots$

11.4.3 The Set SSENT and the Set SENT

In this section, we will show that the set SSENT is also a subset of the set SENT in [36]. Tables 11.4 and 11.5 will show vividly, in another way, that there is no any redundant tree in the set SSENT.

Tables 11.4 and 11.5 just show the difference between the set SSENT and the set SENT, in the extended elementary differentials, while the coefficients of the trees

Table 11.5 Trees in SSENT_m, $m \leq 5$, and the corresponding SENT trees—(b)

trees in SSENT_m	α		trees in $\mathbb{T}_m$	α	$\mathscr{F}$
	6	$f^{(3)}(y', y', f)$		6	$f^{(3)}(y', y', f)$
		$+ f^{(2)}(-My, f)$		6	$f''(-My, f)$
	3	$f^{(2)}(f, f)$		3	$f''(f, f)$
	4	$f^{(2)}(y', f^{(1)}y')$		4	$f''(y', f'y')$
	1	$f^{(1)}f^{(2)}(y', y')$		1	$f'f''(y', y')$
		$+ f^{(1)}f^{(1)}(-My)$		1	$f'f'(-My)$
	1	$f^{(1)}f^{(1)}f$		1	$f'f'f$
	1	$f^{(1)}(-M)f$		1	$f'(-M)f$

are ignored. The extended elementary differential for each tree in the set SSENT is given as a function of $\{f^{(i)}, -M, y, y'\}$. It can be observed that an extended elementary differential is not just one function but a sum of the traditional elementary differentials.

The set SSENT is a significant improvement on the original one in [36] since all the redundant trees disappear. This makes the order conditions of the numerical integrator much clearer and simpler.

11.5 B-series and Order Conditions

In Sect. 11.4, we have established the SSEN-trees and the corresponding six mappings. With all these preliminaries, motivated by the concept of B-series [2, 3, 9–12], we will give a totally different approach from the one in [36] to derive the order conditions.

11.5.1 B-series

The B-series theory has its origin in the paper [2] by Butcher in 1972, although series expansions were not used there. Then series expansions were introduced by Hairer and Wanner [11] in 1974.

We now present the following two elementary theorems.

Theorem 11.1 *With Definition 11.5,* $h^2 f\big(y(t+h)\big)$ *is a B-series,*

$$h^2 f\big(y(t+h)\big) = \sum_{\tau \in SSENT} \frac{h^{\rho(\tau)+1}}{(\rho(\tau)-1)!} \alpha(\tau) \mathscr{F}(\tau)(y).$$

Proof The underlying idea for obtaining the expression is, in fact, quite straightforward. We just insert the series into itself. From the matrix-variation-of-constants formula, the first one of (11.3) with $\mu = 1$, and the Taylor series expansion (11.6), we obtain

$$f\big(y(t+h)\big) = \sum_{n \geq 0} \frac{1}{n!} f^{(n)}\Big|_{\hat{y}} \left(h^2 \int_0^1 (1-z)\phi_1\big((1-z)^2 V\big) f\big(y(t+hz)\big)dz\right)^{\otimes n},$$

(11.21)

where $f\big(y(t+h)\big)$ is expanded at point $\hat{y} = \phi_0(V)y(t) + \phi_1(V)hy'(t)$. Definition 11.5 ensures that $f\big(y(t+h)\big)$ is a B-series. In fact, if $f\big(y(t+h)\big)$ is a B-series, the second term in (11.21) is

$$h^2 \int_0^1 (1-z)\phi_1\big((1-z)^2 V\big) f\big(y(t+hz)\big)dz$$

$$= \sum_{\tau \in SSENT} \int_0^1 (1-z)\phi_1\big((1-z)^2 V\big) \frac{z^{\rho(\tau)-1}}{(\rho(\tau)-1)!}dz \cdot \Big(h^{\rho(\tau)+1}\alpha(\tau)\mathscr{F}(\tau)(y)\Big)$$

$$= \sum_{\tau \in SSENT} \phi_{\rho(\tau)+1}(V) \cdot h^{\rho(\tau)+1}\alpha(\tau)\mathscr{F}(\tau)(y)$$

$$= \sum_{\tau \in SSENT} \sum_{p \geq 0} \frac{(-1)^p V^p}{(\rho(\tau)+1+2p)!}h^{\rho(\tau)+1}\alpha(\tau)\mathscr{F}(\tau)(y).$$

(11.22)

The equality (11.22) follows from (4.7) and (11.4). Then, from (11.13) and (11.22), (11.21) becomes

$$f\big(y(t+h)\big) = \sum_{n,m} \sum_{\tau \in SSENT} \frac{h^s}{n!m!} D^m f^{(n)}\left(\frac{(-M)^{p_1}\alpha(\tau_1)\mathscr{F}(\tau_1)(y)}{(\rho(\tau_1)+1+2p_1)!}, \ldots, \frac{(-M)^{p_n}\alpha(\tau_n)\mathscr{F}(\tau_n)(y)}{(\rho(\tau_n)+1+2p_n)!}\right),$$

(11.23)

where $s = m + \sum_{k=1}^n \big(2p_k + \rho(\tau_k)+1\big)$, $p_i \geq 0$, $(i = 1, 2, \ldots, n)$. By Definition 11.5, we complete the proof. $\qquad\square$

Theorem 11.2 *Given an ERKN scheme (11.5), by Definition 11.5, $h^2 f(Y_i)$ are B-series,*

$$h^2 f(Y_i) = \sum_{\tau \in SSENT} \frac{h^{\rho(\tau)+1}}{\rho(\tau)!} a_i(\tau),$$

where $a_i(\tau) = \Phi_i(\tau) \cdot \kappa(\tau) \cdot \gamma(\tau) \cdot \alpha(\tau) \cdot \mathscr{F}(\tau)(y_0)$.

Proof In a similar way, for an ERKN method (11.5), we expand $f(Y_i)$ at

$$\tilde{y} = \phi_0(c_i^2 V)y_0 + \phi_1(c_i^2 V)c_i h y_0'$$

and obtain the Taylor series expansion as follows:

$$f(Y_i) = \sum_{n \geq 0} \frac{1}{n!} f^{(n)}\Big|_{\tilde{y}} \Big(h^2 \sum_j \bar{a}_{ij}(V) f(Y_j)\Big)^{\otimes n}. \tag{11.24}$$

With (11.14), $f^{(n)}\big|_{\tilde{y}}$ in (11.24) is exactly $g(c_i h)$. Then the Taylor series expansion of $f^{(n)}\big|_{\tilde{y}}$ at $h = 0$ is given by

$$f^{(n)}\Big|_{\tilde{y}} = \sum_{m \geq 0} \frac{c_i^m}{m!} h^m D_h^m f^{(n)}. \tag{11.25}$$

Definition 11.5 ensures that $f(Y_i)$ $(i = 1, \ldots, s)$ are B-series. In fact, $h^2 \sum_j \bar{a}_{ij}(V) f(Y_j)$ in (11.24) is given by

$$h^2 \sum_j \bar{a}_{ij}(V) f(Y_j) = \sum_{\tau \in SSENT} \frac{\sum_j \bar{a}_{ij}(V)}{\rho(\tau)!} h^{\rho(\tau)+1} a_j(\tau)$$

$$= \sum_{\tau \in SSENT} \sum_{p \geq 0} \frac{\sum_j \bar{a}_{ij}^{(2p)}}{\rho(\tau)!} \frac{V^p}{(2p)!} h^{\rho(\tau)+1} a_j(\tau). \tag{11.26}$$

Inserting Eqs. (11.25) and (11.26) into (11.24), we obtain

$$f(Y_i) = \sum_{n,m} \sum_{\tau \in SSENT} \frac{c_i^m h^s}{n!m!} D_h^m f^{(n)} \Big(\frac{\sum_j \bar{a}_{ij}^{(2p_1)}}{\rho(\tau_1)!} \frac{M^{p_1}}{(2p_1)!} a_j(\tau_1), \ldots, \frac{\sum_j \bar{a}_{ij}^{(2p_n)}}{\rho(\tau_n)!} \frac{M^{p_n}}{(2p_n)!} a_j(\tau_n)\Big), \tag{11.27}$$

where $s = m + \sum_{k=1}^n \big(2p_k + \rho(\tau_k) + 1\big)$, $p_i \geq 0$, $(i = 1, 2, \ldots, n)$. This completes the proof. $\square$

11.5.2 Order Conditions

We are now in a position to present the order conditions for the ERKN scheme (11.5).

Theorem 11.3 *The scheme (11.5) has order r if and only if the following conditions*

$$
\begin{cases}
\displaystyle\sum_{i=1}^{s} \bar{b}_i(V)\kappa(\tau)\gamma(\tau)\Phi_i(\tau) = \rho(\tau)!\phi_{\rho(\tau)+1} + \mathcal{O}(h^{r-\rho(\tau)}), & \forall \tau \in SSENT_m, \ m \leq r-1, \\[4mm]
\displaystyle\sum_{i=1}^{s} b_i(V)\kappa(\tau)\gamma(\tau)\Phi_i(\tau) = \rho(\tau)!\phi_{\rho(\tau)} + \mathcal{O}(h^{r-\rho(\tau)+1}), & \forall \tau \in SSENT_m, \ m \leq r,
\end{cases}
$$

are satisfied.

Proof From the matrix-variation-of-constants formula (11.3) and the scheme (11.5), using Theorems 11.1 and 11.2, we have

$$
\begin{aligned}
\boldsymbol{y}_1 &= \phi_0(V)\boldsymbol{y}_0 + h\phi_1(V)\boldsymbol{y}_0' \\
&\quad + \sum_{\tau \in SSENT} \frac{h^{\rho(\tau)+1}}{\rho(\tau)!} \sum_{i=1}^{s} \bar{b}_i(V)\Phi_i(\tau)\kappa(\tau)\gamma(\tau)\alpha(\tau)\mathscr{F}(\tau)(\boldsymbol{y}_0),
\end{aligned}
\tag{11.28}
$$

$$
\begin{aligned}
\boldsymbol{y}(t+h) &= \phi_0(V)\boldsymbol{y} + h\phi_1(V)\boldsymbol{y}' \\
&\quad + \sum_{\tau \in EMT} h^{\rho(\tau)+1}\alpha(\tau)\mathscr{F}(\tau)(\boldsymbol{y}) \int_0^1 (1-z)\frac{z^{\rho(\tau)-1}}{(\rho(\tau)-1)!}\phi_1((1-z)V)\,\mathrm{d}z.
\end{aligned}
\tag{11.29}
$$

Comparing (11.28) with (11.29) and using (11.4), we obtain the first result of Theorem 11.3. Likewise, we can get the second result of Theorem 11.3. ☐

All the rooted trees of SSENT up to order 5, together with the extended differentials $\mathscr{F}(\cdot)(\boldsymbol{y})$ and the expressions of $\Phi_i(\cdot)$, $\rho(\cdot)$, $\alpha(\cdot)$, $\kappa(\cdot)$ and $\gamma(\cdot)$, which are needed for the order conditions, are given in Table 11.2. For sixth-order SSEN-trees and seventh-order SSEN-trees, we do not plot all of them but the newly added trees in Table 11.3.

The tri-coloured tree set in this theory can be regarded as an optimal extension of the classical SN-tree set. When M is the null matrix, from the fact that the set SSENT is exactly the same as that of the classical SN-trees (see, Table 14.3 on the p. 292 in [10]), the order conditions of Theorem 11.3 reduce to those for classical RKN methods when applied to $y'' = f(y)$.

It seems that the results of the order conditions are formally the same as those in [36], but the contents are different. Moreover, the proof here is much clearer and simpler.

Table 11.6 Fifth-order conditions for five-stage explicit ERKN methods

$\tau \in$ SSENT	$\tau \in$ SSENT$_m$, $m \le 4$	$\tau \in$ SSENT$_m$, $m \le 5$
	$\sum\limits_{i=1}^{5} \bar{b}_i(V) = \phi_2 + \mathcal{O}(h^4)$	$\sum\limits_{i=1}^{5} b_i(V) = \phi_1 + \mathcal{O}(h^5)$
	$\sum\limits_{i=1}^{5} \bar{b}_i(V)c_i = \phi_3 + \mathcal{O}(h^3)$	$\sum\limits_{i=1}^{5} b_i(V)c_i = \phi_2 + \mathcal{O}(h^4)$
	$\sum\limits_{i=1}^{5} \bar{b}_i(V)c_i^2 = 2\phi_4 + \mathcal{O}(h^2)$	$\sum\limits_{i=1}^{5} b_i(V)c_i^2 = 2\phi_3 + \mathcal{O}(h^3)$
	$\sum\limits_{i=1}^{5} \bar{b}_i(V)\sum_j \bar{a}_{ij}^{(0)} = \phi_4 + \mathcal{O}(h^2)$	$\sum\limits_{i=1}^{5} b_i(V)\sum_j \bar{a}_{ij}^{(0)} = \phi_3 + \mathcal{O}(h^3)$
	$\sum\limits_{i=1}^{5} \bar{b}_i(V)c_i^3 = 6\phi_5 + \mathcal{O}(h)$	$\sum\limits_{i=1}^{5} b_i(V)c_i^3 = 6\phi_4 + \mathcal{O}(h^2)$
	$\sum\limits_{i=1}^{5} \bar{b}_i(V)c_i\sum_j \bar{a}_{ij}^{(0)} = 3\phi_5 + \mathcal{O}(h)$	$\sum\limits_{i=1}^{5} b_i(V)c_i\sum_j \bar{a}_{ij}^{(0)} = 3\phi_4 + \mathcal{O}(h^2)$
	$\sum\limits_{i=1}^{5} \bar{b}_i(V)\sum_j \bar{a}_{ij}^{(0)}c_j = \phi_5 + \mathcal{O}(h)$	$\sum\limits_{i=1}^{5} b_i(V)\sum_j \bar{a}_{ij}^{(0)}c_j = \phi_4 + \mathcal{O}(h^2)$
	$--$	$\sum\limits_{i=1}^{5} b_i(V)c_i^4 = 24\phi_5 + \mathcal{O}(h)$
	$--$	$\sum\limits_{i=1}^{5} b_i(V)c_i^2\sum_j \bar{a}_{ij}^{(0)} = 12\phi_5 + \mathcal{O}(h)$
	$--$	$\sum\limits_{i=1}^{5} b_i(V)\sum_{j,k} \bar{a}_{ij}^{(0)}\bar{a}_{ik}^{(0)} = 6\phi_5 + \mathcal{O}(h)$
	$--$	$\sum\limits_{i=1}^{5} b_i(V)c_i\sum_j \bar{a}_{ij}^{(0)}c_j = 4\phi_5 + \mathcal{O}(h)$
	$--$	$\sum\limits_{i=1}^{5} b_i(V)\sum_j \bar{a}_{ij}^{(0)}c_j^2 = 2\phi_5 + \mathcal{O}(h)$
	$--$	$\sum\limits_{i=1}^{5} b_i(V)\sum_{j,k} \bar{a}_{ij}^{(0)}\bar{a}_{jk}^{(0)} = \phi_5 + \mathcal{O}(h)$
	$--$	$\sum\limits_{i=1}^{5} b_i(V)\sum_j \bar{a}_{ij}^{(2)} = -2\phi_5 + \mathcal{O}(h)$

Before the end of this section, we give an example of the order conditions derived from the novel theory of SSENT trees. We consider the five-stage explicit ERKN method denoted by the following Butcher tableau:

$$
\begin{array}{c|c}
c & \bar{A}(V) \\
\hline
& \bar{b}^{\mathsf{T}}(V) \\
\hline
& b^{\mathsf{T}}(V)
\end{array}
=
\begin{array}{c|ccccc}
c_1 & \mathbf{0}_{d\times d} & \mathbf{0}_{d\times d} & \mathbf{0}_{d\times d} & \mathbf{0}_{d\times d} & \mathbf{0}_{d\times d} \\
c_2 & \bar{a}_{21}(V) & \mathbf{0}_{d\times d} & \mathbf{0}_{d\times d} & \mathbf{0}_{d\times d} & \mathbf{0}_{d\times d} \\
c_3 & \bar{a}_{31}(V) & \bar{a}_{32}(V) & \mathbf{0}_{d\times d} & \mathbf{0}_{d\times d} & \mathbf{0}_{d\times d} \\
c_4 & \bar{a}_{41}(V) & \bar{a}_{42}(V) & \bar{a}_{43}(V) & \mathbf{0}_{d\times d} & \mathbf{0}_{d\times d} \\
c_5 & \bar{a}_{51}(V) & \bar{a}_{52}(V) & \bar{a}_{53}(V) & \bar{a}_{54}(V) & \mathbf{0}_{d\times d} \\
\hline
& \bar{b}_1(V) & \bar{b}_2(V) & \bar{b}_3(V) & \bar{b}_4(V) & \bar{b}_5(V) \\
\hline
& b_1(V) & b_2(V) & b_3(V) & b_4(V) & b_5(V)
\end{array}
$$

In light of Theorem 11.3, the necessary and sufficient conditions for this five-stage explicit ERKN method of order five are listed by Table 11.6.

The set of SENT trees originally developed in [36] contains enormous redundant trees which result in inconvenience and inefficiency to use the original tri-coloured tree theory for order conditions. However, with the simplified tri-coloured tree theory presented in this chapter, it can be observed clearly from Table 11.6 that each different tree $\tau \in$ SSENT corresponds to a different order condition uniquely. Redundant trees really disappear now.

11.6   Conclusions and Discussions

In this chapter, we have established a compact theory of the order conditions for ERKN methods which are designed especially for multi-frequency oscillatory systems, and were presented initially in [36]. The original tri-coloured tree theory and the order conditions for ERKN methods proposed in the paper [36] are not satisfied with the enormous number of redundant trees. Hence, the authors had been keeping their focus on the simplification of the tri-coloured tree theory and the order conditions for ERKN methods after the publication of the paper [36]. This chapter has succeeded in the simplification by defining two special mappings, namely, the extended elementary differential and the sign mapping. This successful simplification makes the construction of ERKN methods much simpler and more efficient for the multi-frequency and multidimensional oscillatory systems (11.1). In light of the simplified tree theory, almost one half of the algebraic conditions in [36] can be eliminated.

Likewise, we can create a similar compact theory of the order conditions for ERKN methods (1.46) for solving the general multi-frequency oscillatory second-order initial value problem (1.2) in Chap. 1.

This chapter is based on the recent work by Yang et al. [37].

References

1. Boik RJ (2006) Lecture notes: statistics, vol 550. Spring. http://www.math.montana.edu/~rjboik/classes/550/notes.550.06.pdf, pp 33–35
2. Butcher JC (1972) An algebraic theory of integration methods. Math Comp 26:79–106
3. Butcher JC (2008) Numerical methods for ordinary differential equations, 2nd edn. Wiley, Chichester
4. Coleman JP, Ixaru L Gr (1996) P-stability and exponential-fitting methods for $y'' = f(x, y)$. IMA J Numer Anal 16:179–199
5. Deuflhard P (1979) A study of extrapolation methods based on multistep schemes without parasitic solutions. Z Angew Math Phys 30:177–189
6. Fang Y, Wu X (2008) A trigonometrically fitted explicit Numerov-type method for second-order initial value problems with oscillating solutions. Appl Numer Math 58:341–351
7. Gautschi W (1961) Numerical integration of ordinary differential equations based on trigonometric polynomials. Numer Math 3:381–397
8. Hairer E (1976/1977) Méthodes de Nyström pour l'équation différentielle $y'' = f(x, y)$. Numer Math 27:283–300
9. Hairer E, Lubich C, Wanner G (2006) Geometric numerical integration, structure-preserving algorithms for ordinary differential equations, 2nd edn. Springer, Berlin
10. Hairer E, Nørsett SP, Wanner G (1993) Solving ordinary differnetial equations I, nonstiff problems. Springer series in computational mathematics, 2nd edn. Springer, Berlin
11. Hairer E, Wanner G (1974) On the Butcher group and general nulti-value methods. Computing 13:1–15
12. Hairer E, Wanner G (1975/1976) A theory for Nyström methods. Numer Math 25:383–400
13. Hochbruck M, Lubich C (1999) A Gsutschi-type method for oscillatory second-order differential equations. Numer Math 83:403–426
14. Hochbruck M, Lubich C (1999) Exponential integrators for quantum-classical molecular dynamics. BIT Numer Math 39:620–645
15. Li J, Wang B, You X, Wu X (2011) Two-step extended RKN methods for oscillatory systems. Comput Phys Commun 182:2486–2507
16. Li J, Wu X (2013) Adapted Falkner-type methods solving oscillatory second-order differential equations. Numer Algo 62:355–381
17. Murua A, Sanz-Serna JM (1999) Order conditions for numerical integrators obtained by composing simpler integrators, Philos Trans Royal Soc London, ser A 357:1079–1100
18. Nyström EJ (1925) Ueber die numerische Integration von Differentialgleichungen. Acta Soc Sci Fenn 50:1–54
19. Shi W, Wu X, Xia J (2012) Explicit multi-symplectic extended leap-frog methods for Hamiltonian wave equations. J Comput Phys 231:7671–7694
20. Simos TE (1998) An exponentially-fitted Runge-Kutta method for the numerical integration of initial-value problems with periodic or oscillating solutions. Comput Phys Commun 115:1–8
21. Vanden Berghe G, Daele MV, Vyver HV (2003) Exponential fitted Runge-Kutta methods of collocation type: fixed or variable knot points? J Comput Appl Math 159:217–239
22. Vanden Berghe G, De Meyer H, Daele MV, Hecke TV (1999) Exponentially-fitted Runge-Kutta methods. Comput Phys Commun 123:7–15
23. Vanden Berghe G, De Meyer H, Van Daele M, Van Hecke T (2000) Exponentially-fitted explicit Runge-Kutta methods. J Comput Appl Math 125:107–115
24. Wang B, Wu X (2012) A new high precision energy-preserving integrator for system of oscillatory second-order differential equations. Phys Lett A 376:1185–1190
25. Wang B, Wu X (2013) A Filon-type asymptotic approach to solving highly oscillatory second-order initial value problems. J Comput Phys 243:210–223
26. Wang B, Wu X, Zhao H (2013) Novel improved multidimensional Strömer-Verlet formulas with applications to four aspects in scientific computation. Math Comput Model 57:857–872
27. Wu X (2012) A note on stability of multidimensional adapted Runge-Kutta-Nyström methods for oscillatory systems. Appl Math Model 36:6331–6337

28. Wu X, Wang B, Liu K, Zhao H (2013) ERKN methods for long-term integration of multidimensional orbital problems. Appl Math Model 37:2327–2336
29. Wu X, Wang B, Shi W (2013) Efficient energy-perserving integrators for oscillatory Hamiltonian systems. J Comput Phys 235:587–605
30. Wu X, Wang B, Shi W (2013) Effective integrators for nonlinear second-order oscillatory systems with a time-dependent frequency matrix. Appl Math Model 37:6505–6518
31. Wu X, Wang B, Xia J (2010) Extended symplectic Runge-Kutta-Nyström integrators for separable Hamiltonian systems. In: Vigo Aguiar J (ed) Proceedings of the 2010 international conference on computational and mathematical methods in science and engineering, vol III. Spain, pp 1016–1020
32. Wu X, Wang B, Xia J (2012) Explicit symplectic multidimensional exponential fitting modified Runge-Kutta-Nyström methods. BIT Numer Math 52:773–795
33. Wu X, You X, Wang B (2013) Structure-preserving algorithms for oscillatory differential equations. Springer, Berlin
34. Wu X, You X, Shi W, Wang B (2010) ERKN integrators for systems of oscillatory second-order differential equations. Comput Phys Commun 181:1873–1887
35. Yang H, Wu X (2008) Trigonometrically-fitted ARKN methods for perturbed oscillators. Appl Numer Math 58:1375–1395
36. Yang H, Wu X, You X, Fang Y (2009) Extended RKN-type methods for numerical integration of perturbed oscillators. Comput Phys Commun 180:1777–1794
37. Yang H, Zeng X, Wu X, Ru Z (2014) A simplified Nyström-tree theory for extended Runge-Kutta-Nyström integrators solving multi-frequency oscillatory systems. Accepted Manuscript by Comput Phys Commun 185:2841–2850

Chapter 12
General Local Energy-Preserving Integrators for Multi-symplectic Hamiltonian PDEs

In this chapter we present a general approach to constructing local energy-preserving algorithms which can be of arbitrarily high order in time for multi-symplectic Hamiltonian PDEs. This approach is based on the temporal discretization using continuous Runge–Kutta–type methods, and the spatial discretization using pseudospectral methods or Gauss-Legendre collocation methods. The local energy conservation law of the new schemes is analysed in detail. The effectiveness of the novel local energy-preserving integrators is demonstrated by coupled nonlinear Schrödinger equations and 2D nonlinear Schrödinger equations with external fields. Meanwhile, the new schemes are compared with some classical multi-symplectic and symplectic schemes in numerical experiments. The numerical results show the remarkable *long-term behaviour* of the new schemes in preserving the qualitative properties.

12.1 Motivation

Since the development of multi-symplectic structure by Bridges [2] and Marsden et al. [28] for a class of PDEs, the construction and analysis of multi-symplectic numerical integrators conserving the discrete multi-symplectic structure has become one of the central topics in PDE algorithms. Many multi-symplectic schemes have been proposed such as multi-symplectic RK/PRK/RKN methods, finite volume methods, spectral/pseudospectral methods, splitting methods and wavelet collocation methods (see, e.g. [3, 4, 11, 18, 19, 27, 30, 31, 33]). All of these methods focus on the preservation of some kind of discrete multi-symplecticity. However, multi-symplectic PDEs have many other important properties such as the local energy conservation law (ECL) and the local momentum conservation law (MCL). In general, multi-symplectic integrators can only preserve quadratic conservation laws and invariants exactly. In the paper [30], Reich first proposed two methods that preserve the discrete ECL and MCL respectively. In [32], Wang et al. generalized Reich's work.

X. Wu et al., *Structure-Preserving Algorithms for Oscillatory Differential Equations II*, DOI 10.1007/978-3-662-48156-1_12

In Cai et al. [6, 7] and Chen et al. [9], the authors constructed some local structure-preserving schemes for special multi-symplectic PDEs. In [14], Gong et al. developed a general approach to constructing local structure-preserving algorithms. Local energy-preserving algorithms preserve the discrete global energy under suitable boundary conditions. Thus in the case of multi-symplectic PDEs, they cover the traditional global energy-preserving algorithms (see, e.g. [8, 13, 16, 21, 24]). However, most of the local and global energy-preserving methods are based on the discrete gradient for the temporal discretization. Therefore, they can have only second-order accuracy in time. However, it is noted that Hairer [17] developed a family of energy-preserving continuous Runge–Kutta–type methods which can have arbitrarily high order for Hamiltonian ODEs. On the basis of Hairer's work, in this chapter, we consider general local energy-preserving methods for multi-symplectic Hamiltonian PDEs, and we are hopeful of obtaining new high-order schemes which exactly preserve the discrete ECL.

Most of the existing local energy-preserving algorithms are based on the spatial discretization using the implicit midpoint rule. Although the authors in [8, 14] mentioned a class of global energy-preserving schemes based on the (pseudo) spectral discretization for the spatial derivative, it seems that there is little work investigating the local energy-preserving property of these schemes in the literature. In this chapter, we analyse the preservation of the discrete ECL for the underlying schemes which are based on the pseudospectral spatial discretization. Meanwhile, we also derive a class of local energy-preserving schemes based on the general Gauss-Legendre collocation for the spatial discretization.

12.2 Multi-symplectic PDEs and Energy-Preserving Continuous Runge–Kutta Methods

A multi-symplectic PDE with one temporal variable and two spatial variables can be written in the form:

$$M z_t + K z_x + L z_y = \nabla_z S(z), \quad z \in \mathbb{R}^d, \tag{12.1}$$

where M, K, and L are skew-symmetric $d \times d$ real matrices, $S : \mathbb{R}^d \to \mathbb{R}$ is a smooth scalar-valued function of the state variable z and ∇_z is the gradient operator. Three differential 2-forms are defined by

$$\omega = \frac{1}{2} dz \wedge M dz, \quad \kappa = \frac{1}{2} dz \wedge K dz, \quad \tau = \frac{1}{2} dz \wedge L dz.$$

Then Eq. (12.1) has the multi-symplectic conservation law (MSCL):

$$\partial_t \omega + \partial_x \kappa + \partial_y \tau = 0. \tag{12.2}$$

Another important local conservation law is the ECL:

$$\partial_t E + \partial_x F + \partial_y G = 0,$$

where

$$E = S(z) - \frac{1}{2}z^\mathsf{T}Kz_x - \frac{1}{2}z^\mathsf{T}Lz_y, \quad F = \frac{1}{2}z^\mathsf{T}Kz_t, \quad G = \frac{1}{2}z^\mathsf{T}Lz_t.$$

When $L = 0$, Eq. (12.1) reduces to the case of one spatial dimension:

$$Mz_t + Kz_x = \nabla_z S(z). \tag{12.3}$$

Correspondingly, the ECL reduces to:

$$\partial_t E + \partial_x F = 0,$$

where

$$E = S(z) - \frac{1}{2}z^\mathsf{T}Kz_x, \quad F = \frac{1}{2}z^\mathsf{T}Kz_t.$$

We notice that the energy density E is related to the gradient of S. If one is interested in constructing schemes which can preserve the discrete ECL, a natural idea is to replace $\nabla_z S$ by the discrete gradient (DG) $\bar{\nabla}_z S$. For details of the discrete gradient, readers are referred to [15, 26].

A limitation of the DG method is that it can only achieve second-order accuracy in general. Therefore, classical local energy-preserving methods based on the DG cannot reach an order higher than 2 in temporal direction unless the composition technique is applied, which is not our interest in this chapter.

In contrast to the DG method, Hairer's seminal work makes it possible to overcome the order barrier. The following is a summary of this approach.

Consider autonomous ODEs:

$$\begin{cases} y'(t) = f(y(t)), \quad y \in \mathbb{R}^d, \\ y(t_0) = y_0. \end{cases} \tag{12.4}$$

Hairer's approach can be regarded as a continuous Runge–Kutta method:

$$\begin{cases} y_\tau = y_0 + h \displaystyle\int_0^1 A_{\tau,\sigma} f(y_\sigma)d\sigma, \\[2mm] y_1 = y_0 + h \displaystyle\int_0^1 B_\sigma f(y_\sigma)d\sigma, \\[2mm] B_\sigma \equiv 1, \; A_{\tau,\sigma} = \displaystyle\sum_{i=1}^s \frac{1}{b_i}\int_0^\tau l_i(\alpha)d\alpha l_i(\sigma), \end{cases} \tag{12.5}$$

where h is the stepsize, $\{l_i(\tau)\}_{i=1}^{s}$ are Lagrange interpolating polynomials based on the s distinct points $c_1, c_2, \ldots, c_s$, $b_i = \int_0^1 l_i(\tau)d\tau$ for $i = 1, 2, \ldots, s$, and y_τ approximates the value of $y(t_0 + \tau h)$ for $\tau \in [0, 1]$. The continuous RK method can be expressed in a Buchter tableau as

$$\frac{C_\tau \, | \, A_{\tau,\sigma}}{\quad \, | \, B_\tau}$$

with

$$C_\tau = \int_0^1 A_{\tau,\sigma} \, d\sigma = \tau.$$

If $f(y) = J^{-1}\nabla H(y)$, and J is a constant skew-symmetric matrix, then this method preserves the Hamiltonian: $H(y_1) = H(y_0)$. Let r be the order of the quadrature formula $(b_i, c_i)_{i=1}^{s}$. Then this continuous method has order p:

$$p = \begin{cases} 2s & \text{for } r \geq 2s - 1, \\ 2r - 2s + 2 & \text{for } r \leq 2s - 2. \end{cases} \tag{12.6}$$

Moreover, if the quadrature nodes are symmetric, i.e. $c_i = 1 - c_{s+1-i}$ for $i = 1, 2, \ldots, s$, then the method (12.5) is also symmetric. Clearly, by choosing the s-point Gauss-Legendre quadrature formula, we gain a symmetric continuous RK method of order $2s$. Besides, although this method is not symplectic, it is conjugate-symplectic up to at least order $2s + 2$ (its long-term behavior is almost the same as that of a symplectic method). The proof can be found in [17]. In view of these prominent properties, we select (12.5) as the elementary method for the time integration of Hamiltonian PDEs. We denote this method by CRK and call $(b_i, c_i)_{i=1}^{s}$ as the generating quadrature formula in the remainder of this chapter.

12.3 Construction of Local Energy-Preserving Algorithms for Hamiltonian PDEs

12.3.1 Pseudospectral Spatial Discretization

For simplicity, we first consider the following PDE with one spatial variable:

$$M z_t + K z_x = \nabla_z S(z, x). \tag{12.7}$$

In the classical multi-symplectic PDE (12.3), the Hamiltonian S is independent of the variable x. It should be noted that (12.7) does not have the MSCL and the MCL, but the local energy conservation law still holds:

$$\partial_t E + \partial_x F = 0, \tag{12.8}$$

where

$$E = S(z, x) - \frac{1}{2} z^\mathsf{T} K z_x, \quad F = \frac{1}{2} z^\mathsf{T} K z_t.$$

Thus local energy-preserving methods can be more widely used than classical multi-symplectic methods. Most of multi-symplectic methods can be constructed by concatenating two ODE methods in time and space, respectively. The temporal method is always symplectic, while the spatial one may not. However, in the underlying new schemes, we use the CRK method instead of the symplectic method for the time integration. In this section, we consider a class of convenient methods for the spatial discretization under the periodic boundary condition. They are the Fourier spectral, the pseudospectral, and the wavelet collocation methods (see, e.g. [4, 11, 33]). A common characteristic of the three methods is the substitution of a skew-symmetric differential matrix D for the operator ∂_x. For example, assuming $z(x_0, t) = z(x_0 + L, t)$, (12.7) becomes a system of ODEs in time after the pseudospectral spatial discretization:

$$M \frac{\mathrm{d}}{\mathrm{d}t} z_j + K \sum_{k=0}^{N-1} D_{jk} z_k = \nabla_z S(z_j, x_j), \tag{12.9}$$

for $j = 0, 1, \ldots, N - 1$, where N is an even integer, $x_j = x_0 + j \Delta x$, $\Delta x = \frac{L}{N}$, $z_j \approx z(x_j, t)$, and D is a skew-symmetric matrix whose entries are determined by (see, e.g. [11])

$$D_{jk} = \begin{cases} \frac{\pi}{L}(-1)^{j+k} \cot(\pi \frac{x_j - x_k}{L}), & j \neq k, \\ 0, & j = k. \end{cases}$$

Multiplying both sides of (12.9) by $\frac{\mathrm{d}}{\mathrm{d}t} z_j^\mathsf{T}$, we get N semi-discrete ECLs (see, e.g. [23, 33]):

$$\frac{\mathrm{d}}{\mathrm{d}t} E_j + \sum_{k=0}^{N-1} D_{jk} F_{jk} = 0, \tag{12.10}$$

for $j = 0, 1, \ldots, N - 1$, where

$$E_j = S(z_j, x_j) - \frac{1}{2} z_j^\mathsf{T} K \sum_{k=0}^{N-1} D_{jk} z_k,$$

$$F_{jk} = \frac{1}{2} z_k^\mathsf{T} K \frac{\mathrm{d}}{\mathrm{d}t} z_j + \frac{1}{2} z_j^\mathsf{T} K \frac{\mathrm{d}}{\mathrm{d}t} z_k.$$

The term $\sum_{k=0}^{N-1} D_{jk} F_{jk}$ can be considered as the discrete $\partial_x F(z(x_j, t))$:

$$
\begin{aligned}
\sum_{k=0}^{N-1} D_{jk} F_{jk} &= \frac{1}{2} \delta_x z_j^\mathsf{T} K \frac{\mathrm{d}}{\mathrm{d}t} z_j + \frac{1}{2} z_j^\mathsf{T} K \frac{\mathrm{d}}{\mathrm{d}t} \delta_x z_j \\
&\approx \frac{1}{2} \partial_x z(x_j, t)^\mathsf{T} K \frac{\mathrm{d}}{\mathrm{d}t} z(x_j, t) + \frac{1}{2} z(x_j, t)^\mathsf{T} K \frac{\mathrm{d}}{\mathrm{d}t} \partial_x z(x_j, t) = \partial_x F(z(x_j, t)),
\end{aligned}
$$
(12.11)

where $\sum_{k=0}^{N-1} D_{jk} z_k = \delta_x z_j \approx \partial_x z(x_j, t)$.

If S is independent of the variable x, then N semi-discrete MSCLs (see, e.g. [4, 11]) also hold:

$$
\frac{\mathrm{d}}{\mathrm{d}t} \omega_j + \sum_{k=0}^{N-1} D_{jk} \kappa_{jk} = 0,
$$

$$
\omega_j = \frac{1}{2} \mathrm{d}z_j \wedge M \mathrm{d}z_j,
$$

$$
\kappa_{jk} = \frac{1}{2} (\mathrm{d}z_j \wedge K \mathrm{d}z_k + \mathrm{d}z_k \wedge K \mathrm{d}z_j),
$$

for $j = 0, 1, \ldots, N - 1$. Here $\sum_{k=0}^{N-1} D_{jk} \kappa_{jk}$ (the discrete $\partial_x \kappa(z(x_j, t))$) can be comprehended in a way similar to (12.11).

After the temporal discretization using the CRK method (12.5), the full discrete scheme can be written as follows:

$$
\begin{cases}
z_j^\tau = z_j^0 + \Delta t \displaystyle\int_0^1 A_{\tau,\sigma} \delta_t z_j^\sigma \, d\sigma, \\[2mm]
z_j^1 = z_j^0 + \Delta t \displaystyle\int_0^1 \delta_t z_j^\sigma \, d\sigma, \\[2mm]
\delta_x z_j^\tau = \displaystyle\sum_{k=0}^{N-1} D_{jk} z_k^\tau, \\[2mm]
M \delta_t z_j^\tau + K \delta_x z_j^\tau = \nabla_z S(z_j^\tau, x_j),
\end{cases}
$$
(12.12)

for $j = 0, 1, \ldots, N - 1$, where $z_j^\tau \approx z(x_j, t_0 + \tau \Delta t)$, $\delta_t z_j^\tau \approx \partial_t z(x_j, t_0 + \tau \Delta t)$ are polynomials in τ. For the energy-preserving property of the CRK method, we expect this scheme to preserve some discrete ECLs. Firstly, note that $b_i = \int_0^1 l_i(\tau) d\tau$. For convenience, we denote

$$
\langle f \rangle_i = \frac{1}{b_i} \int_0^1 l_i(\tau) f(\tau) d\tau
$$

(i.e. the weighted average of a function f with the weight function $l_i(\tau)$) in the remainder of our paper. Obviously, $\langle \cdot \rangle_i$ is a linear operator.

The next theorem shows the N-discrete local energy conservation law of (12.12).

Theorem 12.1 *The scheme* (12.12) *exactly conserves the* N-*discrete local energy conservation law:*

$$\frac{E_j^1 - E_j^0}{\Delta t} + \sum_{k=0}^{N-1} D_{jk} \bar{F}_{jk} = 0, \tag{12.13}$$

for $j = 0, 1, \ldots, N - 1$, *where*

$$E_j^\alpha = S(z_j^\alpha, x_j) - \frac{1}{2} z_j^{\alpha\mathsf{T}} K \delta_x z_j^\alpha, \alpha = 0, 1,$$

$$\bar{F}_{jk} = \frac{1}{2} \sum_{i=1}^{s} b_i (\langle z_j \rangle_i^\mathsf{T} K \langle \delta_t z_k \rangle_i + \langle z_k \rangle_i^\mathsf{T} K \langle \delta_t z_j \rangle_i).$$

By summing the identities (12.13) from $j = 0$ to $N - 1$, on noticing that $\bar{F}_{jk}$ is symmetric with respect to j, k and D_{jk} is anti-symmetric with respect to j, k, the discrete ECLs lead to the global energy conservation:

$$\Delta x \sum_{j=0}^{N-1} E_j^1 - \Delta x \sum_{j=0}^{N-1} E_j^0 = -\Delta x \Delta t \sum_{j=0}^{N-1} \sum_{k=0}^{N-1} D_{jk} \bar{F}_{jk} = 0. \tag{12.14}$$

If we evaluate the integrals of $\bar{F}_{jk}$ by the generating quadrature formula of the CRK method, we have

$$\bar{F}_{jk} \approx \frac{1}{2} \sum_{i=1}^{s} b_i (z_j^{c_i\mathsf{T}} K \delta_t z_k^{c_i} + z_k^{c_i\mathsf{T}} K \delta_t z_j^{c_i}).$$

Proof First of all, note that the discrete differential operator δ_x is linear, thus it holds that

$$\partial_\tau \delta_x z_j^\tau = \delta_x \partial_\tau z_j^\tau. \tag{12.15}$$

It follows from (12.12) that

$$\partial_\tau z_j^\tau = \Delta t \int_0^1 \sum_{i=1}^{s} \frac{1}{b_i} l_i(\tau) l_i(\sigma) \delta_t z_j^\sigma d\sigma = \Delta t \sum_{i=1}^{s} l_i(\tau) \langle \delta_t z_j \rangle_i. \tag{12.16}$$

We then have

$$S(z_j^1, x_j) - S(z_j^0, x_j)$$
$$= \int_0^1 \partial_\tau z_j^{\tau\mathsf{T}} \nabla_z S(z_j^\tau, x_j) d\tau = \Delta t \sum_{i=1}^{s} b_i \langle \delta_t z_j \rangle_i^\mathsf{T} \langle \nabla_z S_j \rangle_i, \tag{12.17}$$

$$z_j^1 K \delta_x z_j^1 - z_j^0 K \delta_x z_j^0$$

$$= \int_0^1 \partial_\tau (z_j^{\tau\mathsf{T}} K \delta_x z_j^\tau) d\tau = \int_0^1 (\partial_\tau z_j^{\tau\mathsf{T}} K \delta_x z_j^\tau + z_j^{\tau\mathsf{T}} K \delta_x \partial_\tau z_j^\tau) d\tau \qquad (12.18)$$

$$= \Delta t \sum_{i=1}^s b_i \langle \delta_t z_j \rangle_i^\mathsf{T} K \langle \delta_x z_j \rangle_i + \Delta t \sum_{i=1}^s b_i \langle z_j \rangle_i^\mathsf{T} K \langle \delta_x \delta_t z_j \rangle_i.$$

With (12.17) and (12.18), it follows from

$$\mathbf{a}^\mathsf{T} M \mathbf{a} = 0, \quad \mathbf{a}^\mathsf{T} M \mathbf{b} = -\mathbf{b}^\mathsf{T} M \mathbf{a}$$

for $\mathbf{a}, \mathbf{b} \in \mathbb{R}^d$ that

$$(E_j^1 - E_j^0)/\Delta t$$

$$= (S(z_j^1, x_j) - S(z_j^0, x_j) - \frac{1}{2}(z_j^{1\mathsf{T}} K \delta_x z_j^1 - z_j^{0\mathsf{T}} K \delta_x z_j^0))/\Delta t$$

$$= \sum_{i=1}^s b_i \langle \delta_t z_j \rangle_i^\mathsf{T} \langle M \delta_t z_j + K \delta_x z_j \rangle_i - \frac{1}{2} \sum_{i=1}^s b_i \langle \delta_t z_j \rangle_i^\mathsf{T} K \langle \delta_x z_j \rangle_i - \frac{1}{2} \sum_{i=1}^s b_i \langle z_j \rangle_i^\mathsf{T} K \langle \delta_x \delta_t z_j \rangle_i$$

$$= \frac{1}{2} \sum_{i=1}^s b_i \langle \delta_t z_j \rangle_i^\mathsf{T} K \langle \delta_x z_j \rangle_i - \frac{1}{2} \sum_{i=1}^s b_i \langle z_j \rangle_i^\mathsf{T} K \langle \delta_x \delta_t z_j \rangle_i$$

$$= \frac{1}{2} \sum_{k=0}^{N-1} D_{jk} \sum_{i=1}^s b_i \langle \delta_t z_j \rangle_i^\mathsf{T} K \langle z_k \rangle_i - \frac{1}{2} \sum_{k=0}^{N-1} D_{jk} \sum_{i=1}^s b_i \langle z_j \rangle_i^\mathsf{T} K \langle \delta_t z_k \rangle_i$$

$$= -\sum_{k=0}^{N-1} D_{jk} \bar{F}_{jk}.$$

$$(12.19)$$

$\square$

Note that a crucial property of the pseudospectral method is replacing the operator ∂_x with a linear and skew-symmetric differential matrix. Fortunately, this property is shared by spectral methods and wavelet collocation methods, and hence our analysis procedure to construct the local energy-preserving scheme can be extended to them without any trouble.

Our approach can also be easily generalized to high dimensional problems. For example, we consider the following equation:

$$M z_t + K z_x + L z_y = \nabla_z S(z, x, y). \qquad (12.20)$$

The ECL of this equation is:

$$\partial_t E + \partial_x F + \partial_y G = 0, \qquad (12.21)$$

where

$$E = S(z, x, y) - \frac{1}{2}z^{\mathsf{T}}Kz_x - \frac{1}{2}z^{\mathsf{T}}Lz_y, \quad F = \frac{1}{2}z^{\mathsf{T}}Kz_t, \quad G = \frac{1}{2}z^{\mathsf{T}}Lz_t.$$

Applying a CRK method to t-direction and a pseudospectral method to x and y directions (under the periodic boundary condition $z(x_0, y, t) = z(x_0 + L_1, y, t)$, $z(x, y_0, t) = z(x, y_0 + L_2, t)$) gives the following full discrete scheme:

$$\begin{cases} z_{jl}^{\tau} = z_{jl}^0 + \Delta t \int_0^1 A_{\tau,\sigma} \delta_t z_{jl}^{\sigma} d\sigma, \\ \\ z_{jl}^1 = z_{jl}^0 + \Delta t \int_0^1 \delta_t z_{jl}^{\sigma} d\sigma, \\ \\ \delta_x z_{jl}^{\tau} = \sum_{k=0}^{N-1} (D_x)_{jk} z_{kl}^{\tau}, \\ \\ \delta_y z_{jl}^{\tau} = \sum_{m=0}^{M-1} (D_y)_{lm} z_{jm}^{\tau}, \\ \\ M\delta_t z_{jl}^{\tau} + K\delta_x z_{jl}^{\tau} + L\delta_y z_{jl}^{\tau} = \nabla_z S(z_{jl}^{\tau}, x_j, y_l), \end{cases} \quad (12.22)$$

for $j = 0, 1, \ldots, N-1$ and $l = 0, 1, \ldots, M-1$, where $z_{jl}^{\tau} \approx z(x_j, y_l, t_0 + \tau \Delta t)$, $\delta_t z_{jl}^{\tau} \approx \partial_t z(x_j, y_l, t_0 + \tau \Delta t)$ are polynomials in τ, $x_j = x_0 + j\Delta x$, $y_l = y_0 + l\Delta y$, $\Delta x = \frac{L_1}{N}$, $\Delta y = \frac{L_2}{M}$, and D_x and D_y are pseudospectral differential matrices related to x and y directions respectively.

The next theorem presents the discrete local energy conservation laws of (12.22).

Theorem 12.2 *The scheme* (12.22) *exactly conserves the NM-discrete local energy conservation law:*

$$\frac{E_{jl}^1 - E_{jl}^0}{\Delta t} + \sum_{k=0}^{N-1} (D_x)_{jk} \bar{F}_{jk,l} + \sum_{m=0}^{M-1} (D_y)_{lm} \bar{G}_{j,lm} = 0, \quad (12.23)$$

for $j = 0, 1, \ldots, N-1$ and $l = 0, 1, \ldots, M-1$, where

$$E_{jl}^{\alpha} = S(z_{jl}^{\alpha}, x_j, y_l) - \frac{1}{2}z_{jl}^{\alpha\mathsf{T}} K\delta_x z_{jl}^{\alpha} - \frac{1}{2}z_{jl}^{\alpha\mathsf{T}} L\delta_y z_{jl}^{\alpha}, \alpha = 0, 1,$$

$$\bar{F}_{jk,l} = \frac{1}{2} \sum_{i=1}^{s} b_i (\langle z_{jl}\rangle_i^{\mathsf{T}} K \langle \delta_t z_{kl}\rangle_i + \langle z_{kl}\rangle_i^{\mathsf{T}} K \langle \delta_t z_{jl}\rangle_i),$$

$$\bar{G}_{j,lm} = \frac{1}{2} \sum_{i=1}^{s} b_i (\langle z_{jl}\rangle_i^{\mathsf{T}} L \langle \delta_t z_{jm}\rangle_i + \langle z_{jm}\rangle_i^{\mathsf{T}} L \langle \delta_t z_{jl}\rangle_i).$$

Since the proof of Theorem 12.2 is very similar to that of Theorem 12.1, we omit the details here.

Summing the identities (12.23) over all space grid points, on noticing that $\bar{F}_{jk,l}$ is symmetric with respect to j, k, and $(D_x)_{jk}$ is anti-symmetric with respect to j, k, $\bar{G}_{j,lm}$ is symmetric with respect to l, m, and $(D_y)_{lm}$ is anti-symmetric with respect to l, m, again, we obtain the global energy conservation:

$$
\Delta x \Delta y \sum_{j=0}^{N-1}\sum_{l=0}^{M-1} E_{jl}^1 - \Delta x \Delta y \sum_{j=0}^{N-1}\sum_{l=0}^{M-1} E_{jl}^0
$$
$$
= -\Delta t \Delta x \Delta y \sum_{j=0}^{N-1}\sum_{l=0}^{M-1}\sum_{k=0}^{N-1}(D_x)_{jk}\bar{F}_{jk,l} - \Delta t \Delta x \Delta y \sum_{j=0}^{N-1}\sum_{l=0}^{M-1}\sum_{m=0}^{M-1}(D_y)_{lm}\bar{G}_{j,lm} = 0.
$$
$$(12.24)$$

12.3.2 Gauss-Legendre Collocation Spatial Discretization

In multi-symplectic algorithms, another class of methods frequently applied to spatial discretization is the Gauss-Legendre (GL) collocation method. We assume that the Butcher tableau of the GL method is:

$$
\begin{array}{c|ccc}
\tilde{c}_1 & \tilde{a}_{11} & \cdots & \tilde{a}_{1r} \\
\vdots & \vdots & \ddots & \vdots \\
\tilde{c}_r & \tilde{a}_{r1} & \cdots & \tilde{a}_{rr} \\
\hline
 & \tilde{b}_1 & \cdots & \tilde{b}_r
\end{array}
\tag{12.25}
$$

After the spatial discretization using the GL method (12.25) and the temporal discretization using the CRK, we obtain the full discrete scheme of (12.7):

$$
\begin{cases}
z_{n,j}^\tau = z_{n,j}^0 + \Delta t \int_0^1 A_{\tau,\sigma}\delta_t z_{n,j}^\sigma d\sigma, \\
z_{n,j}^1 = z_{n,j}^0 + \Delta t \int_0^1 \delta_t z_{n,j}^\sigma d\sigma, \\
z_{n,j}^\tau = z_n^\tau + \Delta x \sum_{k=1}^r \tilde{a}_{jk}\delta_x z_{n,k}^\tau, \\
z_{n+1}^\tau = z_n^\tau + \Delta x \sum_{j=1}^r \tilde{b}_j\delta_x z_{n,j}^\tau, \\
M\delta_t z_{n,j}^\tau + K\delta_x z_{n,j}^\tau = \nabla_z S(z_{n,j}^\tau, x_n + \tilde{c}_j \Delta x),
\end{cases}
\tag{12.26}
$$

for $j = 1, 2, \ldots, r$, where $z_n^\tau \approx z(x_n, t_0 + \tau \Delta t)$, $z_{n+1}^\tau \approx z(x_n + \Delta x, t_0 + \tau \Delta t)$, $z_{n,j}^\tau$ $\approx z(x_n + \tilde{c}_j \Delta x, t_0 + \tau \Delta t)$, $\delta_t z_{n,j}^\tau \approx \partial_t z(x_n + \tilde{c}_j \Delta x, t_0 + \tau \Delta t)$, $\delta_x z_{n,j}^\tau \approx \partial_x z(x_n + \tilde{c}_j \Delta x, t_0 + \tau \Delta t)$ are polynomials in τ. This is a local scheme on the box $[x_n, x_n + \Delta x] \times [t_0, t_0 + \Delta t]$. To show that (12.26) exactly conserves the discrete ECL, we should make sure that there is some commutation law between δ_t and δ_x. To this end we introduce the following auxiliary system:

$$
\begin{cases}
\delta_x z_{n,j}^\tau = \delta_x z_{n,j}^0 + \Delta t \displaystyle\int_0^1 A_{\tau,\sigma} \delta_t \delta_x z_{n,j}^\sigma \, d\sigma, \\[2ex]
\delta_x z_{n,j}^1 = \delta_x z_{n,j}^0 + \Delta t \displaystyle\int_0^1 \delta_t \delta_x z_{n,j}^\sigma \, d\sigma, \\[2ex]
\delta_t z_{n,j}^\tau = \delta_t z_n^\tau + \Delta x \displaystyle\sum_{k=1}^r \tilde{a}_{jk} \delta_x \delta_t z_{n,k}^\tau, \\[2ex]
\delta_t z_{n+1}^\tau = \delta_t z_n^\tau + \Delta x \displaystyle\sum_{j=1}^r \tilde{b}_j \delta_x \delta_t z_{n,j}^\tau,
\end{cases}
\tag{12.27}
$$

for $j = 1, 2, \ldots, r$, where $\delta_t \delta_x z_{n,j}^\sigma \approx \partial_t \partial_x z(x_n + \tilde{c}_j \Delta x, t_0 + \sigma t)$, $\delta_x \delta_t z_{n,j}^\sigma \approx \partial_x \partial_t z(x_n + \tilde{c}_j \Delta x, t_0 + \sigma t)$. Then

$$
\begin{aligned}
z_{n,j}^\tau &= z_{n,j}^0 + \Delta t \int_0^1 A_{\tau,\sigma} \delta_t z_{n,j}^\sigma \, d\sigma \\
&= z_{n,j}^0 + \Delta t \int_0^1 A_{\tau,\sigma} \left(\delta_t z_n^\sigma + \Delta x \sum_{k=1}^r \tilde{a}_{jk} \delta_x \delta_t z_{n,k}^\sigma \right) d\sigma \\
&= z_{n,j}^0 + z_n^\tau - z_n^0 + \Delta t \Delta x \sum_{k=1}^r \tilde{a}_{jk} \int_0^1 A_{\tau,\sigma} \delta_x \delta_t z_{n,k}^\sigma \, d\sigma.
\end{aligned}
\tag{12.28}
$$

Likewise,

$$
z_{n,j}^\tau = z_n^\tau + z_{n,j}^0 - z_n^0 + \Delta x \Delta t \sum_{k=1}^r \tilde{a}_{jk} \int_0^1 A_{\tau,\sigma} \delta_t \delta_x z_{n,k}^\sigma \, d\sigma.
\tag{12.29}
$$

Equations (12.28) and (12.29) lead to

$$
\sum_{k=1}^r \tilde{a}_{jk} \int_0^1 A_{\tau,\sigma} \delta_x \delta_t z_{n,k}^\sigma \, d\sigma = \sum_{k=1}^r \tilde{a}_{jk} \int_0^1 A_{\tau,\sigma} \delta_t \delta_x z_{n,k}^\sigma \, d\sigma.
$$

Since the matrix $(\tilde{a}_{jk})_{1 \le j, \, k \le r}$ is invertible, we have

$$
\int_0^1 A_{\tau,\sigma} \delta_x \delta_t z_{n,k}^\sigma \, d\sigma = \int_0^1 A_{\tau,\sigma} \delta_t \delta_x z_{n,k}^\sigma \, d\sigma, \quad \tau \in [0, 1],
\tag{12.30}
$$

for $k = 1, 2, \ldots, r$. Taking derivatives with respect to τ on both sides of Eq. (12.30), we arrive at

$$\int_0^1 (\sum_{i=1}^s \frac{1}{b_i} l_i(\tau) l_i(\sigma)) \delta_x \delta_t z_{n,k}^\sigma d\sigma = \int_0^1 (\sum_{i=1}^s \frac{1}{b_i} l_i(\tau) l_i(\sigma)) \delta_t \delta_x z_{n,k}^\sigma d\sigma, \quad \tau \in [0, 1],$$

for $k = 1, 2, \ldots, r$. Finally, setting $\tau = c_1, \ldots, c_s$, we have the following lemma:

Lemma 12.1 *The following discrete commutability between δ_t and δ_x holds:*

$$\langle \delta_x \delta_t z_{n,j} \rangle_i = \langle \delta_t \delta_x z_{n,j} \rangle_i, \tag{12.31}$$

for $i = 1, \ldots, s$ and $j = 1, 2, \ldots, r$.

Theorem 12.3 *The scheme* (12.26) *conserves the following discrete local energy conservation law:*

$$\Delta x \sum_{j=1}^r \tilde{b}_j (E_{n,j}^1 - E_{n,j}^0) + \Delta t (\bar{F}_{n+1} - \bar{F}_n) = 0, \tag{12.32}$$

where

$$E_{n,j}^\alpha = S(z_{n,j}^\alpha, x_n + \tilde{c}_j \Delta x) - \frac{1}{2} z_{n,j}^{\alpha \mathsf{T}} K \delta_x z_{n,j}^\alpha, \alpha = 0, 1,$$

$$\bar{F}_\beta = \frac{1}{2} \sum_{i=1}^s b_i \langle z_\beta \rangle_i^\mathsf{T} K \langle \delta_t z_\beta \rangle_i, \beta = n, n+1.$$

Proof It follows from the first equations of (12.26) and (12.27) that,

$$\partial_\tau z_{n,j}^\tau = \Delta t \sum_{i=1}^s l_i(\tau) \langle \delta_t z_{n,j} \rangle_i, \quad \partial_\tau \delta_x z_{n,j}^\tau = \Delta t \sum_{i=1}^s l_i(\tau) \langle \delta_t \delta_x z_{n,j} \rangle_i. \tag{12.33}$$

The result in the temporal direction is similar to the pseudospectral case:

$$S(z_{n,j}^1, x_j + \tilde{c}_j \Delta x) - S(z_{n,j}^0, x_j + \tilde{c}_j \Delta x)$$

$$= \int_0^1 \partial_\tau z_{n,j}^{\tau \mathsf{T}} \nabla_z S(z_{n,j}^\tau, x_j + \tilde{c}_j \Delta x) d\tau = \Delta t \sum_{i=1}^s b_i \langle \delta_t z_{n,j} \rangle_i^\mathsf{T} \langle \nabla_z S_{n,j} \rangle_i,$$

$$z_{n,j}^1 K \delta_x z_{n,j}^1 - z_{n,j}^0 K \delta_x z_{n,j}^0 = \int_0^1 \partial_\tau (z_{n,j}^{\tau \mathsf{T}} K \delta_x z_{n,j}^\tau) d\tau$$

$$= \int_0^1 (\partial_\tau z_{n,j}^{\tau \mathsf{T}} K \delta_x z_{n,j}^\tau + z_{n,j}^{\tau \mathsf{T}} K \partial_\tau \delta_x z_{n,j}^\tau) d\tau$$

$$= \Delta t \sum_{i=1}^s b_i \langle \delta_t z_{n,j} \rangle_i^\mathsf{T} K \langle \delta_x z_{n,j} \rangle_i + \Delta t \sum_{i=1}^s b_i \langle z_{n,j} \rangle_i^\mathsf{T} K \langle \delta_t \delta_x z_{n,j} \rangle_i.$$

Hence

$$
E_{n,j}^1 - E_{n,j}^0
$$

$$
= S(z_{n,j}^1, x_n + \tilde{c}_j \Delta x) - S(z_{n,j}^0, x_n + \tilde{c}_j \Delta x) - \frac{1}{2}(z_{n,j}^{1\,\mathsf{T}} K \delta_x z_{n,j}^1 - z_{n,j}^{0\,\mathsf{T}} K \delta_x z_{n,j}^0)
$$

$$
= \Delta t \sum_{i=1}^{s} b_i \langle \delta_t z_{n,j} \rangle_i^{\mathsf{T}} \langle M \delta_t z_{n,j} + K \delta_x z_{n,j} \rangle_i - \frac{1}{2} \Delta t \sum_{i=1}^{s} b_i \langle \delta_t z_{n,j} \rangle_i^{\mathsf{T}} K \langle \delta_x z_{n,j} \rangle_i
$$

$$
- \frac{1}{2} \Delta t \sum_{i=1}^{s} b_i \langle z_{n,j} \rangle_i^{\mathsf{T}} K \langle \delta_t \delta_x z_{n,j} \rangle_i
$$

$$
= \frac{1}{2} \Delta t \sum_{i=1}^{s} b_i \langle \delta_t z_{n,j} \rangle_i^{\mathsf{T}} K \langle \delta_x z_{n,j} \rangle_i - \frac{1}{2} \Delta t \sum_{i=1}^{s} b_i \langle z_{n,j} \rangle_i^{\mathsf{T}} K \langle \delta_t \delta_x z_{n,j} \rangle_i .
$$

$$(12.34)$$

On the other hand, we have

$$
z_{n+1}^{\mathsf{T}\,\mathsf{T}} K \delta_t z_{n+1}^{\sigma} - z_n^{\mathsf{T}\,\mathsf{T}} K \delta_t z_n^{\sigma}
$$

$$
= (z_n^{\mathsf{T}\,\mathsf{T}} + \Delta x \sum_{j=1}^{r} \tilde{b}_j \delta_x z_{n,j}^{\mathsf{T}\,\mathsf{T}}) K (\delta_t z_n^{\sigma} + \Delta x \sum_{j=1}^{r} \tilde{b}_j \delta_x \delta_t z_{n,j}^{\sigma}) - z_n^{\mathsf{T}\,\mathsf{T}} K \delta_t z_n^{\sigma}
$$

$$
= \Delta x \sum_{j=1}^{r} \tilde{b}_j z_n^{\mathsf{T}\,\mathsf{T}} K \delta_x \delta_t z_{n,j}^{\sigma} + \Delta x \sum_{j=1}^{r} \tilde{b}_j \delta_x z_{n,j}^{\mathsf{T}\,\mathsf{T}} K \delta_t z_n^{\sigma} + \Delta x^2 \sum_{j,k=1}^{r} \tilde{b}_j \tilde{b}_k \delta_x z_{n,j}^{\mathsf{T}\,\mathsf{T}} K \delta_x \delta_t z_{n,k}^{\sigma}
$$

$$
= \Delta x \sum_{j=1}^{r} \tilde{b}_j (z_{n,j}^{\mathsf{T}\,\mathsf{T}} - \Delta x \sum_{k=1}^{r} \tilde{a}_{jk} \delta_x z_{n,k}^{\mathsf{T}\,\mathsf{T}}) K \delta_x \delta_t z_{n,j}^{\sigma} + \Delta x \sum_{j=1}^{r} \tilde{b}_j \delta_x z_{n,j}^{\mathsf{T}\,\mathsf{T}} K (\delta_t z_{n,j}^{\sigma} - \Delta x \sum_{k=1}^{r} \tilde{a}_{jk} \delta_x \delta_t z_{n,k}^{\sigma})
$$

$$
+ \Delta x^2 \sum_{j,k=1}^{r} \tilde{b}_j \tilde{b}_k \delta_x z_{n,j}^{\mathsf{T}\,\mathsf{T}} K \delta_x \delta_t z_{n,k}^{\sigma}
$$

$$
= \Delta x \sum_{j=1}^{r} \tilde{b}_j z_{n,j}^{\mathsf{T}\,\mathsf{T}} K \delta_x \delta_t z_{n,j}^{\sigma} + \Delta x \sum_{j=1}^{r} \tilde{b}_j \delta_x z_{n,j}^{\mathsf{T}\,\mathsf{T}} K \delta_t z_{n,j}^{\sigma} + \Delta x^2 \sum_{j,k=1}^{r} (\tilde{b}_j \tilde{b}_k - \tilde{b}_j \tilde{a}_{jk} - \tilde{b}_k \tilde{a}_{kj}) \delta_x z_{n,j}^{\mathsf{T}\,\mathsf{T}} K \delta_x \delta_t z_{n,k}^{\sigma}
$$

$$
= \Delta x \sum_{j=1}^{r} \tilde{b}_j z_{n,j}^{\mathsf{T}\,\mathsf{T}} K \delta_x \delta_t z_{n,j}^{\sigma} + \Delta x \sum_{j=1}^{r} \tilde{b}_j \delta_x z_{n,j}^{\mathsf{T}\,\mathsf{T}} K \delta_t z_{n,j}^{\sigma} .
$$

$$(12.35)$$

It follows from (12.35) that

$$
\bar{F}_{n+1} - \bar{F}_n
$$

$$
= \frac{1}{2} \sum_{i=1}^{s} b_i (\langle z_{n+1} \rangle_i^{\mathsf{T}} K \langle \delta_t z_{n+1} \rangle_i - \langle z_n \rangle_i^{\mathsf{T}} K \langle \delta_t z_n \rangle_i)
$$

$$(12.36)$$

$$
= \frac{1}{2} \Delta x \sum_{j=1}^{r} \sum_{i=1}^{s} \tilde{b}_j b_i (\langle z_{n,j} \rangle_i^{\mathsf{T}} K \langle \delta_x \delta_t z_{n,j} \rangle_i + \langle \delta_x z_{n,j} \rangle_i^{\mathsf{T}} K \langle \delta_t z_{n,j} \rangle_i) .
$$

From (12.34) and (12.36), using Lemma 12.1, we obtain

$$
\Delta x \sum_{j=1}^{r} \tilde{b}_j (E_{n,j}^1 - E_{n,j}^0) + \Delta t (\bar{F}_{n+1} - \bar{F}_n)
$$

$$
= \frac{1}{2} \Delta t \Delta x \sum_{i=1}^{s} \sum_{j=1}^{r} \tilde{b}_j b_i (\langle \delta_t z_{n,j} \rangle_i^\mathsf{T} K \langle \delta_x z_{n,j} \rangle_i - \langle z_{n,j} \rangle_i^\mathsf{T} K \langle \delta_t \delta_x z_{n,j} \rangle_i
$$

$$
+ \langle z_{n,j} \rangle_i^\mathsf{T} K \langle \delta_x \delta_t z_{n,j} \rangle_i + \langle \delta_x z_{n,j} \rangle_i^\mathsf{T} K \langle \delta_t z_{n,j} \rangle_i)
$$

$$
= \frac{1}{2} \Delta x \Delta t \sum_{j=1}^{r} \sum_{i=1}^{s} \tilde{b}_j b_i \langle z_{n,j} \rangle_i^\mathsf{T} K (\langle \delta_x \delta_t z_{n,j} \rangle_i - \langle \delta_t \delta_x z_{n,j} \rangle_i) = 0. \qquad \Box
$$

(12.37)

Assume that the spatial domain is divided equally into N intervals and the corresponding grids are $x_0, x_1, \dots, x_N$. By summing the identities (12.32) from $n = 0$ to $N - 1$, we obtain the global energy conservation of the scheme (12.26) under the periodic boundary condition:

$$
\Delta x \sum_{n=0}^{N-1} \sum_{j=1}^{r} \tilde{b}_j E_{n,j}^1 - \Delta x \sum_{n=0}^{N-1} \sum_{j=1}^{r} \tilde{b}_j E_{n,j}^0 = -\Delta t \sum_{n=0}^{N-1} (\bar{F}_{n+1} - \bar{F}_n) = 0. \quad (12.38)
$$

The GL spatial discretization is not restricted to the periodic boundary condition (PBC). Thus the discrete ECL (12.32) is superior to the discrete global energy conservation (12.38). However, discretizing space by high-order GL methods may lead to singular and massive ODE systems which are expensive to solve (see, e.g. [27, 31]). For this reason, we will not include the scheme (12.26) in our numerical experiments in Sects. 12.6 and 12.7.

12.4 Local Energy-Preserving Schemes for Coupled Nonlinear Schrödinger Equations

An important class of multi-symplectic PDEs is the (coupled) nonlinear Schrödinger equation ((C)NLS). A great number of them have polynomial nonlinear terms, and hence we can calculate the integrals exactly in our method (for example, by symbol calculations). Here we summarize the multi-symplectic structure of the 2-coupled NLS:

$$
\begin{cases}
i u_t + i \alpha u_x + \dfrac{1}{2} u_{xx} + (|u|^2 + \beta |v|^2) u = 0, \\[2mm]
i v_t - i \alpha v_x + \dfrac{1}{2} v_{xx} + (\beta |u|^2 + |v|^2) v = 0,
\end{cases}
\qquad (12.39)
$$

where u, v are complex variables, and i is the imaginary unit. Assuming $u = q_1 + iq_2$ and $v = q_3 + iq_4$, $\partial_x q_j = 2p_j$, q_j are real variables for $j = 1, \ldots, 4$, we transform this equation into the multi-symplectic form (see, e.g. [10]):

$$\begin{pmatrix} \mathbf{J}_1 & \mathbf{O} \\ \mathbf{O} & \mathbf{O} \end{pmatrix} z_t + \begin{pmatrix} \mathbf{J}_2 & \mathbf{I} \\ -\mathbf{I} & \mathbf{O} \end{pmatrix} z_x = \nabla_z S(z),$$

where

$$\mathbf{J}_1 = \begin{pmatrix} 0 & -1 & 0 & 0 \\ 1 & 0 & 0 & 0 \\ 0 & 0 & 0 & -1 \\ 0 & 0 & 1 & 0 \end{pmatrix}, \mathbf{J}_2 = \begin{pmatrix} 0 & -\alpha & 0 & 0 \\ \alpha & 0 & 0 & 0 \\ 0 & 0 & 0 & \alpha \\ 0 & 0 & -\alpha & 0 \end{pmatrix}, \mathbf{O} = \begin{pmatrix} 0 & 0 & 0 & 0 \\ 0 & 0 & 0 & 0 \\ 0 & 0 & 0 & 0 \\ 0 & 0 & 0 & 0 \end{pmatrix}, \mathbf{I} = \begin{pmatrix} 1 & 0 & 0 & 0 \\ 0 & 1 & 0 & 0 \\ 0 & 0 & 1 & 0 \\ 0 & 0 & 0 & 1 \end{pmatrix},$$

and

$$z = (q_1, q_2, q_3, q_4, p_1, p_2, p_3, p_4)^\mathsf{T},$$
$$S = -\frac{1}{4}(q_1^2 + q_2^2)^2 - \frac{1}{4}(q_3^2 + q_4^2)^2 - \frac{1}{2}\beta(q_1^2 + q_2^2)(q_3^2 + q_4^2) - (p_1^2 + p_2^2 + p_3^2 + p_4^2).$$

The corresponding energy density E and flux F in the ECL (12.8) are:

$$E = S - \alpha(q_2 p_1 - q_1 p_2 + q_3 p_4 - q_4 p_3) - \frac{1}{2}\sum_{i=1}^{4}(q_i \partial_x p_i - 2p_i^2),$$

$$F = \frac{1}{2}(\alpha(q_2 \partial_t q_1 - q_1 \partial_t q_2 + q_3 \partial_t q_4 - q_4 \partial_t q_3) + \sum_{i=1}^{4}(q_i \partial_t p_i - p_i \partial_t q_i)).$$

The corresponding MCL of this equation is:

$$\partial_t I + \partial_x G = 0,$$

where

$$I = q_2 p_1 - q_1 p_2 + q_4 p_3 - q_3 p_4,$$
$$G = S - \frac{1}{2}(q_2 \partial_t q_1 - q_1 \partial_t q_2 + q_4 \partial_t q_3 - q_3 \partial_t q_4).$$

Integrating the ECL and MCL with respect to the variable x under the PBC leads to the global energy and the momentum conservation:

$$\int_{x_0}^{x_0+L} E(x, t)\mathrm{d}x = \int_{x_0}^{x_0+L} E(x, 0)\mathrm{d}x, \quad \int_{x_0}^{x_0+L} I(x, t)\mathrm{d}x = \int_{x_0}^{x_0+L} I(x, 0)\mathrm{d}x.$$

Besides, the global charges of u and v are constant under the PBC:

$$\int_{x_0}^{x_0+L} |u(x,t)|^2 dx = \int_{x_0}^{x_0+L} |u(x,0)|^2 dx, \quad \int_{x_0}^{x_0+L} |v(x,t)|^2 dx = \int_{x_0}^{x_0+L} |v(x,0)|^2 dx.$$

Applying the discrete procedure (12.12) to (12.39) gives the following scheme in a vector form:

$$
\begin{cases}
q_1^\tau = q_1^0 + \Delta t \int_0^1 A_{\tau,\sigma}(-\alpha Dq_1^\sigma - Dp_2^\sigma - (((q_1^\sigma)^{\cdot 2} + (q_2^\sigma)^{\cdot 2}) + \beta((q_3^\sigma)^{\cdot 2} + (q_4^\sigma)^{\cdot 2})) \cdot q_2^\sigma) d\sigma, \\[4pt]
q_2^\tau = q_2^0 + \Delta t \int_0^1 A_{\tau,\sigma}(-\alpha Dq_2^\sigma + Dp_1^\sigma + (((q_1^\sigma)^{\cdot 2} + (q_2^\sigma)^{\cdot 2}) + \beta((q_3^\sigma)^{\cdot 2} + (q_4^\sigma)^{\cdot 2})) \cdot q_1^\sigma) d\sigma, \\[4pt]
q_3^\tau = q_3^0 + \Delta t \int_0^1 A_{\tau,\sigma}(\alpha Dq_3^\sigma - Dp_4^\sigma - (\beta((q_1^\sigma)^{\cdot 2} + (q_2^\sigma)^{\cdot 2}) + ((q_3^\sigma)^{\cdot 2} + (q_4^\sigma)^{\cdot 2})) \cdot q_4^\sigma) d\sigma, \\[4pt]
q_4^\tau = q_4^0 + \Delta t \int_0^1 A_{\tau,\sigma}(\alpha Dq_4^\sigma + Dp_3^\sigma + (\beta((q_1^\sigma)^{\cdot 2} + (q_2^\sigma)^{\cdot 2}) + ((q_3^\sigma)^{\cdot 2} + (q_4^\sigma)^{\cdot 2})) \cdot q_3^\sigma) d\sigma, \\[4pt]
q_1^1 = q_1^0 + \Delta t \int_0^1 (-\alpha Dq_1^\sigma - Dp_2^\sigma - (((q_1^\sigma)^{\cdot 2} + (q_2^\sigma)^{\cdot 2}) + \beta((q_3^\sigma)^{\cdot 2} + (q_4^\sigma)^{\cdot 2})) \cdot q_2^\sigma) d\sigma, \\[4pt]
q_2^1 = q_2^0 + \Delta t \int_0^1 (-\alpha Dq_2^\sigma + Dp_{1,k}^\sigma + (((q_1^\sigma)^{\cdot 2} + (q_2^\sigma)^{\cdot 2}) + \beta((q_3^\sigma)^{\cdot 2} + (q_4^\sigma)^{\cdot 2})) \cdot q_1^\sigma) d\sigma, \\[4pt]
q_3^1 = q_3^0 + \Delta t \int_0^1 (\alpha Dq_3^\sigma - Dp_4^\sigma - (\beta((q_1^\sigma)^{\cdot 2} + (q_2^\sigma)^{\cdot 2}) + ((q_3^\sigma)^{\cdot 2} + (q_4^\sigma)^{\cdot 2})) \cdot q_4^\sigma) d\sigma, \\[4pt]
q_4^1 = q_4^0 + \Delta t \int_0^1 (\alpha Dq_4^\sigma + Dp_3^\sigma + (\beta((q_1^\sigma)^{\cdot 2} + (q_2^\sigma)^{\cdot 2}) + ((q_3^\sigma)^{\cdot 2} + (q_4^\sigma)^{\cdot 2})) \cdot q_3^\sigma) d\sigma, \\[4pt]
\delta_x q_i^\sigma = Dq_i^\sigma = 2p_i^\sigma, i = 1, \ldots, 4,
\end{cases}
$$

$$(12.40)$$

where $q_i^a = (q_{i,0}^a, q_{i,1}^a, \ldots, q_{i,N-1}^a)^\mathsf{T}$, $p_i^a = (p_{i,0}^a, p_{i,1}^a, \ldots, p_{i,N-1}^a)^\mathsf{T}$, $a = 0, 1, \tau$ for $i = 1, \ldots, 4$ and $j = 1, \ldots, N-1$. Here $q_{i,j}^\tau$, $p_{i,j}^\tau$ are polynomials in τ. The symbols "$\cdot 2$" and "$\cdot$" indicate the entrywise square operation and the entrywise multiplication operation, respectively.

It can be observed that p_i^τ can be eliminated from (12.40). If the generating quadrature formula has s nodes, then $A_{\tau,\sigma}$ is a polynomial of degree s in variable τ, and so are $q_{i,j}^\tau$ for $i = 1, \ldots, 4$ and $j = 0, 1, \ldots, N-1$. These polynomials are uniquely determined by their values at $s+1$ points. For convenience, we choose $0, \frac{1}{s}, \frac{2}{s}, \ldots, 1$. Then $q_{i,j}^\tau$ can be expressed by Lagrange interpolating polynomials based on these $s+1$ points. Fixing τ at $\frac{1}{s}, \frac{2}{s}, \ldots, \frac{s-1}{s}$, we get a system of algebraic equations in $q_{i,j}^c$ with $c = 0, \frac{1}{s}, \frac{2}{s}, \ldots, 1$. The polynomial integrals in this system can be calculated accurately. Solving the algebraic system by an iteration method, we finally obtain the numerical solution $q_{i,j}^1$.

For example, if we select the CRK method generated by the 2-point GL quadrature formula, then q_i^σ, p_i^σ are vectors whose entries are polynomials of the second degree. Thus we have

$$q_i^\sigma = q_i^0 \tilde{l}_1(\sigma) + q_i^{\frac{1}{2}} \tilde{l}_2(\sigma) + q_i^1 \tilde{l}_3(\sigma),$$

where $\tilde{l}_1(\sigma), \tilde{l}_2(\sigma), \tilde{l}_3(\sigma)$ are Lagrange interpolating polynomials based on the nodes $0, \frac{1}{2}, 1$. Let $\tau = \frac{1}{2}$. Then the first four equations of (12.40) can be written in practical forms:

$$q_1^{\frac{1}{2}} = q_1^0 + \Delta t \int_0^1 A_{\frac{1}{2},\sigma}(-\alpha D q_1^\sigma - D p_2^\sigma - (((q_1^\sigma)^2 + (q_2^\sigma)^2) + \beta((q_3^\sigma)^2 + (q_4^\sigma)^2)) \cdot q_2^\sigma) d\sigma,$$

$$q_2^{\frac{1}{2}} = q_2^0 + \Delta t \int_0^1 A_{\frac{1}{2},\sigma}(-\alpha D q_2^\sigma + D p_1^\sigma + (((q_1^\sigma)^2 + (q_2^\sigma)^2) + \beta((q_3^\sigma)^2 + (q_4^\sigma)^2)) \cdot q_1^\sigma) d\sigma,$$

$$q_3^{\frac{1}{2}} = q_3^0 + \Delta t \int_0^1 A_{\frac{1}{2},\sigma}(\alpha D q_3^\sigma - D p_4^\sigma - (\beta((q_1^\sigma)^2 + (q_2^\sigma)^2) + ((q_3^\sigma)^2 + (q_4^\sigma)^2)) \cdot q_4^\sigma) d\sigma,$$

$$q_4^{\frac{1}{2}} = q_4^0 + \Delta t \int_0^1 A_{\frac{1}{2},\sigma}(\alpha D q_4^\sigma + D p_3^\sigma + (\beta((q_1^\sigma)^2 + (q_2^\sigma)^2) + ((q_3^\sigma)^2 + (q_4^\sigma)^2)) \cdot q_3^\sigma) d\sigma.$$

$$(12.41)$$

After integrating the linear and nonlinear terms about σ, (12.41) becomes an undetermined system of equations in unknown vectors $q_1^{\frac{1}{2}}, q_2^{\frac{1}{2}}, q_3^{\frac{1}{2}}, q_4^{\frac{1}{2}}, q_1^1, q_2^1, q_3^1, q_4^1$. By combining them with the 5, 6, 7 and 8th equations of (12.40), we obtain an entirely determined algebraic system about them which can be easily solved by a fixed-point iteration in practical computations. If the generating quadrature formula has only one node, for example, the implicit midpoint rule, then $q_i^\sigma = (1 - \sigma) q_i^0 + \sigma q_i^1$. In this particular case, the first four equations are not necessary to be taken into account.

According to Theorem 12.1, the scheme (12.40) preserves the discrete ECLs:

$$\frac{E_j^1 - E_j^0}{\Delta t} + \sum_{k=0}^{N-1} D_{jk} \bar{F}_{jk} = 0, \qquad (12.42)$$

for $j = 0, 1, \ldots, N - 1$, where

$$E_j^a = S_j^a - \alpha(q_{2,j}^a p_{1,j}^a - q_1^a p_{2,j}^a + q_{3,j}^a p_{4,j}^a - q_{4,j}^a p_{3,j}^a)$$

$$- \frac{1}{2} \sum_{i=1}^4 (q_{i,j}^a \sum_{k=0}^{N-1} D_{jk} p_{i,k}^a - 2(p_{i,j}^a)^2), \quad a = 0, 1,$$

$$\bar{F}_{jk} = \frac{1}{2} \sum_{i=1}^s b_i (\alpha(\langle q_{2,j}\rangle_i \langle \delta_t q_{1,k}\rangle_i - \langle q_{1,j}\rangle_i \langle \delta_t q_{2,k}\rangle_i + \langle q_{3,j}\rangle_i \langle \delta_t q_{4,k}\rangle_i - \langle q_{4,j}\rangle_i \langle \delta_t q_{3,k}\rangle_i)$$

$$+ \sum_{\gamma=1}^4 (\langle q_{\gamma,j}\rangle_i \langle \delta_t p_{\gamma,k}\rangle_i - \langle p_{\gamma,j}\rangle_i \langle \delta_t q_{\gamma,k}\rangle_i) + \alpha(\langle q_{2,k}\rangle_i \langle \delta_t q_{1,j}\rangle_i - \langle q_{1,k}\rangle_i \langle \delta_t q_{2,j}\rangle_i$$

$$+ \langle q_{3,k}\rangle_i \langle \delta_t q_{4,j}\rangle_i - \langle q_{4,k}\rangle_i \langle \delta_t q_{3,j}\rangle_i) + \sum_{\gamma=1}^4 (\langle q_{\gamma,k}\rangle_i \langle \delta_t p_{\gamma,j}\rangle_i - \langle p_{\gamma,k}\rangle_i \langle \delta_t q_{\gamma,j}\rangle_i)).$$

12.5   Local Energy-Preserving Schemes for 2D Nonlinear Schrödinger Equations

Another PDE to which we pay attention is the NLS with two spatial variables:

$$i\psi_t + \alpha(\psi_{xx} + \psi_{yy}) + V'(|\psi|^2, x, y)\psi = 0. \tag{12.43}$$

Here the symbol $'$ indicates the derivative of V with respect to the first variable. Let $\psi = p + iq$, where p and q are real and imaginary parts of ψ, respectively. Introducing $v = \partial_x p$, $w = \partial_x q$, $a = \partial_y p$, $b = \partial_y q$, we transform this equation into the compact form (12.20), where

$$
M = \begin{pmatrix} 0 & 1 & 0 & 0 & 0 & 0 \\ -1 & 0 & 0 & 0 & 0 & 0 \\ 0 & 0 & 0 & 0 & 0 & 0 \\ 0 & 0 & 0 & 0 & 0 & 0 \\ 0 & 0 & 0 & 0 & 0 & 0 \\ 0 & 0 & 0 & 0 & 0 & 0 \end{pmatrix},\quad
K = \begin{pmatrix} 0 & 0 & -\alpha & 0 & 0 & 0 \\ 0 & 0 & 0 & -\alpha & 0 & 0 \\ \alpha & 0 & 0 & 0 & 0 & 0 \\ 0 & \alpha & 0 & 0 & 0 & 0 \\ 0 & 0 & 0 & 0 & 0 & 0 \\ 0 & 0 & 0 & 0 & 0 & 0 \end{pmatrix},\quad
L = \begin{pmatrix} 0 & 0 & 0 & 0 & -\alpha & 0 \\ 0 & 0 & 0 & 0 & 0 & -\alpha \\ 0 & 0 & 0 & 0 & 0 & 0 \\ 0 & 0 & 0 & 0 & 0 & 0 \\ \alpha & 0 & 0 & 0 & 0 & 0 \\ 0 & \alpha & 0 & 0 & 0 & 0 \end{pmatrix},
$$

and

$$z = (p, q, v, w, a, b)^{\mathsf{T}}, \quad S = \frac{1}{2}V(p^2 + q^2, x, y) + \frac{\alpha}{2}(v^2 + w^2 + a^2 + b^2).$$

According to (12.21), the ECL of Eq. (12.43) reads:

$$\partial_t E + \partial_x F + \partial_y G = 0, \tag{12.44}$$

where

$$E = \frac{1}{2}V(p^2 + q^2, x, y) + \frac{\alpha}{2}(pv_x + qw_x + pa_y + qb_y),$$
$$F = \frac{\alpha}{2}(-pv_t - qw_t + vp_t + wq_t), \quad G = \frac{\alpha}{2}(-pa_t - qb_t + ap_t + bq_t).$$

Equation (12.43) also has the local charge conservation law:

$$\partial_t C + \partial_x P + \partial_y Q = 0,$$

where

$$C = \frac{1}{2}(p^2 + q^2), \ P = \alpha(-vq + wp), \ Q = \alpha(-aq + bp).$$

If V is independent of the variables x, y, then (12.43) is a multi-symplectic PDE. According to (12.2), the MSCL is:

$$\partial_t(dp \wedge dq) + \partial_x(-\alpha dp \wedge dv - \alpha dq \wedge dw) + \partial_y(-\alpha dp \wedge da - \alpha dq \wedge db) = 0.$$

All of these conservation laws lead to corresponding global invariants under the PBC. The full discretized scheme of (12.43) in vector form derived from our discrete procedure (12.22) is:

$$
\begin{cases}
p^\tau = p^0 + \Delta t \displaystyle\int_0^1 A_{\tau,\sigma}(-(D_x \otimes I_M)\alpha w^\sigma - (I_N \otimes D_y)\alpha b^\sigma \\
\quad - V'((p^\sigma)^{\cdot 2} + (q^\sigma)^{\cdot 2}, x \otimes e_M, e_N \otimes y) \cdot q^\sigma)d\sigma, \\
q^\tau = q^0 + \Delta t \displaystyle\int_0^1 A_{\tau,\sigma}((D_x \otimes I_M)\alpha v^\sigma + (I_N \otimes D_y)\alpha a^\sigma \\
\quad + V'((p^\sigma)^{\cdot 2} + (q^\sigma)^{\cdot 2}, x \otimes e_M, e_N \otimes y) \cdot p^\sigma)d\sigma, \\
p^1 = p^0 + \Delta t \displaystyle\int_0^1 (-(D_x \otimes I_M)\alpha w^\sigma - (I_N \otimes D_y)\alpha b^\sigma \\
\quad - V'((p^\sigma)^{\cdot 2} + (q^\sigma)^{\cdot 2}, x \otimes e_M, e_N \otimes y) \cdot q^\sigma)d\sigma, \\
q^1 = q^0 + \Delta t \displaystyle\int_0^1 (D_x \otimes I_M)\alpha v^\sigma + (I_N \otimes D_y)\alpha a^\sigma \\
\quad + V'((p^\sigma)^{\cdot 2} + (q^\sigma)^{\cdot 2}, x \otimes e_M, e_N \otimes y) \cdot p^\sigma)d\sigma, \\
\delta_x p^\sigma = (D_x \otimes I_M)p^\sigma = v^\sigma, \delta_x q^\sigma = (D_x \otimes I_M)q^\sigma = w^\sigma, \\
\delta_y p^\sigma = (I_N \otimes D_y)p^\sigma = a^\sigma, \delta_y q^\sigma = (I_N \otimes D_y)q^\sigma = b^\sigma,
\end{cases}
\tag{12.45}
$$

where the entries p_{jl}, q_{jl} of vectors p, q for $0 \le j \le N - 1$ and $0 \le l \le M - 1$ are arranged according to lexicographical order:

$$(j, l) \prec (k, m), \text{ when } j \prec k, \quad (j, l) \prec (j, m), \text{ when } l \prec m,$$

$x = (x_0, x_1, \ldots, x_{N-1})^\mathsf{T}$, $y = (y_0, y_1, \ldots, y_{M-1})^\mathsf{T}$, I_N, I_M, e_N, e_M are Nth and Mth order identity matrices, N length and M length identity vectors, respectively. If the potential V is a polynomial in the first variable, then the scheme (12.45) can be implemented in a way similar to (12.40).

By Theorem 12.2, (12.45) preserves the discrete ECLs:

$$\frac{E_{jl}^1 - E_{jl}^0}{\Delta t} + \sum_{k=0}^{N-1}(D_x)_{jk}\bar{F}_{jk,l} + \sum_{m=0}^{M-1}(D_y)_{lm}\bar{G}_{j,lm} = 0, \tag{12.46}$$

for $j = 0, 1, \ldots, N - 1$ and $l = 0, 1, \ldots, M - 1$, where

$$E_{jl}^c = \frac{1}{2}V((p_{jl}^c)^2 + (q_{jl}^c)^2, x_j, y_l) + \frac{\alpha}{2}(p_{jl}^c \delta_x v_{jl}^c + q_{jl}^c \delta_x w_{jl}^c + p_{jl}^c \delta_y a_{jl}^c + q_{jl}^c \delta_y b_{jl}^c), c = 0, 1,$$

$$\bar{F}_{jk,l} = \frac{\alpha}{2}\sum_{i=1}^{s} b_i(-\langle p_{jl}\rangle_i \langle \delta_t v_{kl}\rangle_i - \langle q_{jl}\rangle_i \langle \delta_t w_{kl}\rangle_i + \langle v_{jl}\rangle_i \langle \delta_t p_{kl}\rangle_i + \langle w_{jl}\rangle_i \langle \delta_t q_{kl}\rangle_i$$

$$+ \frac{\alpha}{2}\sum_{i=1}^{s} b_i(-\langle p_{kl}\rangle_i \langle \delta_t v_{jl}\rangle_i - \langle q_{kl}\rangle_i \langle \delta_t w_{jl}\rangle_i + \langle v_{kl}\rangle_i \langle \delta_t p_{jl}\rangle_i + \langle w_{kl}\rangle_i \langle \delta_t q_{jl}\rangle_i),$$

$$\bar{G}_{j,lm} = \frac{\alpha}{2}\sum_{i=1}^{s} b_i(-\langle p_{jl}\rangle_i \langle \delta_t a_{jm}\rangle_i - \langle q_{jl}\rangle_i \langle \delta_t b_{jm}\rangle_i + \langle a_{jl}\rangle_i \langle \delta_t p_{jm}\rangle_i + \langle b_{jl}\rangle_i \langle \delta_t q_{jm}\rangle_i)$$

$$+ \frac{\alpha}{2}\sum_{i=1}^{s} b_i(-\langle p_{jm}\rangle_i \langle \delta_t a_{jl}\rangle_i - \langle q_{jm}\rangle_i \langle \delta_t b_{jl}\rangle_i + \langle a_{jm}\rangle_i \langle \delta_t p_{jl}\rangle_i + \langle b_{jm}\rangle_i \langle \delta_t q_{jl}\rangle_i).$$

However, the expressions of E_{jl}^c, $\bar{F}_{jk,l}$, $\bar{G}_{j,lm}$ above are lengthy and difficult to be calculated. We thus rewrite them as:

$$E_{jl}^c = \frac{1}{2}V((p_{jl}^c)^2 + (q_{jl}^c)^2, x_j, y_l) - \frac{\alpha}{2}((v_{jl}^c)^2 + (w_{jl}^c)^2 + (a_{jl}^c)^2 + (b_{jl}^c)^2) + \tilde{E}_{jl}^c, c = 0, 1,$$

$$\bar{F}_{jk,l} = \alpha \sum_{i=1}^{s} b_i(\langle v_{jl}\rangle_i \langle \delta_t p_{kl}\rangle_i + \langle w_{jl}\rangle_i \langle \delta_t q_{kl}\rangle_i + \langle v_{kl}\rangle_i \langle \delta_t p_{jl}\rangle_i + \langle w_{kl}\rangle_i \langle \delta_t q_{jl}\rangle_i) + \tilde{F}_{jk,l},$$

$$\bar{G}_{j,lm} = \alpha \sum_{i=1}^{s} b_i(\langle a_{jl}\rangle_i \langle \delta_t p_{jm}\rangle_i + \langle b_{jl}\rangle_i \langle \delta_t q_{jm}\rangle_i + \langle a_{jm}\rangle_i \langle \delta_t p_{jl}\rangle_i + \langle b_{jm}\rangle_i \langle \delta_t q_{jl}\rangle_i) + \tilde{G}_{j,lm},$$

$$(12.47)$$

where $\tilde{E}_{jl}^c$, $\tilde{F}_{jk,l}$ $\tilde{G}_{j,lm}$ are the corresponding residuals.

Taking derivatives with respect to τ on both sides of

$$\delta_x p_{jl}^\tau = v_{jl}^\tau = v_{jl}^0 + \Delta t \int_0^1 A_{\tau,\sigma} \delta_t v_{jl}^\sigma d\sigma,$$

and setting $\tau = c_1, \ldots, c_s$, we have

$$\langle \delta_x \delta_t p_{jl}\rangle_i = \langle \delta_t v_{jl}\rangle_i,$$

for $i = 1, 2, \ldots, s$. By using this law of commutation and following the standard proof procedure of Theorem 12.1, the terms involving v_{jl} can be eliminated from $\tilde{E}_{jl}^c$, $\tilde{F}_{jk,l}$ $\tilde{G}_{j,lm}$. The terms involving w_{jl}, a_{jl}, b_{jl} can be dealt with in the same way.

Therefore, we arrive at

$$\frac{\tilde{E}_{jl}^1 - \tilde{E}_{jl}^0}{\Delta t} + \sum_{k=0}^{N-1}(D_x)_{jk}\tilde{F}_{jk,l} + \sum_{m=0}^{M-1}(D_y)_{lm}\tilde{G}_{j,lm} = 0. \qquad (12.48)$$

Subtracting (12.48) from (12.46), we obtain the new discrete ECLs of (12.45):

$$\frac{E_{jl}^1 - E_{jl}^0}{\Delta t} + \sum_{k=0}^{N-1}(D_x)_{jk}\bar{F}_{jk,l} + \sum_{m=0}^{M-1}(D_y)_{lm}\bar{G}_{j,lm} = 0, \qquad (12.49)$$

for $j = 0, 1, \ldots, N-1$ and $l = 0, 1, \ldots, M-1$, where

$$E_{jl} = \frac{1}{2}V(p_{jl}^2 + q_{jl}^2, x_j, y_l) - \frac{\alpha}{2}(v_{jl}^2 + w_{jl}^2 + a_{jl}^2 + b_{jl}^2),$$

$$\bar{F}_{jk,l} = \alpha\sum_{i=1}^{s}b_i(\langle v_{jl}\rangle_i\langle\delta_t p_{kl}\rangle_i + \langle w_{jl}\rangle_i\langle\delta_t q_{kl}\rangle_i + \langle v_{kl}\rangle_i\langle\delta_t p_{jl}\rangle_i + \langle w_{kl}\rangle_i\langle\delta_t q_{jl}\rangle_i),$$

$$\bar{G}_{j,lm} = \alpha\sum_{i=1}^{s}b_i(\langle a_{jl}\rangle_i\langle\delta_t p_{jm}\rangle_i + \langle b_{jl}\rangle_i\langle\delta_t q_{jm}\rangle_i + \langle a_{jm}\rangle_i\langle\delta_t p_{jl}\rangle_i + \langle b_{jm}\rangle_i\langle\delta_t q_{jl}\rangle_i).$$

Equation (12.49) can be thought of as a discrete version of:

$$\partial_t(\frac{1}{2}V(p^2 + q^2, x, y) - \frac{\alpha}{2}(v^2 + w^2 + a^2 + b^2)) + \partial_x(\alpha v p_t + \alpha w q_t)$$
$$+ \partial_y(\alpha a p_t + \alpha b q_t) = 0,$$

which is a more common ECL of Eq. (12.43) than Eq. (12.44).

Equation (12.49) involves less discrete derivatives than Eq. (12.46), thus can be easily calculated.

12.6 Numerical Experiments for Coupled Nonlinear Schrödingers Equations

If we choose the 2-point Gauss-Legendre quadrature formula:

$$b_1 = \frac{1}{2}, b_2 = \frac{1}{2},$$

$$c_1 = \frac{1}{2} - \frac{\sqrt{3}}{6}, c_2 = \frac{1}{2} + \frac{\sqrt{3}}{6},$$

for the CRK method, then

$$A_{\tau,\sigma} = \tau((4 - 3\tau) - 6(1 - \tau)\sigma).$$

This CRK method is of order four by (12.6). In this section, we use it for the temporal discretization while the spatial direction is discretized by the pseudospectral method. The corresponding local energy-preserving method for the CNLS is denoted by ET4.

Throughout the experiments in this section we always take the periodic boundary condition $u(x_0, t) = u(x_0 + L, t)$, $v(x_0, t) = v(x_0 + L, t)$ and set the initial time $t_0 = 0$. Besides the discrete global energy which has been mentioned in (12.14), we define these discrete global quantities as follows:

1. The discrete global charges of u and v at time $n\Delta t$:

$$\begin{cases} CH_U^n = \Delta x \sum_{j=0}^{N-1} ((q_{1,j}^n)^2 + (q_{2,j}^n)^2), \\[4mm] CH_V^n = \Delta x \sum_{j=0}^{N-1} ((q_{3,j}^n)^2 + (q_{4,j}^n)^2). \end{cases}$$

2. The discrete global momentum at time $n\Delta t$:

$$I^n = \Delta x \sum_{j=0}^{N-1} (q_{2,j}^n p_{1,j}^n - q_{1,j}^n p_{2,j}^n + q_{4,j}^n p_{3,j}^n - q_{3,j}^n p_{4,j}^n).$$

The (relative) global energy error (GEE^n), global momentum error (GIE^n), global charge errors of u (GCE_U^n) and v (GCE_V^n) at time $n\Delta t$ will be calculated by the following formulae:

$$GEE^n = \frac{E^n - E^0}{|E^0|}, \quad GIE^n = \frac{I^n - I^0}{|I^0|},$$

$$GCE_U^n = \frac{CH_U^n - CH_U^0}{|CH_U^0|}, \quad GCE_V^n = \frac{CH_V^n - CH_V^0}{|CH_V^0|},$$

respectively.

Experiment 12.1 We first consider to set the constants $\alpha, \beta = 0$ in (12.39). Then the CNLS can be decomposed into two independent NLSs:

$$\begin{cases} iu_t + \dfrac{1}{2} u_{xx} + |u|^2 u = 0, \\[4mm] iv_t + \dfrac{1}{2} v_{xx} + |v|^2 v = 0. \end{cases} \tag{12.50}$$

Given the initial conditions:

$$\begin{cases} u(x, 0) = \mathrm{sech}(x), \\[3mm] v(x, 0) = \mathrm{sech}(x)\exp(i\dfrac{x}{\sqrt{10}}), \end{cases}$$

the analytic expressions of u and v are given by

$$
\begin{cases}
u(x,t) = \operatorname{sech}(x)\exp(i\frac{t}{2}), \\
v(x,t) = \operatorname{sech}(x - \frac{t}{\sqrt{10}})\exp(i(\frac{x}{\sqrt{10}} + \frac{9}{20}t)).
\end{cases}
\tag{12.51}
$$

In this experiment, we compute the difference between the numerical solution and the exact solution of u. Since u decays exponentially away from the point $(0, t)$, we can take the boundary condition $u(-30, 0) = u(30, 0)$, $v(-30, 0) = v(30, 0)$ with little loss of accuracy on u. We also compare our local energy-preserving method ET4 with a classical multi-symplectic scheme (MST4) which is obtained by concatenating the two-point Gauss–Legendre symplectic Runge–Kutta method in time and the pseudospectral method in space. Note that ET4 and MST4 are of the same order. Let $N = 300$, $\Delta t = 0.4$, 0.8 and set $\varepsilon = 10^{-14}$ as the error tolerance for iteration solutions. The numerical results over the time interval $[0, 1200]$, which is about 100 multiples of the period of u, are plotted in Figs. 12.1, 12.2, 12.3, 12.4, 12.5 and 12.6.

Figures 12.1 and 12.3 illustrate that ET4 conserves the discrete global energy exactly (regardless of round-off errors). Although ET4 cannot preserve discrete global charges, its global charge errors show reasonable oscillation in magnitude 10^{-10} ($\Delta t = 0.4$) and 10^{-4} ($\Delta t = 0.8$), respectively. We attribute this behaviour to the conjugate-symplecticity of the CRK method.

On the contrary, Figs. 12.2 and 12.4 show that MST4 conserves global charges exactly (regardless of round-off errors) while its global energy errors oscillate in magnitude 10^{-8} ($\Delta t = 0.4$) and 10^{-3} ($\Delta t = 0.8$). This is a character of symplectic integrators.

According to Figs. 12.1, 12.2, 12.3 and 12.4, MST4 preserves the discrete global momentum better than ET4 in this experiment.

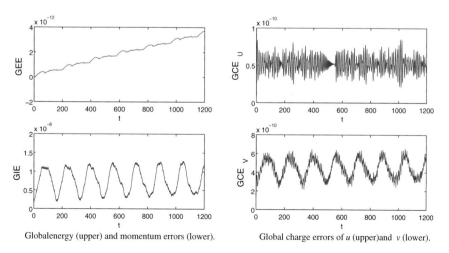

Globalenergy (upper) and momentum errors (lower).   Global charge errors of u (upper)and v (lower).

Fig. 12.1 Errors obtained by ET4, $\Delta t = 0.4$

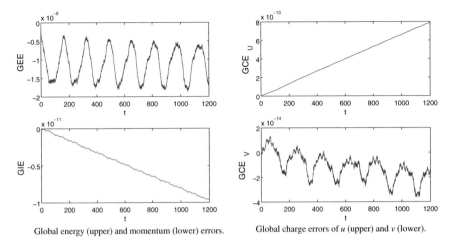

Global energy (upper) and momentum (lower) errors.   Global charge errors of u (upper) and v (lower).

Fig. 12.2   Errors obtained by MST4, $\Delta t = 0.4$

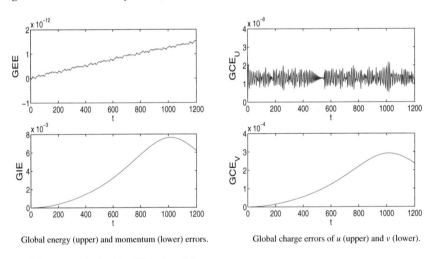

Global energy (upper) and momentum (lower) errors.   Global charge errors of u (upper) and v (lower).

Fig. 12.3   Errors obtained by ET4, $\Delta t = 0.8$

It can be observed from (12.51) that the amplitudes of u and v are both 1. Figure 12.5 shows that ET4 and MST4 both have excellent long-term behaviours. The relative maximum global errors do not exceed 1.5 % ($\Delta t = 0.4$) and 25 % ($\Delta t = 0.8$) over the time interval [0,1200].

Here we point out that ET4 and MST4 have the same iteration cost with the same Δt and ε. In the case $\Delta t = 0.4$, both of them need 19 iterations per step. This phenomenon also occurs in the following experiments.

Experiment 12.2   We then start to simulate the collision of double solitons for (12.39) with the initial conditions:

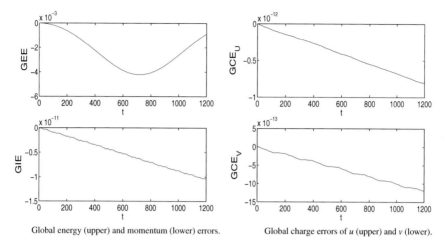

Global energy (upper) and momentum (lower) errors.　　Global charge errors of u (upper) and v (lower).

Fig. 12.4 Errors obtained by MST4, $\Delta t = 0.8$

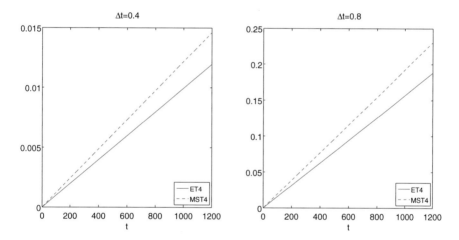

Fig. 12.5 Maximum global errors of ET4 and MST4

$$\begin{cases} u(x,0) = \sum_{j=1}^{2} \sqrt{\frac{2a_j}{1+\beta}} \operatorname{sech}(\sqrt{2a_j}(x-x_j)) \exp(i(\gamma_j - \alpha)(x-x_j)), \\[4mm] v(x,0) = \sum_{j=1}^{2} \sqrt{\frac{2a_j}{1+\beta}} \operatorname{sech}(\sqrt{2a_j}(x-x_j)) \exp(i(\gamma_j + \alpha)(x-x_j)). \end{cases}$$

This is an initial condition resulting in a collision of two separate single solitons. Here we choose $x_0 = 0, L = 100, \alpha = 0.8, \beta = \frac{2}{3}, a_1 = 1, a_2 = 0.8, \gamma_1 = 1.5,$ $\gamma_2 = -1.5, x_1 = 20, x_2 = 80$. Take the temporal stepsize $\Delta t = 0.2$ and spatial grid number $N = 450$. The numerical results are shown in Figs. 12.7 and 12.8.

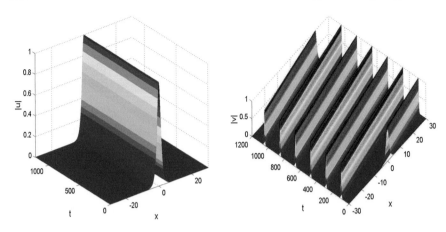

Fig. 12.6 Numerical shapes of u (*left*) and v (*right*), obtained by ET4

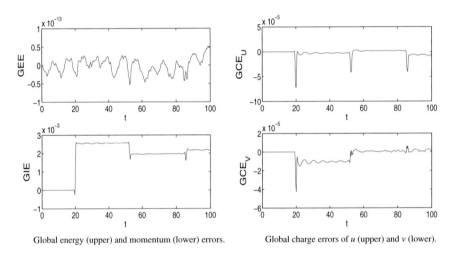

Global energy (upper) and momentum (lower) errors.   Global charge errors of u (upper) and v (lower).

Fig. 12.7 Errors obtained by ET4, $\Delta t = 0.2$, $N = 450$

Obviously, ET4 successfully simulates the collision of two solitons and the effects of boundaries on bisolitons. It exactly preserves the discrete energy and conserves the discrete charges and momentum very well.

Experiment 12.3 The last experiment on the CNLS (12.39) is the simulation of the interaction among triple solitons with the initial conditions:

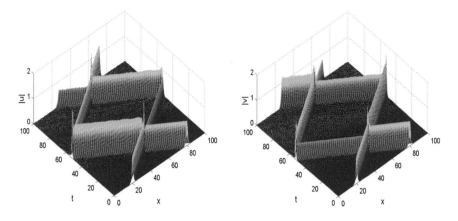

The shape of bisoliton u.

The shape of bisoliton v.

Fig. 12.8 Numerical shapes of u (*left*) and v (*right*) obtained from ET4

$$
\begin{cases}
u(x,0) = \sum_{j=1}^{3} \sqrt{\frac{2a_j}{1+\beta}}\, \mathrm{sech}(\sqrt{2a_j}(x-x_j))\exp(i(\gamma_j-\alpha)(x-x_j)), \\[2ex]
v(x,0) = \sum_{j=1}^{3} \sqrt{\frac{2a_j}{1+\beta}}\, \mathrm{sech}(\sqrt{2a_j}(x-x_j))\exp(i(\gamma_j+\alpha)(x-x_j)).
\end{cases}
$$

Here we also test another scheme associated with ET4. The only difference between it and ET4 is that we evaluate the nonlinear integrals in ET4 not by symbol calculation, but by a high-order quadrature formula. In the case of ET4, the polynomials are of degrees 6. Hence we can calculate them exactly by the 4-point GL quadrature formula. To illustrate the alternative scheme, we evaluate the nonlinear integrals by the 3-point GL quadrature formula:

$$
b_1 = \frac{5}{18},\; b_2 = \frac{4}{9},\; b_3 = \frac{5}{18},
$$
$$
c_1 = \frac{1}{2} - \frac{\sqrt{15}}{10},\; c_2 = \frac{1}{2},\; c_3 = \frac{1}{2} + \frac{\sqrt{15}}{10}.
$$

For example, the first nonlinear integral of (12.40) is approximated by

$$
\int_0^1 (((q_1^\sigma)^{\cdot 2} + (q_2^\sigma)^{\cdot 2}) + \beta((q_3^\sigma)^{\cdot 2} + (q_4^\sigma)^{\cdot 2})) \cdot q_2^\sigma \, d\sigma
$$
$$
\approx \sum_{i=1}^{3} b_i (((q_1^{c_i})^{\cdot 2} + (q_2^{c_i})^{\cdot 2}) + \beta((q_3^{c_i})^{\cdot 2} + (q_4^{c_i})^{\cdot 2})) \cdot q_2^{c_i}.
$$

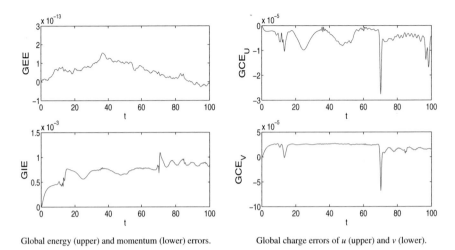

Global energy (upper) and momentum (lower) errors.   Global charge errors of u (upper) and v (lower).

Fig. 12.9 Errors obtained by ET4, $\Delta t = 0.2$, $N = 360$

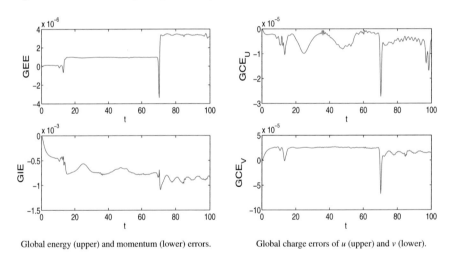

Global energy (upper) and momentum (lower) errors.   Global charge errors of u (upper) and v (lower).

Fig. 12.10 Errors obtained by ET4GL6, $\Delta t = 0.2$, $N = 360$

For convenience, we denote the scheme by ET4GL6. Setting $\Delta t = 0.2$, $N = 360$, $x_0 = 0$, $L = 80$, $\alpha = 0.5$, $\beta = \frac{2}{3}$, $\gamma_1 = 1.5$, $\gamma_2 = 0.1$, $\gamma_3 = -1.2$, $a_1 = 0.75$, $a_2 = 1$, $a_3 = 0.5$, $x_1 = 20$, $x_2 = 40$, $x_3 = 60$, we compute it over the time interval $[0,100]$. Numerical results are presented in Figs. 12.9, 12.10, 12.11 and 12.12. The behaviours of ET4, ET4GL6 are very similar in conserving momentum and charges. Unsurprisingly, ET4 and MST4 preserve exactly the discrete global energy and charges, respectively. However, ET4GL6 can conserve the discrete energy in magnitude 10^{-6}, whereas MST4 only preserves the energy in magnitude 10^{-4}. Hence if we give more

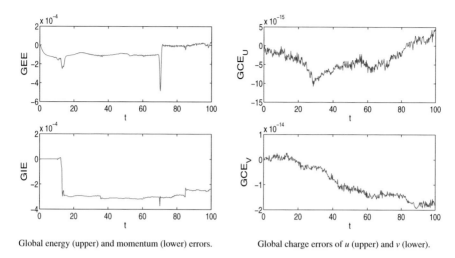

Global energy (upper) and momentum (lower) errors.   Global charge errors of u (upper) and v (lower).

Fig. 12.11 Errors obtained by MST4, $\Delta t = 0.2$, $N = 360$

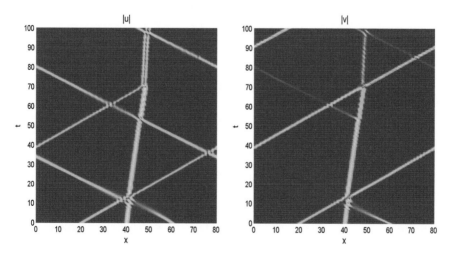

Fig. 12.12 Numerical solitons of u and v obtained by ET4

weight on the discrete energy, ET4GL6 is a favourable scheme. In fact, when the nonlinear integrals cannot be calculated exactly or have to be integrated in very complicated forms, ETGL6 is a reasonable alternative scheme.

12.7   Numerical Experiments for 2D Nonlinear Schrödinger Equations

In this section, we apply the CRK method of second order (i.e. average vector field method) to t-direction and the pseudospectral method to x and y directions. This scheme is denoted by ET2. To illustrate the method, we will compare it with another prominent traditional scheme which is obtained by the implicit midpoint temporal discretization and the pseudospectral spatial discretization (ST2). If (12.43) is linear, the scheme ET2 is the same as ST2. Hence we will not give numerical examples of 2D linear Schrödinger equations.

The boundary condition is always taken to be periodic:

$$u(x_l, y, t) = u(x_r, y, t), u(x, y_l, t) = u(x, y_r, t). \tag{12.52}$$

The discrete global charge CH will still be taken into account:

$$CH^n = \Delta x \Delta y \sum_{j=0}^{N-1} \sum_{l=0}^{M-1} ((p_{jl}^n)^2 + (q_{jl}^n)^2),$$

where

$$CH^n \approx \int_{x_l}^{x_r} \int_{y_l}^{y_r} (p(x, y, n\Delta t)^2 + q(x, y, n\Delta t)^2) dxdy.$$

Besides, the residuals in the ECL (12.49) are defined as:

$$RES_{jl}^n = \frac{E_{jl}^{n+1} - E_{jl}^n}{\Delta t} + \sum_{k=0}^{N-1} (D_x)_{jk} \bar{F}_{jk,l} + \sum_{m=0}^{M-1} (D_y)_{lm} \bar{G}_{j,lm},$$

for $j = 0, 1, \ldots, N - 1$ and $l = 0, 1, \ldots, M - 1$.

In this section, we calculate RES^n: the residual with the maximum absolute value at the time level $n\Delta t$.

Experiment 12.4 Let $\alpha = \frac{1}{2}$, $V(\xi, x, y) = V_1(x, y)\xi + \frac{1}{2}\beta\xi^2$. Then (12.43) becomes the Gross–Pitaevskii (GP) equation:

$$i\psi_t + \frac{1}{2}(\psi_{xx} + \psi_{yy}) + V_1(x, y)\psi + \beta|\psi|^2\psi = 0. \tag{12.53}$$

This equation is an important mean-field model for the dynamics of a dilute gas Bose-Einstein condensate (BEC) (see, e.g. [12]). The parameter β determines whether (12.53) is attractive ($\beta > 0$) or repulsive ($\beta < 0$).

We should note that Eq. (12.53) is no longer multi-symplectic and the scheme ST2 is only symplectic in time. We first consider the attractive case $\beta = 1$. The external potential V_1 is:

$$V_1(x, y) = -\frac{1}{2}(x^2 + y^2) - 2\exp(-(x^2 + y^2)).$$

The initial condition is given by:

$$\psi(x, y, 0) = \sqrt{2}\exp(-\frac{1}{2}(x^2 + y^2)).$$

This IVP has the exact solution (see, e.g. [1]):

$$\psi(x, y, t) = \sqrt{2}\exp(-\frac{1}{2}(x^2 + y^2))\exp(-it).$$

For the same reason in Experiment 12.1, we set the spatial domain as $x_l = -6$, $x_r = 6$, $y_l = -6$, $y_r = 6$. The temporal stepsize is chosen as $\Delta t = 0.10, 0.08, 0.05$, respectively. Fixing the number of spatial grids $N = M = 42$, we compute the numerical solution over the time interval $[0, 45]$. The numerical results are shown in Figs. 12.13, 12.14, 12.15, 12.16 and 12.17.

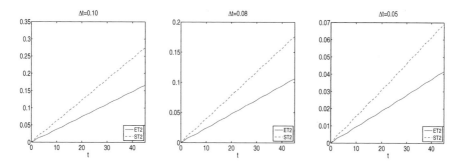

Fig. 12.13 Maximum global errors

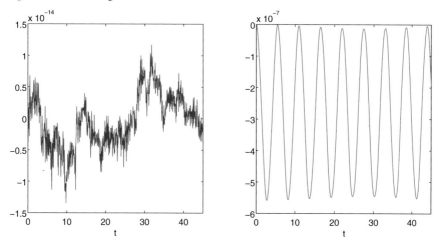

Fig. 12.14 Global energy errors of ET2 (*left*) and ST2 (*right*), $\Delta t = 0.05$

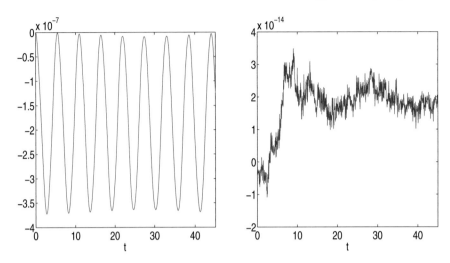

Fig. 12.15 Global charge errors of ET2 (*left*) and ST2 (*right*), $\Delta t = 0.05$

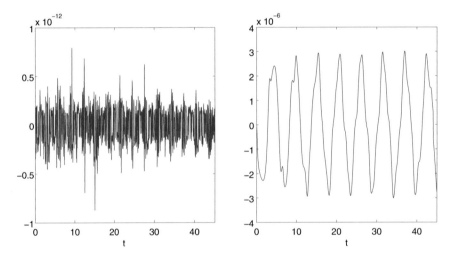

Fig. 12.16 Maximum residuals (RES) of ET2 (*left*) and ST2 (*right*) in the ECL, $\Delta t = 0.05$

From the results, we can see that ET2 conserves both the global energy and the ECL exactly while its global charge errors oscillates in magnitude 10^{-7}. On the other hand, ST2 preserves the global charge accurately whereas its global energy errors oscillates in magnitude 10^{-7} and its maximum residuals in the ECL oscillates in magnitude 10^{-6}. However, the maximum global errors of ST2 are twice as large as that of ET2 under the three different Δt.

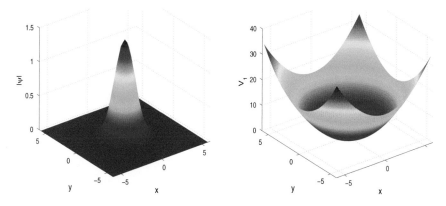

Fig. 12.17 Shapes of the solution (*left*) and the potential V_1 (*right*)

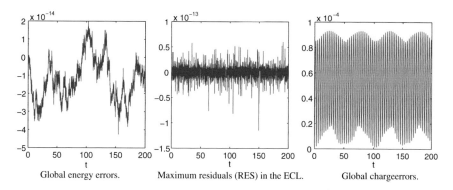

| Global energy errors. | Maximum residuals (RES) in the ECL. | Global chargeerrors. |

Fig. 12.18 Errors obtained by ET2

Experiment 12.5 Let $\alpha = \frac{1}{2}$, $V_1(x, y) = -\frac{1}{2}(x^2 + y^2)$, $\beta = -2$. Given the initial condition

$$\psi(x, y, 0) = \frac{1}{\sqrt{\pi}} \exp(-\frac{1}{2}(x^2 + y^2)),$$

we now consider the repulsive GP equation in space $[-8, 8] \times [-8, 8]$ (see, e.g. [22]). Let $N = M = 36$, $\Delta t = 0.1$, we compute the numerical solution over the time interval $[0, 200]$. The results are plotted in Figs. 12.18 and 12.19. Obviously, ET2 still show the eminent long-term behaviour dealing with high dimensional problems.

Experiment 12.6 We then consider the 2DNLS with quintic nonlinearity:

$$i\psi_t + \psi_{xx} + \psi_{yy} + V_1(x, y)\psi + |\psi|^4 \psi = 0, \tag{12.54}$$

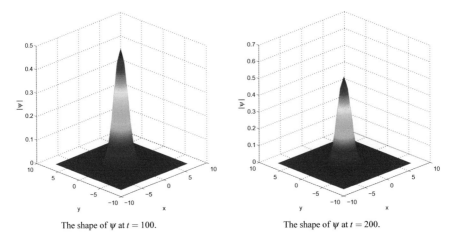

The shape of ψ at $t = 100$.   The shape of ψ at $t = 200$.

Fig. 12.19 The numerical shapes of ψ

where

$$V_1(x, y) = -\frac{A^4}{4}(A^4 x^2 + A^4 y^2) - A^4 \exp(-A^4 x^2 - A^4 y^2)$$

is an external field, and A is a constant. Its potential is:

$$V(\xi, x, y) = V_1(x, y)\xi + \frac{1}{3}\xi^3.$$

This equation admits the solution:

$$\psi(x, y, t) = A\exp(-\frac{1}{4}A^4(x^2 + y^2))\exp(-i A^4 t).$$

Its period is $\frac{2\pi}{A^4}$. Setting $A = 1.5$, $x_l = -4$, $x_r = 4$, $y_l = -4$, $y_r = 4$, $\Delta t = 0.01$, $N = M = 42$, we integrate (12.54) over a very long interval [0,124] which is about 100 multiples of the period. Since the behaviours of ET2 and ST2 in conserving the global charge and the energy are very similar to those in Experiments 12.4 and 12.5, they are omitted here. The global errors of ET2 and ST2 in l^∞ and $\frac{1}{\sqrt{NM}}l^2$ norms are shown in Fig. 12.20.

Clearly, in the quintic case, the method ET2 again wins over the classical symplectic scheme ST2.

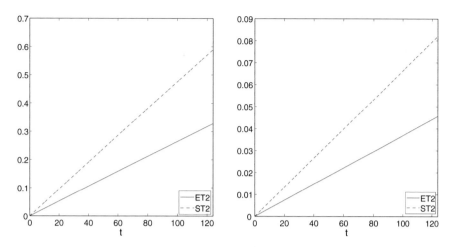

Fig. 12.20 l^∞ global errors (*left*) and $\frac{1}{\sqrt{NM}}l^2$ global errors (*right*)

12.8 Conclusions

"For Hamiltonian differential equations there is a long-standing dispute on the question whether in a numerical simulation it is more important to preserve energy or symplecticity. Many people give more weight on symplecticity, because it is known (by backward error analysis arguments) that symplectic integrators conserve a modified Hamiltonian" (quote from Hairer's paper [17]).

However, due to the complexity of PDEs, the theory on multi-symplectic integrators is still far from being satisfactory. There are only a few results on some simple schemes (e.g. the Preissman and the Euler box scheme) and on special PDEs (e.g. the nonlinear wave equation and the nonlinear Schrödinger equation) based on backward error analysis (see, e.g. [5, 20, 29]). These theories show that a class of box schemes conserves the modified ECL and MCL(see, e.g. [20]). Besides, it seems there is no robust theoretical results for the multi-symplectic (pseudo) spectral scheme. Therefore, the local energy-preserving algorithms may play a much more important role in PDEs than their counterparts in ODEs.

In this chapter, we presented general local energy-preserving schemes which can have arbitrarily high orders for solving multi-symplectic Hamiltonian PDEs. In these schemes, time is discretized by a continuous Runge–Kutta method and space is discretized by a pseudospectral method or a Gauss-Legendre collocation method. It should be noted that more Hamiltonian PDEs admit a local energy conservation law than do a multi-symplectic conservation law. Hence the local energy-preserving methods can be more widely applied to Hamiltonian PDEs than multi-symplectic methods in the literature. The numerical results accompanied in this chapter are promising. In the experiments on CNLSs, the local energy-preserving methods and the associated methods behave similarly to the multi-symplectic methods of the same order. In the experiments on 2D NLSs with external fields, the local energy-preserving methods behave better than symplectic methods in both cubic and quintic nonlinear problems. The numerical results show the excellent qualitative properties of the new schemes in *long-term numerical simulation*.

This chapter is based on the very recent work by Li and Wu [25].

References

1. Antar N, Pamuk N (2013) Exact solutions of two dimensional nonlinear Schrödinger equations with external potentials. Appl Comput Math 2:152–158
2. Bridges TJ (1997) Multi-symplectic structures and wave propagation. Math Proc Camb Philos Soc 121:147–190
3. Bridges TJ, Reich S (2001) Multi-symplectic integrators: numerical schemes for Hamiltonian PDEs that conserve symplecticity. Phys Lett A 284:184–193
4. Bridges TJ, Reich S (2001) Multi-symplectic spectral discretizations for the Zakhakarov-Kuznetsov and shallow water equations. Physica D 152–153:491–504
5. Bridges TJ, Reich S (2006) Numerical methods for Hamiltonian PDEs. J Phys A: Math Gen 39:5287
6. Cai J, Wang Y, Liang H (2013) Local energy-preserving and momentum-preserving algorithms for coupled nonlinear Schrödinger system. J Comput Phys 239:30–50
7. Cai J, Wang Y (2013) Local structure-preserving algorithms for the "good" Boussinesq equation. J Comput Phys 239:72–89
8. Celledoni E, Grimm V, Mclachlan RI, Maclaren DI, O'Neale D, Owren B, Quispel GRW (2012) Preserving energy resp. dissipation in numerical PDEs using the 'Average Vector Field' method. J Comput Phys 231:6770–6789
9. Chen Y, Sun Y, Tang Y (2011) Energy-preserving numerical methods for Landau-Lifshitz equation. J Phys A: Math Theor 44:295207–295222
10. Chen Y, Zhu H, Song S (2010) Multi-symplectic splitting method for the coupled nonlinear Schrödinger equation. Comput Phys Comm 181:1231–1241
11. Chen JB, Qin MZ (2001) Multisymplectic Fourier pseudospectral method for the nonlinear Schrödinger equation. Electon Trans Numer Anal 12:193–204
12. Deconinck B, Frigyik BA, Kutz JN (2001) Stability of exact solutions of the defocusing nonlinear Schrodinger equation with periodic potential in two dimensions. Phys Lett A 283:177–184
13. Fei Z, Vázquez L (1991) Two energy-conserving numerical schemes for the sine-Gordon equation. Appl Math Comput 45:17–30
14. Gong Y, Cai J, Wang Y (2014) Some new structure-preserving algorithms for general multi-symplectic formulations of Hamiltonian PDEs. J Comput Phys 279:80–102
15. Gonzalez O (1996) Time integration and discrete hamiltonian systems. J Nonlinear Sci 6:449–467
16. Guo BY, Vázquez L (1983) A numerical scheme for nonlinear Klein-Gordon equation. J Appl Sci 1:25–32
17. Hairer E (2010) Energy-preserving variant of collocation methods. J Numer Anal Ind Appl Math 5:73–84
18. Hong J, Liu H, Sun G (2005) The multi-symplecticity of partitioned Runge-Kutta methods for Hamiltonian PDEs. Math Comp 75:167–181
19. Hong J, Liu XY, Li C (2007) Multi-symplectic Runge-Kutta-Nyström methods for Schrödinger equations with variable coefficients. J Comput Phys 226:1968–1984
20. Islas AL, Schober CM, Li C (2005) Backward error analysis for multisymplectic discretizations of Hamiltonian PDEs. Math Comput Simul 69:290–303

21. Karasözen B, Simsek G (2013) Energy preserving integration of bi-Hamiltonian partial differential equations. TWMS J App Eng Math 3:75–86
22. Kong L, Hong J, Fu F, Chen J (2011) Symplectic structure-preserving integrators for the two-dimensional Gross-Pitaevskii equation for BEC. J Comput Appl Math 235:4937–4948
23. Kong L, Wang L, Jiang S, Duan Y (2013) Multisymplectic Fourier pseudo-spectral integrators for Klein-Gordon-Schrödinger equations. Sci China Math 56:915–932
24. Li S, Vu-Quoc L (1995) Finite difference calculus invariant structure of a class of algorithms for the nonlinear Klein-Gordon equation. SIAM J Numer Anal 32:1839–1875
25. Li YW, Wu X (2015) General local energy-preserving integrators for solving multi-symplectic Hamiltonian PDEs. J Comput Phys 301:141–166
26. Mclachlan RI, Quispel GRW, Robidoux N (1999) Geometric integration using discrete gradients. Philos Trans R Soc A 357:1021–1046
27. Mclachlan RI, Ryland BN, Sun Y (2014) High order multisymplectic Runge-Kutta methods. SIAM J Sci Comput 36:A2199–A2226
28. Marsden JE, Patrick GP, Shkoller S (1999) Multi-symplectic, variational integrators, and nonlinear PDEs. Comm Math Phys 4:351–395
29. Moore BE, Reich S (2003) Backward error analysis for multi-symplectic integration methods. Numerische Mathematik 95:625–652
30. Reich S (2000) Multi-Symplectic Runge-Kutta Collocation Methods for Hamiltonian Wave Equation. J Comput Phys 157:473–499
31. Ryland BN, Mclachlan RI, Franco J (2007) On multi-symplecticity of partitioned Runge-Kutta and splitting methods. Int J Comput Math 84:847–869
32. Wang Y, Wang B, Qin MZ (2008) Local structure-preserving algorithms for partial differential equations. Sci. China Series A: Math. 51:2115–2136
33. Zhu H, Song S, Tang Y (2011) Multi-symplectic wavelet collocation method for the nonlinear Schrödinger equation and the Camassa-Holm equation. Comput. Phys. Comm. 182:616–627

Conference Photo (Appendix)

3rd International Conference on Numerical Analysis of Differential Equations
(Nanjing, March 29-31, 2013)

© Springer-Verlag Berlin Heidelberg and Science Press, Beijing, China 2015
X. Wu et al., *Structure-Preserving Algorithms for Oscillatory
Differential Equations II*, DOI 10.1007/978-3-662-48156-1

4th International Conference on Numerical Analysis of Differential Equations
(Nanjing, May 8-10, 2014)

Index

© Springer-Verlag Berlin Heidelberg and Science Press, Beijing, China 2015
X. Wu et al., *Structure-Preserving Algorithms for Oscillatory Differential Equations II*, DOI 10.1007/978-3-662-48156-1